# Current Topics in Microbiology and Immunology

## Volume 328

James L. Van Etten

Editor

# Lesser Known Large dsDNA Viruses

 Springer

*Editor*
James L. Van Etten
Department of Plant Pathology
   and Nebraska Center for Virology
University of Nebraska
Lincoln, NE 68583-0722
USA

ISBN 978-3-540-68617-0          e-ISBN 978-3-540-68618-7

DOI 10.1007/978-3-540-68618-7

Current Topics in Microbiology and Immunology ISSN 0070-217x

Library of Congress Catalog Number: 2008930849

*Cover design*: WMX Design GmbH, Heidelberg, Germany

Printed on acid-free paper

9 8 7 6 5 4 3 2 1

springer.com

# Preface

Several large dsDNA-containing viruses such as poxviruses (smallpox) and herpes viruses are well known among the scientific community, as well as the general populace, because they cause human diseases. The large dsDNA insect-infecting baculoviruses are also well known in the scientific community because they are used both as biological control agents and as protein expression systems. However, there are other large dsDNA-containing viruses, including the giant 1.2-Mb mimivirus, which are less well known even though all of them play important roles in everyday life. Seven of these virus families are reviewed in this book.

Examples of their importance include the virus that causes white spot disease of shrimp (WSSV) (a single representative in the family *Whispoviridae*). WSSV is responsible for the loss of millions of dollars in the commercial shrimp farming industry every year. Likewise, some iridoviruses, such as members in the genus *Megalocytivirus*, cause serious diseases in commercial fish farms throughout Asia. Other Iridovirus members in the genus *Ranavirus* are the causative agent for approximately 50% of the documented cases of amphibian mortality reported in the United States between 1996 and 2001.

African swine fever virus (ASFV) (a single representative in the family *Asfarviridae*), which is vectored by argasid ticks and may actually be a tick virus, is usually lethal to domestic swine and outbreaks of the virus have led to large swine kills in Southern Europe, primarily in Spain.

Large dsDNA viruses that infect algae, family *Phycodnaviridae*, are also just beginning to be appreciated by the scientific community because of their influence on the global environment. That is, more than 50% of the $CO_2$ fixed on the planet is by marine microorganisms, called phytoplankton. The majority of these microorganisms are photosynthetic cyanobacteria. However, a significant number of these microorganisms are eukaryotic algae. Studies in the past 10 years indicate that approximately 20% or more of these photosynthetic microorganisms are infected with a virus at any one time; thus viruses are playing a large role in the turnover of these microorganisms and their role in global $CO_2$ fixation is only beginning to be appreciated by many marine scientists. Phycodnaviruses also contribute to the disappearance of some massive algal blooms, often referred to as red tides or brown tides.

The huge *Mimivirus*, which was originally reported to only infect protozoans belonging to the *Acanthamoeba* (amoebe) genus, may be involved in some human pneumonia-like diseases, although this role is still subject to verification. Currently, *Mimivirus* is the only member in the family *Mimiviridae*, but other relatives have been discovered recently and their genomes are being sequenced. Also, the massive genomic sequencing projects that are occurring, e.g., DNA from the Sargasso Sea, are revealing many genes that are related to *Mimivirus genes*, indicating that these large viruses are probably more common than expected.

Members of the virus family *Ascoviridae* infect insects. Ascoviruses primarily infect lepidopterous insects and they have a fascinating life style.

Some viruses reviewed in this book, specifically ASFV, *Mimivirus*, the iridoviruses, and the phycodnaviruses, along with the poxviruses, probably have a common evolutionary ancestry. Collectively, these viruses are referred to as nuclear, cytoplasmic, large dsDNA viruses, abbreviated as NCLDV. As mentioned in the chapter on the *Ascoviridae*, the ascoviruses have a strong evolutionary connection to the iridoviruses and so eventually they will probably be included as a member of the NCLDVs.

The NCLDVs are gaining the attention of some scientists interested in evolution for two reasons. First, accumulating evidence indicates that the NCLDVs may be ancient viruses; in fact, there is a suggestion that the predecessor of the NCLDVs may have existed at about the time eukaryotes and prokaryotes diverged and that these viruses might even represent the fourth domain of life. A separate cladistic study suggests that the *Phycodnaviridae* δ DNA polymerases may be near the origin of all eukaryotic δ DNA polymerases. Second, other investigators have suggested that an NCLDV predecessor may have been the origin of the nucleus in eukaryotic cells or that the nucleus is the origin of the NCLDVs.

The evolutionary origin of some of these other large dsDNA viruses is a complete mystery. For example, currently WSSV does not seem to be related to any other family of viruses. Of its predicted 181 open reading frames (ORFs), only approximately 25% match proteins in the public databases; however, another 40 of these proteins are structural proteins associated with the virus particles.

The large bacteriophage (~500 kb genome) Phage G, which infects Bacillus megaterium, was always considered to be an abnormality. Although, the phage was first described roughly 40 years ago, it was essentially ignored until recently. However, scientists have recently isolated more of these giant, plaque-forming bacteriophage from nature. Ironically, many of these large dsDNA bacteriophage were missed because the agar concentrations used to plaque the phage were so high that the viruses did not have time to diffuse in the soft agar on the plates.

It should also be noted that many of these large viruses, such as *Mimivirus*, are trapped in filters commonly used to remove bacteria from natural samples. Consequently, they can be easily missed unless special precautions are taken. Without doubt many more large dsDNA-containing viruses remain to be discovered.

Finally, as editor of this volume, I want to thank each of the authors for their contributions. However, only one, Greg Chinchar and his co-authors, actually met the original deadline for their chapter and so he receives a gold medal. A special thank you goes to Chu-Fang Lo and her co-authors for agreeing at the last minute to write the chapter on the *Whispovirus*, after the original author backed out of the assignment. Finally, I want to thank Anne Clauss and her assistants at Springer for making this book happen.

Lincoln, NE, USA                                                    Jim Van Etten

# Contents

# Contributors

C. Abergel
Structural & Genomic Information Laboratory, Parc Scientifique de Luminy,
Case 934, 13288, Marseille cedex 09, France

M.J. Allen
Plymouth Marine Laboratory, Prospect Place, The Hoe, Plymouth PL1 3DH, UK

D.K. Bideshi
Department of Entomology and Interdepartmental Programs in Microbiology,
Genetics, and Molecular Biology, University of California, Riverside,
Riverside, CA 92507, USA

Y. Bigot
Université François Rabelais, U.F.R. des Sciences et Techniques, Laboratoire
d'Etude des Parasites Génétiques, FRE-CNRS 2535, Parc Grandmont,
37200 Tours, France

V.G. Chinchar
Department of Microbiology, University of Mississippi Medical Center,
Jackson, MS, USA
vchinchar@microbio.umsmed.edu

J.-M. Claverie
Structural & Genomic Information Laboratory, Parc Scientifique de Luminy,
Case 934, 13288 Marseille cedex 09, France
Jean-Michel.Claverie@univmed.fr

G.A. Delhon
Department of Pathobiology, College of Veterinary Medicine, University of
Illinois, Urbana, IL 61802, USA

J.L. Van Etten
Department of Plant Pathology and Nebraska Center for Virology,
University of Nebraska, Lincoln, NE 68583-0722, USA

B.A. Federici
Department of Entomology and Interdepartmental Programs in Microbiology,
Genetics, and Molecular Biology, University of California, Riverside, Riverside,
CA 92507, USA
brian.federici@ucr.edu

R.W. Hendrix
Pittsburgh Bacteriophage Institute and Department of Biological Sciences,
University of Pittsburgh, Pittsburgh, PA 15260, USA
rhx@pitt.edu

A. Hyatt
Australian Animal Health Laboratory, Geelong, VIC, Australia

G.-H. Kou
Institute of Zoology, National Taiwan University, Taipei 106,
Taiwan, ROC

B.K. Ku
National Veterinary Research and Quarantine Service,
Anyang Kyonggido 430-016, Korea

J.-H. Leu
Lo Institute of Zoology, National Taiwan University, Taipei 106,
Taiwan, ROC

C.-F. Lo
Institute of Zoology, National Taiwan University, Taipei 106,
Taiwan, ROC
gracelow@ntu.edu.tw

T. Miyazaki
Graduate School of Bioresources, Mie University, Mie, Japan

H. Ogata
Structural & Genomic Information Laboratory, Parc Scientifique de Luminy,
Case 934, 13288, Marseille cedex 09, France

D.L. Rock
Department of Pathobiology, College of Veterinary Medicine,
University of Illinois, Urbana, IL 61802, USA

T. Spears
Department of Entomology and Interdepartmental Programs in Microbiology,
Genetics, and Molecular Biology, University of California, Riverside,
Riverside, CA 92507, USA

Y. Tan
Department of Entomology and Interdepartmental Programs in Microbiology,
Genetics, and Molecular Biology, University of California, Riverside,
Riverside, CA 92507, USA

E.R. Tulman
Department of Pathobiology and Veterinary Science, and Center of
Excellence for Vaccine Research, University of Connecticut, Storrs,
CT 06269, USA
edan.tulman@uconn.edu

T. Williams
Instituto de Ecologia AC, Xalapa, Veracruz, Mexico

W.H. Wilson
Bigelow Laboratory for Ocean Sciences, 180 McKown Point, P.O. Box 475,
West Boothbay Harbor, ME 04575-0475, USA
wwilson@bigelow.org

X. Xu
Key Laboratory of Marine Biogenetic Resources, Third Institute of Oceanography,
SOA, Xiamen, PR China

F. Yang
Key Laboratory of Marine Biogenetic Resources, Third Institute of Oceanography,
SOA, Xiamen, PR China

X. Zhang
Key Laboratory of Marine Biogenetic Resources, Third Institute of Oceanography,
SOA, Xiamen, PR China

# The *Phycodnaviridae*: The Story of How Tiny Giants Rule the World

W.H. Wilson(✉), J.L. Van Etten, M.J. Allen

## Contents

**Abstract** The family *Phycodnaviridae* encompasses a diverse and rapidly expanding collection of large icosahedral, dsDNA viruses that infect algae. These lytic and lysogenic viruses have genomes ranging from 160 to 560 kb. The family consists of six genera based initially on host range and supported by sequence comparisons. The family is monophyletic with branches for each genus, but the phycodnaviruses have evolutionary roots that connect them with several other families of large DNA viruses, referred to as the nucleocytoplasmic large DNA viruses (NCLDV).

W.H. Wilson
Bigelow Laboratory for Ocean Sciences, 180 McKown Point, P.O. Box 475, West Boothbay Harbor, ME 04575-0475, USA
wwilson@bigelow.org

James L. Van Etten (ed.) *Lesser Known Large dsDNA Viruses.*
Current Topics in Microbiology and Immunology 328.
© Springer-Verlag Berlin Heidelberg 2009

The phycodnaviruses have diverse genome structures, some with large regions of noncoding sequence and others with regions of ssDNA. The genomes of members in three genera in the *Phycodnaviridae* have been sequenced. The genome analyses have revealed more than 1000 unique genes, with only 14 homologous genes in common among the three genera of phycodnaviruses sequenced to date. Thus, their gene diversity far exceeds the number of so-called core genes. Not much is known about the replication of these viruses, but the consequences of these infections on phytoplankton have global affects, including influencing geochemical cycling and weather patterns.

## Introduction

The illuminated region is only a small part of the 3.7-km mean depth of the ocean, yet it houses several of the great engines of planetary control (Tett 1990). Absorption of heat energy by the ocean and light energy by tiny floating marine plants, known as phytoplankton, are the engines that help regulate many aspects of the global environment. Phytoplankton (microalgae) form the base of the marine food web and their photosynthetic activities provide an important carbon sink that influences the global carbon cycle and even climate (Charlson et al. 1987). Conservative estimates suggest there is somewhere between 100,000 to several million species of algae and that only approximately 40,000 have been identified. The *Phycodnaviridae* consists of a family of viruses that infect these globally important players. A literal translation of *Phycodnaviridae* is, DNA viruses that infect algae. Given the number of potential hosts, it is incredible so few phycodnaviruses have been isolated to date. With only approximately 150 formal identifications (Wilson et al. 2005b) and at least 100 others mentioned in the literature, it is certain that most phycodnaviruses, containing an almost infinite reservoir of genetic diversity, remain to be discovered.

The *Phycodnaviridae* comprise a genetically diverse (Dunigan et al. 2006), yet morphologically similar, family of large icosahedral viruses that infect marine or freshwater eukaryotic algae with dsDNA genomes ranging from 160 to 560 kb (Van Etten et al. 2002). Members of the *Phycodnaviridae* are currently grouped into six genera (named after the hosts they infect): *Chlorovirus*, *Coccolithovirus*, *Prasinovirus*, *Prymnesiovirus*, *Phaeovirus* and *Raphidovirus* (Wilson et al. 2005b). Nomenclature anomalies have already crept into the system with the formation of the genus *Coccolithovirus*, a group of viruses that infect *Emiliania huxleyi* (an alga species in the class *Prymnesiophyceae*), which logically should have been classified within the genus *Prymnesiovirus*. However, phylogenetic analysis of the DNA polymerase gene from these viruses indicated they belong to a new genus (Schroeder et al. 2002). It is likely the *Phycodnaviridae* will eventually be split into several subfamilies and numerous genera as more isolates are characterized and the current nomenclature will probably become obsolete. Complete genome sequences have been obtained from representatives of the *Chlorovirus*, *Coccolithovirus* and *Phaeovirus* genera (Dunigan et al. 2006) and evolutionary analysis of their genomes places them within a major, monophyletic assemblage of large eukaryotic dsDNA viruses termed the Nucleo-Cytoplasmic Large DNA Viruses (NCLDVs) (Iyer et al. 2001, 2006; Allen et al. 2006c; Raoult et al. 2004). There are five families in the NCLDV clade that include *Poxviridae*, *Iridoviridae*, *Asfarviridae*, *Phycodnaviridae* and *Mimiviridae*. Probably, the virus family *Ascoviridae* should also be included in this grouping because these viruses have clearly evolved from the Iridoviruses (see the chapter by V.G. Chinchar et al., this volume). The grouping of the NCLDVs is significant for two reasons: (i) as the name suggests, it implies a likely propagation mechanism for members of the *Phycodnaviridae* where replication is probably initiated in the nucleus and is completed in the cytoplasm; (ii) the NCLDVs are proposed to have an ancient evolutionary lineage (e.g., Villarreal and DeFilippis 2000; Iyer et al. 2006; Raoult et al. 2004). Phylogeny of the NCLDVs constructed by cladistic analysis indicates that the major families may have diverged prior to the divergence of the major eukaryotic lineages 2–3 billion years ago (Iyer et al. 2006; Raoult et al. 2004). The finding that there are only 14 genes in common, from a pool of approximately 1000 genes, between three genomes from different genera of the *Phycodnaviridae* (Allen et al. 2006c) supports the idea that the *Chlorovirus*, *Coccolithovirus* and *Phaeovirus* genera diverged a long time ago. As more *Phycodnaviridae* genomes are sequenced, there is likely to be an explosion of exciting gene discoveries with novel functions.

Reports of virus-like particles (VLPs) in at least 44 taxa of eukaryotic algae have appeared since the early 1970s (Van Etten et al. 1991). These include incidental observations of VLPs in electron micrographs, for example the first description of VLPs in *E. huxleyi* was reported in 1974 (Manton and Leadbeater 1974). Although these investigators did not publish an electron micrograph showing VLPs in *E. huxleyi*, they mentioned that VLPs in *Chrysochromulina mantoniae* (plates 65 and 66, Manton and Leadbeater 1974) resembled those of VLPs commonly found in moribund or dead *Coccolithus huxleyi* cells (now referred to as *E. huxleyi*). Many images containing VLPs are probably labeled miscellaneous and filed into obscurity in laboratories around the world (e.g., Fig. 1). These old images could be a valuable resource to help identify new viruses and susceptible host strains. It was not until 1979 that a phycodnavirus was even isolated; the virus infected the marine unicellular alga *Micromonas pusilla* (Mayer and Taylor 1979). However, this report was largely ignored until the early 1990s (Cottrell and Suttle 1991) when high concentrations of viruses in aquatic environments were being described (Bergh et al. 1989). Perhaps most significantly, in the early 1980s a group of viruses were characterized that infect freshwater unicellular, eukaryotic, exsymbiotic chlorella-like green algae, called chloroviruses (Meints et al. 1981; Van Etten et al. 1981, 1982, 1983a). These reports were followed in the early 1990s by research into marine filamentous brown algal viruses (Muller et al. 1990; Henry and Meints 1992; Muller and Stache 1992). Thereafter, use of genetic markers such as virus-encoded DNA polymerases, a core gene present in all NCLDVs (Iyer et al. 2001, 2006; Allen et al. 2006c), revealed that phycoviruses are a diverse and ubiquitous component of aquatic environments (Chen and Suttle 1995; Chen et al. 1996; Short and Suttle 2002). The field of phycodnavirology is now well established and

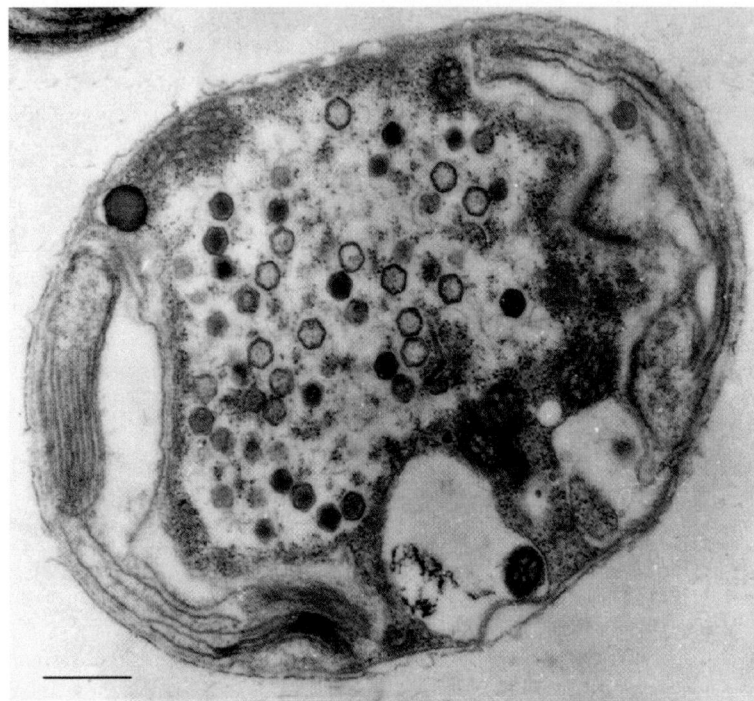

**Fig. 1** Final stages of infection in the marine phytoplankton *Pavlova virescens*. Note the different stages of virus assembly in the cell cytoplasm. Although never characterized, these VLPs have the hallmarks of a phycodnavirus. Samples were prepared by thin-sectioning in 1978! Image was given to the author when the investigator, John Green (formerly Marine Biological Association, Plymouth), was clearing out his office prior to retirement. No information was available on the size of the scale bar

expanding rapidly; indeed there have been four international meetings dedicated to the subject (Bergen, Norway 1998; Galway, Ireland 2000; Hiroshima, Japan 2002; and Amsterdam, Netherlands 2005). Previous algal virus reviews have focused primarily on the chloroviruses (Van Etten and Meints 1999; Van Etten et al. 2002, 2003; Kang et al. 2005; Dunigan et al. 2006; Yamada et al. 2006) because of the extensive research on these viruses. The ecological aspects of marine phytoplankton viruses were reviewed recently by Brussaard (2004). The current chapter will provide a broad overview of the *Phycodnaviridae*, attempting to describe the novelty and incredible genetic diversity of this ancient group of viruses. Starting with a brief introduction to the algal hosts these viruses infect, the review will then cover virus propagation strategies, genome structure, an analysis of known and novel genes, and finish with a discussion on core genes and their implication in *Phycodnaviridae* evolution.

# Hosts

Eukaryotic algae are a group of oxygen-evolving, photosynthetic organisms that include seaweeds (macroalgae) and a large diverse group of microorganisms generically referred to as microalgae. Photosynthetic prokaryotes such as pico-cyanobacteria (e.g. *Synechococcus* or *Prochlorococcus*) are not included in this group, however, they are infected by bacteriophage-like viruses (Suttle 2000a, 2000b; Clokie and Mann 2006). Eukaryotic algae range in size from the smallest known eukaryote, *Ostreococcus,* at approximately 1 μm in diameter (Derelle et al. 2006; Palenik et al. 2007), through numerous chain-forming and colonial species that are visible to the naked eye to the large kelp (seaweed) forests in coastal regions. Algae consist of at least five distinct evolutionary lineages (plants, cercozoa, alveolates, heterokonts and discicristates) (Baldauf 2003) and they are ubiquitous in marine, freshwater and terrestrial habitats. The number of algal species (mostly microalgae) has been estimated to be as high as several million; hence their overall diversity is probably enormous. It is likely that viruses infect all of these species; furthermore, not all of these viruses will be assigned to the *Phycodnaviridae* family. Indeed, other types of viruses that infect algae are being discovered and characterized (e.g., ssRNA, dsRNA, and ssDNA containing viruses, Tai et al. 2003; Brussaard et al. 2004a; Nagasaki et al. 2004). Thus, algal virology is in its infancy.

## *Chlorella and Chloroviruses*

The genus *Chlorella* consists of small, unicellular, non-motile, asexual green algae with a global distribution (Fig. 2A) (e.g., Shihra and Krauss 1965). They have a simple developmental cycle and reproduce by mitotic division. Vegetative cells increase in size and, depending on the species and environmental conditions, divide into two, four, eight or more progeny, which are released by rupture or enzymatic digestion of the parental walls. Chlorella species are usually free living, although many species have symbiotic relationships with organisms from different classes in the animal kingdom including *Rhizopoda*, *Ciliata*, *Hydrozoa* and *Turbellaria* (Reisser 1992). To date, the only described chloroviruses infect symbiotic chlorella, often referred to as zoochlorellae, such as those associated with the protozoan *Paramecium bursaria*, the coelenterate *Hydrozoa viridis* and the heliozoon *Ancanthocystis turfacea* (Kawakami and Kawakami 1978; Meints et al. 1981; Van Etten et al. 1982; Bubeck and Pfitzner 2005). Fortunately, many of these chlorella strains can be grown in the laboratory, independent of their symbiotic partner, including *P. bursaria*-associated chlorella isolates NC64A and Pbi. The chloro-viruses can be produced in large quantities and assayed by plaque formation. Chloroviruses are ubiquitous in nature and have been isolated from freshwater collected throughout the world (Yamata et al. 2006). Typically, the virus titers in

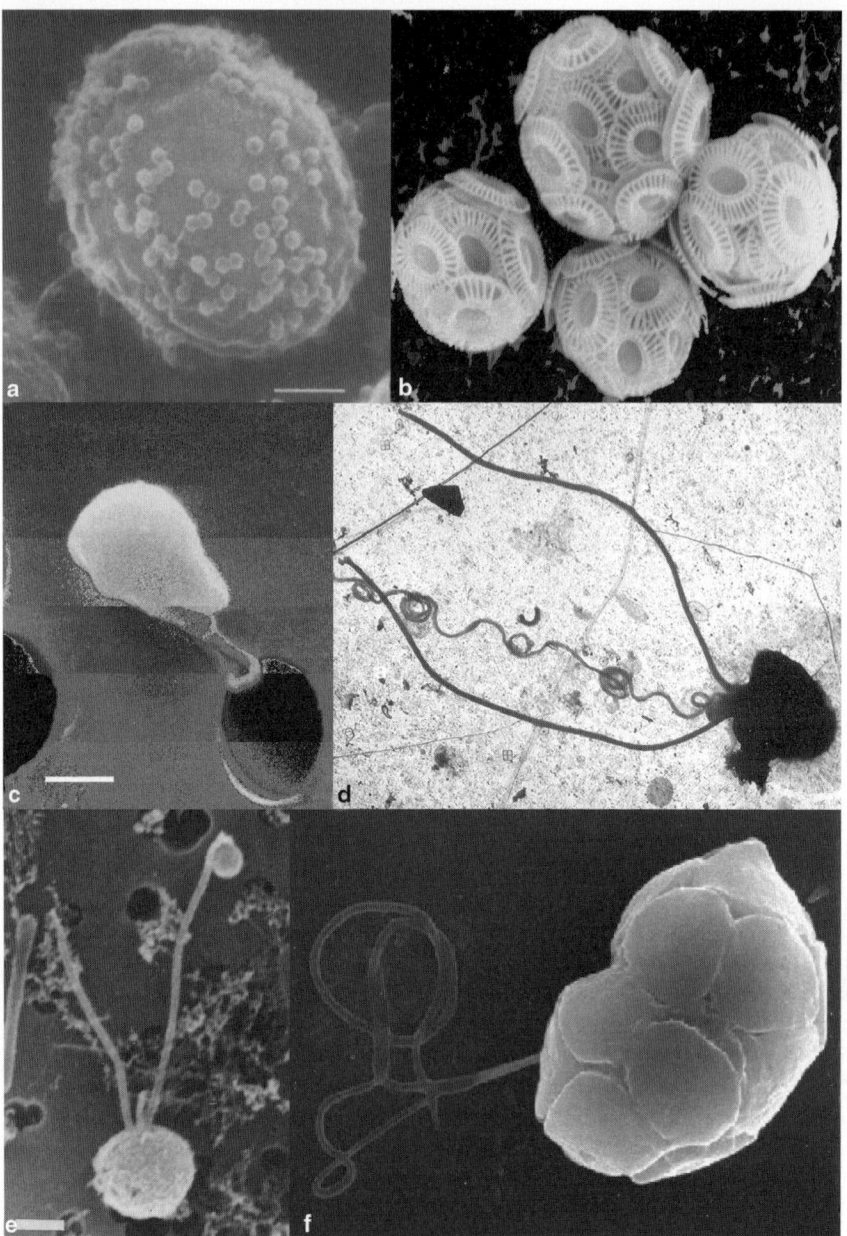

**Fig. 2a–f** Pictures of some representative algae that are hosts for phycodnaviruses. **a** Scanning electron micrograph (SEM) of *Chlorella* NC64A with PBCV-1 particles attached to its surface (scale bar, approx. 500 nm). **b** SEM of four *Emiliania huxleyi* cells (each cell, approx. 5 μm in diameter). **c** SEM *Micromonas pusilla* (scale bar, approx. 500 nm). **d** Transmission electron micrograph (TEM) of a *Chrysochromulina* sp. cell in a seawater sample taken from a Norwegian fjord (cell diameter, approx. 3 μm). **e** SEM of *Phaeocystis* sp. (strain Naples) (scale bar, approx. 1 μm). **f** SEM of *Heterosigma akashiwo* (cell diameter, approx. 12 μm). Images courtesy of (**a**) J.L. Van Etten (Meints et al. 1984); (**b**) W. Wilson (unpublished); (**c**) Bengt Karlson (unpublished) (**d**) G. Bratbak and M. Heldal (unpublished); (**e**) D. Vaulot (Vaulot et al. 1994); (**f**) S. Itakura and K. Nagasaki (unpublished)

native waters are 1–100 plaque-forming units (PFU) per ml, but titers as high as 100,000 PFU/ml of native water have been obtained. Titers fluctuate with the seasons, with the highest titers occurring in the spring. Chlorovirus PBCV-1, which infects *Chlorella* NC64A, is currently the best-studied phycodnavirus (Van Etten 2003; Yamada et al. 2006).

## Emiliania huxleyi *and Coccolithoviruses*

*Emiliania huxleyi* (Lohmann) Hay et Mohler is a representative of the coccolitho-phores, a group of unicellular calcifying marine microalgae that have a global distribution (Brown and Yoder 1994). In open oceanic environments, they consti-tute a significant fraction of the phytoplankton and have an exceptionally rich fossil record spanning approximately 200 million years (Bown 1998), with *E. huxleyi* appearing approximately 260,000 years ago. *E. huxleyi* is an important species with respect to past and present marine primary productivity, sediment formation and climate change. It is the most numerous coccolithophore in our oceans and satellite observations of mesoscale *E. huxleyi* blooms (Holligan et al. 1983, 1993) verified what marine geologists had long known, that calcite derived from these vast blooms is an important component of marine carbon. Ironically, the elaborate calcium carbonate armory of liths covering the surface of these cells was always thought to prevent virus infection (Fig. 2B).

*E. huxleyi* greatly impacts marine ecosystems and, in particular, the global carbon and sulphur cycles (Westbroek et al. 1993; Burkill et al. 2002). Blooms of this ubiqui-tous microalga are known to affect the oceanic carbon pump (Elderfield 2002) and climate (Charlson et al. 1987). Vast coastal and mid-ocean populations of *E. huxleyi* often suddenly disappear, causing substantial fluxes of calcite to the seabed (Ziveri et al. 2000) and cloud-forming dimethyl sulfide to the atmosphere (Malin 1997; Evans et al. 2007). Until recently, the mechanisms of *E. huxleyi* bloom disintegration were poorly understood but it is now commonly accepted that viruses are intrinsically linked to these sudden crashes (Bratbak et al. 1993, 1996; Brussaard et al. 1996; Castberg et al. 2002; Jacquet et al. 2002; Wilson et al. 2002; Schroeder et al. 2003).

## *Prasinophytes and Prasinoviruses*

Prasinophytes belong to a group of marine phytoplankton, generically referred to as photosynthetic picoeukaryotes (PPEs), which consist of small (>3 μm) photosyn-thetic eukaryotic cells. They are a ubiquitous, abundant and highly diverse component of the world's oceans. This abundance, coupled with their high carbon fixation rates (Li 1994; Worden et al. 2004), means that PPEs are important contributors to bio-mass production in the open ocean, despite usually being numerically outnumbered by photosynthetic prokaryotes (e.g., *Synechococcus* or *Prochlorococcus*). The incredibly high diversity of PPEs is only now being appreciated (Diez et al. 2001; Moon-van der Staay et al. 2001; Zeidner et al. 2003; Fuller et al. 2006; Worden 2006). Among the many classes included in the PPEs, members of the class Prasinophyceae are ubiquitous in clone libraries collected from oceanic samples and they often dominate the PPE community (Not et al. 2004; Worden 2006). These include members in the genera *Ostreococcus*, the smallest know eukaryote (Chrétiennot-Dinet et al. 1995), *Bathycoccus* (Eikrem and Throndsen 1990), and *Micromonas* (Fig. 2C), a major component in many oceanic and coastal regions (Not et al. 2005).

Despite the obvious ubiquity and abundance of these algae, relatively little research has been conducted on prasinoviruses and only a few viruses have been described. This is perhaps ironic, given that the first characterized virus assigned to the *Phycodnaviridae* was a prasinovirus that infects *Micromonas pusilla* (Mayer and Taylor 1979). *M. pusilla*-specific viruses are the only prasinoviruses that have been described. These viruses are easy to isolate, simply adding a small volume of filtered coastal seawater to an exponentially growing culture of *M. pusilla* usually results in virus-induced lysis. Research has focused largely on their ecological role and their genetic diversity (Cottrell and Suttle 1991, 1995b; Sahlsten 1998; Sahlsten and Karlson 1998; Zingone et al. 1999, 2006). *M. pusilla* viruses can lyse up to 25% of the daily *Micromonas* population (Evans et al. 2003); however, the high growth rates of the hosts, coupled with the high diversity of both host (Worden 2006) and virus (Chen et al. 1996), allows them to propagate in a stable co-existence (Cottrell and Suttle 1995a) as compared to the bloom-bust scenario observed with *Coccolithovirus* infection of *E. huxleyi* blooms (Bratbak et al. 1993; Jacquet et al. 2002; Wilson et al. 2002).

Except for a report of a transient *Ostreococcus* bloom off the eastern coast of the United States, whose rapid decline was partially attributed to viruses (O'Kelly et al. 2003), no other reports exit of viruses infecting prasinophytes. However, the authors are aware of an unpublished investigation on the isolation of prasinoviruses that infect the smallest known eukaryotic cell, *Ostreococcus tauri*. These dsDNA viruses are readily isolated by plaque assay and are morphologically similar to the *M. pusilla* viruses (K. Weynberg, N. Grimsley, H. Moreau, personal communication). Of the six *Phycodnaviridae* genera, the prasinoviruses and viruses of PPEs are the least represented in the literature despite their global and ecological importance.

## *Prymnesiophytes and Prymnesioviruses*

Algae assigned to the class Prymnesiophyceae, from the division Haptophyta (and generally referred to as haptophytes), occur predominantly in marine environments. These algae have a global distribution and they are often associated with large-scale microalgal blooms. The principle characteristic of the haptophytes is the haptonema, a filiform organelle or flagella whose length ranges from barely detectable (Green and Pienaar 1977) to many times that of the mother cell (Fig. 2D) (Parke et al. 1959). The presence of two-layered microfibrillar scales on the cell surface is also a characteristic and it is traditionally considered an important taxonomic character at the species level. In addition, the scales allow easy identification in mixed natural assemblages of microalgae (Leadbeater 1972). To date, prymnesioviruses have been isolated from members of the genera *Chrysochromulina* (Suttle and Chan 1995; Sandaa et al. 2001) and *Phaeocystis* (Jacobsen et al. 1996; Brussaard et al. 2004b; Wilson et al. 2006). One of the earliest reports of a phycodnavirus was that of a putative prymnesiovirus observed in thin sections of a *Chysochromulina* sp. (Manton and Leadbeater 1974).

*Chrysochromulina* (Fig. 2D) is considered a cosmopolitan genus; however, information on the distribution and abundance of *Chrysochromulina* at the species level is limited. Although abundance over an annual cycle is highly variable and generally low, high N:P ratios, stratified conditions with a low salinity surface layer during summer, together with low phosphate concentrations favor growth of *Chrysochromulina* spp. in coastal waters (Dahl et al. 2005). On the rare occasions when they do bloom, they can produce toxins that cause catastrophic damage to fisheries: for example, in 1988 a large *C. polylepis* bloom (~100,000/ml) off southern Norway killed large numbers of both wild and caged fish (Dundas et al. 1989). However, large monospecific blooms of *Chrysochromulina* spp. are rare; typically they are present in low concentrations (Estep and MacIntyre 1989). This property has led to speculation that *Chrysochromulina* spp. infecting viruses only require a low host density for propagation (Sandaa et al. 2001) and that the viruses may even prevent bloom formation (Suttle and Chan 1995).

In contrast, a lot of information exists on the biogeochemical impact on the marine ecosystem by members of the genus *Phaeocystis* (Fig. 2E); these interdisciplinary studies are summarized in several recent reviews (Veldhuis and Wassmann 2005; Verity et al. 2007). *Phaeocystis* spp. form dense spatially and temporally extensive monospecific blooms consisting of a mixture of colonial (within a gelatinous matrix) and unicellular cells that collapse suddenly in a virus-induced crash (Brussaard et al. 2005a). This crash leads to a rapid shift in the composition of the bacterial community due to the massive flux of released organic nutrients (Brussaard et al. 2005b). Intense *Phaeocystis* spp. blooms can lead to anoxia and impressive foam formation on beaches during their decline (Lancelot et al. 1987), hence their label as harmful algal blooms (HABs). Similar to *E. huxleyi*, *Phaeocystis* spp. play important roles in $CO_2$ and sulphur cycling, which ultimately have major implications in the global climate (Liss et al. 1994) and infection by viruses can exacerbate this process (Malin et al. 1998).

## *Phaeophyceae and Phaeoviruses*

The class Phaeophyceae is more commonly referred to as the brown algae. They are an important economic resource that includes uses in the cosmetic, food and fertilizer industries as well as a source of important biomolecules (McHugh 1991). As multicellular, filamentous macroalgae, they differ from the other unicellular, microalgal hosts of the *Phycodnaviridae*. Another major difference is that during infection, the virus genomes are integrated into the host genome; the virus genomes are then inherited in a Mendelian fashion and virus particles only appear in the reproductive organs of the infected algae (Müller et al. 1998). Phaeophyceae predominantly inhabit temperate near-shore marine coastal environments around the world. They can be significant components of fouling communities that occur on marine structures such as docks, buoys and the exterior surfaces of ships. Viruses are known to infect members in the genera *Ectocarpus*, *Feldmannia*, *Hincksia*, *Myriotrichia*, and *Pilayella*. The life cycle of the filamentous brown alga, *Ectocarpus* (Fig. 3) is perhaps the best studied and it serves as a model organism for studying the molecular genetics of brown algae (Peters et al. 2004). *Ectocarpus* has a complicated sexual life cycle comprising alternating generations. Male and female gametophytes carry plurilocular zoidangia, which produce swimming gametes. The viruses only infect these free-swimming gametes. Gamete fusion

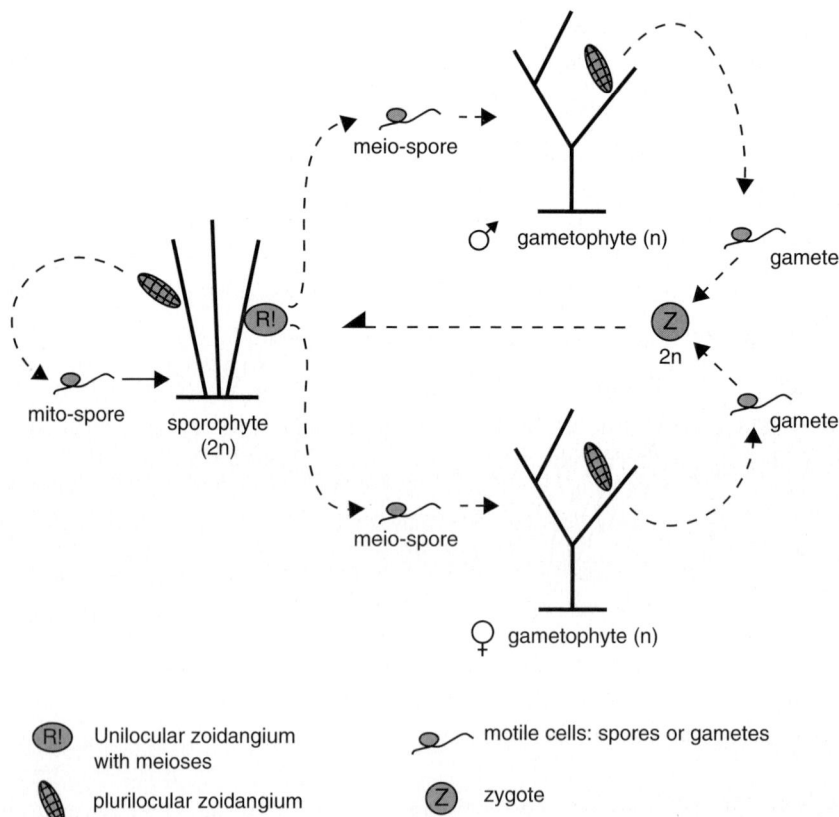

**Fig. 3** Life cycle of *Ectocarpus siliculosus*. Zoidangium: sporangium on sporophytes, gametangium on gametophytes. Note that the sporophyte possesses two types of reproductive organs: (*1*) unilocular sporangia where meioses occur (*R'*) producing taploid (*n*) meiospores, and (*2*) plurilocular sporangia where diploid (2n) mitospores are produced by multiple mitoses. (Reproduced with permission from Müller DG et al. 1998)

results in zygotes that develop into diploid sporophytes. The sporophytes produce unilocular sporangia where meiosis leads to the formation of haploid spores. These spores go on to produce the next gametophyte generation. The sporophytes can also develop plurilocular sporangia by mitotic cell division. The diploid spores can therefore repeat the diploid sporophyte phase (Müller et al. 1998). Virus particles are only formed in gametangia or sporangia (zoidangia) cells of the host (Müller 1996). Since viruses displace the normal reproductive organs, the algae are rendered sterile; however, this does not affect their vegetative growth (Müller et al. 1990).

## *Raphidophytes and Raphidoviruses*

Microalgae in the class Raphidophyceae are important bloom-forming species found in coastal and subartic regions of the world's oceans, although freshwater species also exist (Figueroa and Rengefors 2006). Members are often associated with toxic red tides (Khan et al. 1997) and subsequent fish kills, particularly in the aquaculture industry (Hiroishi et al. 2005). Classical identification of the Raphidophytes relies on morphological differences, but it is extremely difficult to distinguish some of the species due to a pleomorphology that changes shapes and sizes in various environmental conditions (Khan et al. 1995). Consequently, efforts to develop rapid and sensitive polymerase chain reaction (PCR)-based diagnostic techniques are ongoing (Kai et al. 2006). To date, the only viruses reported in this genus infect the red tide forming species *Heterosigma akashiwo* (Fig. 2F); the viruses are referred to as HaV (*Heterosigma akashiwo* virus) (Nagasaki and Yamaguchi 1997, 1998a; Nagasaki et al. 1999b). *H. akashiwo* is a single species belonging to the genus *Heterosigma* (Throndsen 1996) and, although not usually associated with human illness, blooms of *H. akashiwo* have caused massive fish kills, typically of caged fish such as salmon and yellowtail (Honjo 1993; Khan et al. 1997). The susceptibility of *H. akashiwo* to HaV differs among clonal strains in the laboratory (Nagasaki and Yamaguchi 1998b; Nagasaki et al. 1999a; Tarutani et al. 2006). Marine field surveys and cross-reactivity tests between *H. akashiwo* host strains and HaV clones suggest that this strain-dependent infection plays an important role in determining clonal composition and effectively maintaining intraspecies diversity in natural *H. akashiwo* populations (Tarutani et al. 2000; Tomaru et al. 2004). This diversity probably contributes to the success of *H. akashiwo* as a ubiquitous and problematic bloom former in coastal regions.

## Virion Morphology/Structure

Despite the wide host range displayed by the *Phycodnaviridae*, all members have similar structural morphology; this is consistent (independent of genomic information) with a common ancestry. Few have been studied in great detail, but the overlying pattern suggests a general conservation of structure. The lifestyle of viruses, i.e., they enter a cell, hijack the cell metabolic machinery and make hundreds to thousands of new viruses, creates a vital need for a simple and easily assembled virion particle, hence the conservation of structure (Bamford et al. 2002). The virions are large (to contain the large genomes) layered structures 100–220 nm in diameter with a dsDNA-protein core surrounded by a lipid bilayer and an icosahedral capsid (Fig. 4) (Nandhagopal et al. 2002). The capsids have 20 equilateral triangle faces

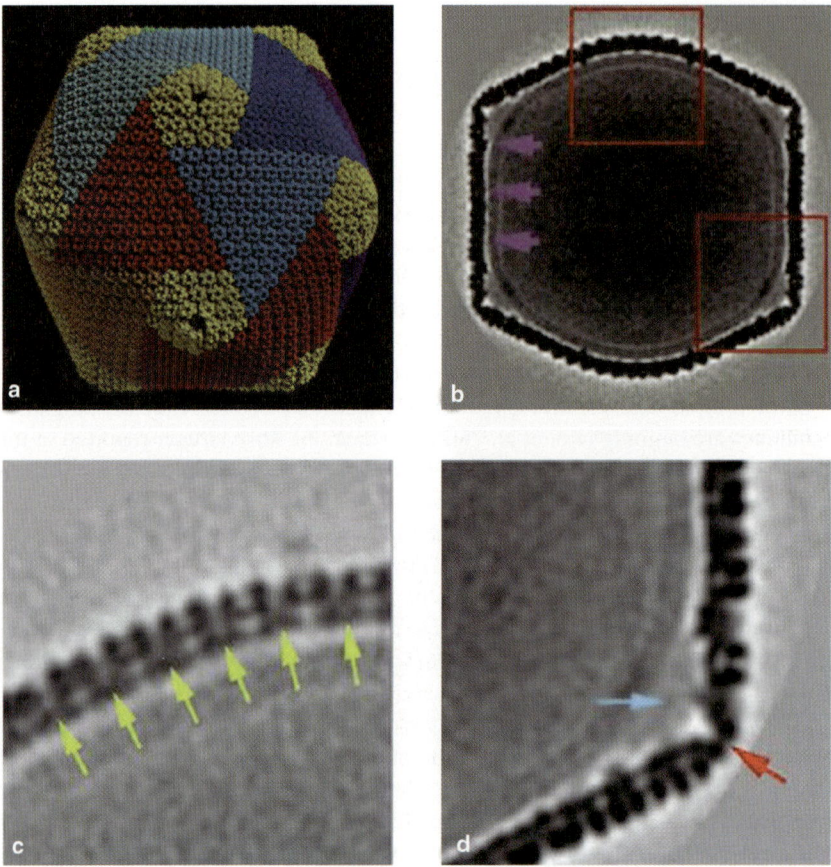

**Fig. 4a–d** Three-dimensional image reconstruction of chlorella virus PBCV-1 from cryoelectron micrographs. **a** The virion capsid consists of 12 pentasymmetrons (highlighted in *yellow*) and 20 trisymmetrons (highlighted in *red* and *two shades of blue*). A pentavalent capsomer lies at the center of each pentasymmetron. Each pentasymmetron consists of one pentamer plus 30 trimers. Eleven capsomers form the edge of each trisymmetron and therefore each trisymmetron has 66 trimers. **b** Cross-sectional view of PBCV-1 along the twofold axis. A lipid bilayered membrane, like a railroad track, exists beneath the capsid shell (*magenta arrows*). Magnified views at two- and fivefold axes (outlines in **b** are shown in **c** and **d**, respectively). **c** The capsomers are interconnected by "cement" proteins (*yellow arrows*). **d** Dense material (*blue arrow*) (cell wall-digesting enzyme(s)?) is present at each vertex (*red arrow*) between the vertex and the membrane. (**a** reproduced with permission from Simpson et al. 2003. **b–d** reproduced with permission from Van Etten 2003)

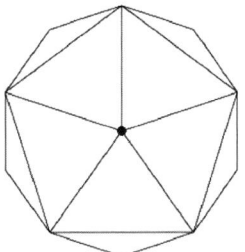 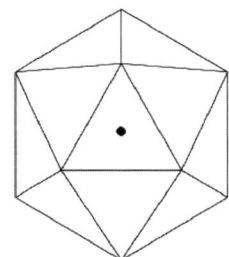 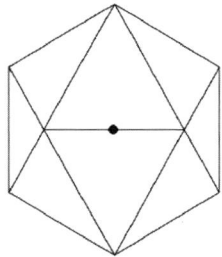

**Fig. 5** The *Phycodnaviridae* capsids have 20 equilateral triangle faces composed of protein subunits and are defined by having a two-, three- and fivefold axis of symmetry. From left to right, there are six fivefold axes of symmetry passing through the vertices, ten threefold axes extending through each face and 15 twofold axes passing through the edges of an icosahedron. (Diagram produced by Thomas Locke)

composed of protein subunits and are defined by having two-, three- and fivefold axis of symmetry (Crick and Watson 1956). There are six fivefold axes of symmetry passing through the vertices, ten threefold axes extending through each face and fifteen twofold axes passing through the edges (Fig. 5). Like other icosahedron virions, the capsids are composed of well-ordered capsid substructures known as trisymmetrons (triangular) and pentasymmetrons (pentagonal) which are centred on the three- and fivefold axes of symmetry (Wrigley 1969; Stoltz 1971). The trisymmetrons do not represent the triangular faces of the icosahedron, but instead curve around the edges allowing the pentasymmetrons to fill the gaps at the fivefold vertices (Fig. 4) (Nandhagopal et al. 2002; Simpson et al. 2003). In all of the *Phycodnaviridae* studied so far, the capsid is composed of 20 trisymmetrons and 12 pentasymmetrons (Yan et al. 2000, 2005). Trisymmetrons and pentasymmetrons are made up of trimeric capsomers, with each doughnut-shaped capsomer composed of three monomers of the major capsid protein.

Most of these data are based on PBCV-1. Cryoelectron microscopy and three-dimensional (3D) image reconstruction of the PBCV-1 virion at 26-Å resolution indicate that the outer capsid is icosahedral and covers a lipid bilayer membrane (Fig. 4A, B) (Yan et al. 2000; Simpson et al. 2003). The membrane is required for infection, because PBCV-1 loses infectivity after exposure to organic solvents. The outer diameter of the viral capsid ranges from 1,650 Å along the two- and threefold axes to 1,900 Å along the fivefold axis. The capsid shell consists of 1,680 donut-shaped trimeric capsomers plus 12 pentameric capsomers at each icosahedral vertex. The trimeric capsomers are arranged into 20 triangular facets (trisymmerons, each containing 66 trimers) and 12 pentagonal facets (pentasymmetrons, each containing 30 trimers and one pentamer at the icosahedral vertices). Assuming all the trimeric capsomers are identical, the outer capsid of the virus contains 5,040 copies of the major capsid protein, Vp54. PBCV-1 has a triangulation number (T number) of 169.

Most of the PBCV-1 trimeric capsomers have a central, concave depression surrounded by three protruding towers. The trimeric capsomers are 72 Å in diameter and 75 Å high. The prominent, cylinder portion of each trimer extends 50 Å above the surface of the capsid shell. The capsomers interconnect at their bases in a contiguous shell that is 20–25 Å thick. Within each trimer, the three monomers connect in the middle; thus each trimer has the appearance of the letter H when viewed in cross-section. This structure agrees with the Vp54 structural studies (Nandhagopal et al. 2002). Cross-sectional views of the reconstructions reveal a "cement" protein(s) that connects the bases of the trimeric capsomers (arrow in Fig. 4C).

Twelve pentamer capsomers, each 70 Å in diameter, exist at the virus fivefold vertices and probably consist of a different protein. Each pentamer is surrounded by radially distributed trimers, and the axis of each of these trimers tilts away from the pentamer. Each pentamer has a cone-shaped, axial channel at its base. One or more proteins appear below the axial channel and outside the inner membrane (Fig. 4D). This protein(s) may be responsible for digesting the host cell wall during infection. Presumably contact between the virus and its host receptor alters the channel sufficiently to release the wall-degrading enzyme(s).

The PBCV-1 virion contains more than 110 different virus-encoded proteins (D.D. Dunigan et al., unpublished results). The PBCV-1 54-kDa major capsid protein (Vp54) is a glycoprotein and comprises approximately 40% of the total virus protein. Vp54 consists of two eight-stranded, antiparallel β-barrel, jelly-roll domains related by pseudo-sixfold rotation (Nandhagopal et al. 2002). The glycan portion of Vp54, which is oriented to the outside of the particle, contains seven neutral sugars, glucose, fucose, rhamnose, galactose, mannose, xylose, and arabinose (Wang et al. 1993). The Vp54 crystal structure revealed six glycans attached to the protein, four N-linked glycans and two O-linked glycans (Nandhagopal et al. 2002). However, the four glycosylated Asn residues are not located in consensus sequences typical of eukaryotic N-linked glycans. This finding, together with other results (Wang et al. 1993; Que et al. 1994; Graves et al. 2001), led to the conclusion that, unlike other viruses, PBCV-1 encodes at least part, if not all, of the machinery required to glycosylate its major capsid protein (Markine-Goriaynoff et al. 2004). Furthermore, glycosylation probably occurs independently of the host endoplasmic reticulum-Golgi system.

The Vp54 structure resembles the major coat proteins from some other, smaller dsDNA viruses representing all three domains of life, including bacteriophage PRD1, human adenoviruses, and a virus infecting the Archaea, *Sulfolobus solfataricus*. This finding led to the suggestion that these three viruses also have a common evolutionary ancestor with the NCLDVs, even though there is no amino acid sequence similarity among their major capsid proteins (Bamford et al. 2002; Benson et al. 2004).

The Prymnesiovirus *Phaeocystis pouchetii* virus (PpV01) has the largest capsid of the *Phycodnaviridae* and has a maximum outer diameter of 2200 Å between opposed vertices (fivefold axes), which drops to 1900 Å along the two- and three-fold axes (Yan et al. 2005). Like PBCV-1, the pentasymmetrons are composed of 31 capsomers (30 trimers and 1 pentamer); however, the trisymmetrons are larger, containing 91 trimers (13 per ~100 nm edge). PpV01 has a T number of 219.

The nature and origin of the lipid bilayer membrane remains relatively unstudied in the phycodnaviruses, although experiments in progress indicate that the PBCV-1 membrane probably originates from the endoplasmic reticulum (J. Heuser et al., unpublished results). A lipid profile of virus EhV-86 is very similar to that of its host *E. huxleyi*, suggesting the direct requisition of host membrane lipids during virion assembly (T. Dunn, personal communication).

Over the last decade we have been increasingly aware that the *Phycodnaviridae* is a highly diverse family whose members share very little else in common other than the nature of their genomic material and their morphology. The following sections discuss this diversity with regards to lifestyle, genomic structure, coding and noncoding potential and infection strategy.

## Propagation Strategies

Members of the *Phycodnaviridae* have a variety of lifestyles. The replication strategies of three of these viruses will be briefly described in this section, PBCV-1 representing the chloroviruses, EsV-1 representing the phaeoviruses and EhV-86 representing the coccolithoviruses.

PBCV-1 initiates infection by attaching rapidly, specifically and irreversibly to the chlorella cell wall (Meints et al. 1984, 1988), probably at a unique virus vertex (Onimatsu et al. 2006); attachment is immediately followed by degradation of the host wall at the point of contact by a virus-packaged enzyme(s) (Fig. 6). The chloroviruses encode several proteins involved in polysaccharide degradation that may be involved in degrading the cell wall (see Sect. 6.3.1). Following host cell wall degradation, the viral internal membrane presumably fuses with the host membrane, resulting in entry of the viral DNA and virion-associated proteins into the cell, leaving an empty virus capsid attached to the cell wall. This process triggers a rapid depolarization of the host membrane (probably triggered by a virus encoded potassium channel located in the virus internal membrane) (Frohns et al. 2006) and the rapid release of potassium ions from the cell (Neupartl et al. 2008); this depolarization may function to prevent infection by a second virus (Mehmel et al. 2003; Frohns et al. 2006). The rapid loss of potassium ions and associated water fluxes from the host also lowers its turgor pressure and this lowering is postulated to aid ejection of DNA from the virions into the host (Neupartl et al. 2008).

Circumstantial evidence indicates that the viral DNA and probably DNA-associated proteins quickly move to the nucleus where early transcription is detected within 5–10 min p.i. (Schuster et al. 1986). Unlike EhV-86 early gene expression (Allen et al. 2006a), the PBCV-1 early expressed genes are scattered throughout the virus genome. Shortly after infection, host chromosomal DNA begins to be degraded (Agarkova et al. 2006), presumably to aid in inhibiting host transcription and to provide the recycling of nucleotides for viral DNA replication. In this immediate-early phase of infection, the host is reprogrammed to transcribe viral RNAs. Very little is known as to how this occurs, but chromatin remodeling may be involved. PBCV-1 encodes a 119 amino acid SET domain containing protein (referred to as vSET) that di-methylates Lys27 in histone 3 (Manzur et al. 2003). vSET is packaged in the PBCV-1 virion and accumulating evidence indicates that vSET is involved in repression of host transcription following PBCV-1 infection (S. Mujtaba et al., unpublished data).

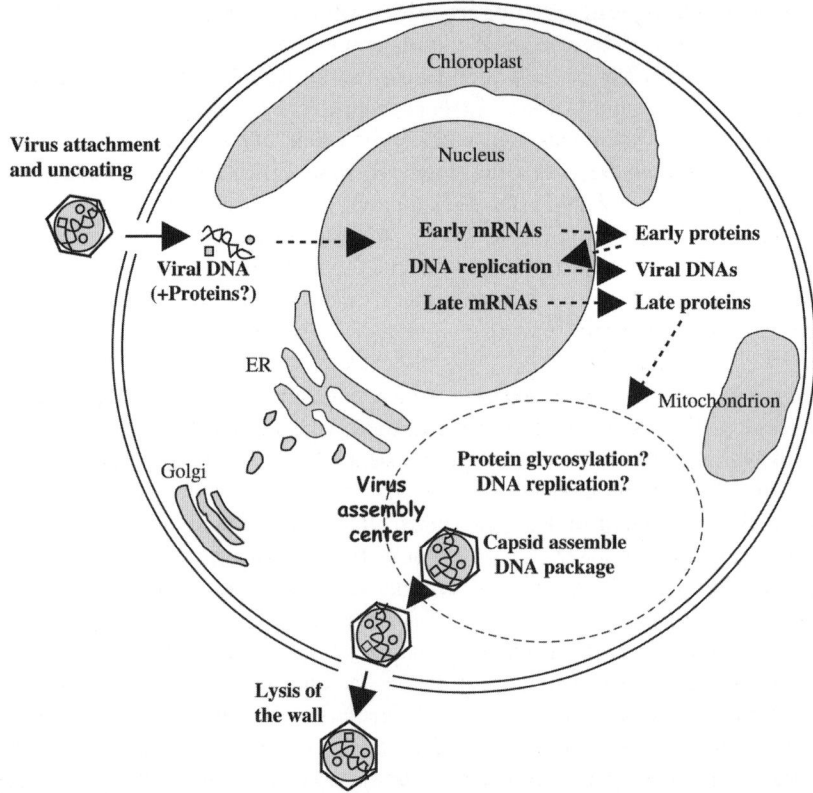

**Fig. 6** Proposed replication cycle of chlorella virus PBCV-1. The virus uncoats at the surface of the alga and the viral DNA, possibly with associated proteins, is assumed to move to the nucleus where early gene transcription begins within 5–10 min postinfection (p.i.) The early mRNAs are transported to the cytoplasm for translation, and the early proteins presumably return to the nucleus to initiate DNA replication, which begins 60–90 min p.i., followed by late gene transcription. Late mRNAs are transported to the cytoplasm for translation, and many of these late proteins are targeted to the cytoplasmically located virus assembly centers, where virus capsids are formed. The algal cell membrane and wall lyses and infectious PBCV-1 progeny viruses are released at 6–8 h p.i. *Solid arrows* are known events; *dotted arrows* are hypothesized events. (Reproduced with permission from Kang et al. 2005)

Viral DNA replication begins 60–90 min after infection (Van Etten et al. 1984) and is followed by transcription of late genes (Schuster et al. 1986). Ultrastructural studies of PBCV-1-infected chlorella suggest that the nuclear membrane remains intact, at least during the early stages of virus replication (Meints et al. 1986). However, a functional host nucleus is not required for virus replication since PBCV-1 can replicate, albeit poorly and with a small burst size, in UV-irradiated cells (Van Etten et al. 1986). Approximately 2–3 h postinfection (p.i.), assembly of virus capsids begins in localized regions in the cytoplasm, called virus assembly centers, which become prominent at 3–4 h p.i. (Meints et al. 1986). By 5– 6 h p.i., the cytoplasm becomes filled with infectious progeny virus particles and by 6–8 h p.i., localized lysis of the host cell releases progeny (Van Etten et al. 1983b). Approximately 1,000 particles are released per cell, of which approximately 30% are infectious.

A different lifestyle occurs in the phaeoviruses: in contrast to the lytic lifestyle of the chloroviruses, the phaeoviruses are lysogenic (Müller et al. 1998; Van Etten et al. 2002). Viruses EsV-1 and EfasV-1 only infect free-swimming, wall-less gametes of their filamentous hosts *Ectocarpus siliculousus* and *Ectocarpus fasciculatus*, respectively. Following virion attachment, the viral genomic DNA and associated protein material is released into the cytoplasm and then immediately moves to the nucleus within 5 min p.i. (Maier et al. 2002). The genomic material is then integrated into the host genome (presumably using a virally encoded integrase like the one in virus EsV-1; R.H. Meints et al., unpublished data). The viral genome is transmitted mitotically to all the cells in the developing host thallus (Brautigam et al. 1995; Delaroque et al. 1999; Müller et al. 1991). The virus genome remains latent in the vegetative filamentous cells until the formation of algal reproductive cells, either sporangia or gametangia, in the mature host. This differentiation triggers virus DNA replication in the sporangia or gametangia, followed by breakdown of the nuclear membrane, virion assembly and release of infectious virions from the cells (Wolf et al. 1998, 2000; Müller et al. 1998).

The coccolithoviruses have a different propagation strategy than either the lytic PBCV-1 or the latent EsV-1 systems, which may be similar to the lifestyle of the ancestral phycodnavirus (Allen et al. 2006c). The host alga *E. huxleyi* is covered with a calcium carbonate shell (Fig. 2B) that would appear to create a physical barrier to virus adsorption. However, despite this barrier, virus adsorption is rapid and appears to be intrinsically linked to the cell cycle of the host (W.H. Wilson, unpublished data). Whereas both PBCV-1 and EsV-1 are completely dependent on the host transcriptional machinery, the coccolithoviruses are unique among the phycodnaviruses in that they have several RNA polymerase encoding genes (Wilson et al. 2005a). These genes suggest a viral replication strategy that could be either entirely or partially independent of the host nucleus. Viral transcription begins immediately after infection, but it is limited to a distinct and localized 100-kb region in the virus genome; this region contains a unique promoter element (Allen et al. 2006a, 2006d). The only genes transcribed during the 1st h after infection have this element. None of these genes match anything in the databases and are referred to as ORFans (Fischer and Eisenberg 1999). Intriguingly, proteomic analysis of EhV-86 virions did not detect any transcriptional machinery packaged in mature virions, therefore a host nuclear RNA polymerase(s) is presumably responsible for this early transcription (M.J. Allen, unpublished data). Between 1 and 2 h p.i., a second transcriptional phase begins with gene expression occurring from the remainder of the genome. Since viral RNA polymerase components are expressed in this second phase, viral replication may no longer be nuclear dependent and transcription may move to the cytoplasm. Unlike the other phycodnavirus systems, where nascent virions accumulate in the cytoplasm prior to release by cell lysis, the coccolithovirus virions are released gradually in a controlled fashion. One possibility is that the virus utilizes the ready-made lith export pathway to maintain this controlled and orderly release. It has also been suggested that the coccolithoviruses use the virally encoded sphingolipid metabolism pathway to manipulate the sphingolipids in the host cell wall or membrane, possibly to prevent programmed cell death and to prolong the length of infection (Han et al. 2006).

## Genome Structure and Properties

By definition, all phycodnaviruses have large dsDNA genomes (Wilson et al. 2005b). These genomes range in size from 100 kb to over 550 kb with G+C contents ranging from 40% to 50% (Van Etten and Meints 1999; Dunigan et al. 2006). Complete genome sequences are available for members in three *Phycodnaviridae* genera: six chloroviruses, PBCV-1, NY-2A, AR158, MT325, FR483 and ATCV-1 (Li et al. 1997; Fitzgerald et al. 2007a, 2007b, 2007c), two phaeoviruses EsV-1 and FirrV (Delaroque et al. 2001, 2003) and the coccolithovirus EhV-86 (Wilson et al. 2005a; Allen et al. 2006b). Partial sequence (approximately 80%) is available for a second coccolithovirus, EhV-163 (Allen et al 2006b). Additional phycodnavirus genomes are being sequenced, but are not yet publicly available.

There is considerable variation in genome structure among the phycodnaviruses. The PBCV-1 genome is a linear 330-kb, nonpermuted dsDNA molecule with covalently closed hairpin termini (Rohozinski et al. 1989). The termini consist of 35 nucleotide-long covalently closed hairpin loops that exist in one of two forms; the two forms are complementary when the 35-nucleotide sequences are inverted (flip-flop) (Zhang et al. 1994). Identical 2221-bp inverted repeats are adjacent to each hairpin end (Strasser et al. 1991). The remainder of the PBCV-1 genome contains primarily single-copy DNA. EsV-1 has a linear dsDNA genome with almost perfect inverted repeats at each end allowing for circularization of the genome (Delaroque et al. 2001). Indeed, prior to sequencing, EsV-1 was thought to have a circular genome (Lanka et al. 1993). It is proposed that the inverted repeats anneal with each other to form a cruciform structure that effectively circularizes the genome (Delaroque et al. 2001).

EhV-86 was originally proposed to have a linear genome (Schroeder et al. 2002). However, PCR amplification over the termini revealed a random A/T single nucleotide overhang (50% A, 50% T), suggesting the virus genome has both linear and circular phases (Wilson et al. 2005a; Allen et al. 2006a, 2006d). The detection of a DNA ligase and four endonucleases in EhV-86 hints that a linear genome may be packaged in the virions that circularize during DNA replication (Allen et al. 2006d).

Viruses typically have compact genomes for replication efficiency and the phycodnaviruses fit this pattern with approximately one gene per 900 to 1000 bp of genomic sequence (based on the predicted 366, 404, 360, 331, 335, 329, 233 and 473 coding sequences (CDS) of PBCV-1, NY-2A, AR158, MT325, FR483, ATCV-1 and EhV-86 (Table 1) (Li et al. 1997; Wilson et al. 2005a; Allen et al. 2006a; Fitzgerald et al. 2007a, 2007b, 2007c). One exception is the 336-kb EsV-1 genome that only has 231 protein encoding genes, which gives it one gene per roughly

**Table 1** Genome data of sequenced phycodnaviruses

| Genus | Virus | Host | Genome size (bp) | %G+C content | ORFs | tRNAs | Nts/gene |
|-------|-------|------|------------------|--------------|------|-------|----------|
| *Chlorovirus* | PBCV–1 | *Chlorella* NC64A | 330,743[a] | 40.0 | 366 | 11 | 904 |
| | NY–2A | *Chlorella* NC64A | 368,683[a] | 40.7 | 404 | 7 | 913 |
| | AR158 | *Chlorella* NC64A | 344,690[a] | 40.8 | 360 | 6 | 957 |
| | MT325 | *Chlorella* Pbi | 314,335[a] | 45.3 | 331 | 10 | 950 |
| | FR483 | *Chlorella* Pbi | 321,240[a] | 44.6 | 335 | 9 | 959 |
| | ATCV–1 | *Chlorella* SAG 3.83 | 288,047[a] | 49.4 | 329 | 11 | 875 |
| *Phaeovirus* | EsV–1 | *Ectocarpus siliculousus* | 335,593 | 51.7 | 231 | 0 | 1453 |
| | FirrV–1 | *Feldmannia irregularis* | 191,667[b] | | 156 | 0 | 1229 |
| *Coccolithovirus* | EhV–86 | *Emiliania huxleyi* | 407,339 | 40.2 | 472 | 5 | 863 |

[a] Does not include the nts at the hairpin ends
[b] Is not one continuous contig

1450 bp. The 366 PBCV-1 protein-encoding genes are evenly distributed on both strands and, with one exception, intergenic space is minimal. In fact, 275 ORFs are separated by fewer than 100 nucleotides. The exception is a 1788-nucleotide sequence near the middle of the genome. This DNA region, which contains many stop codons in all reading frames, encodes 11 tRNA genes.

However, some repetitive DNA occurs in the PBCV-1, EsV-1 and EhV-86 genomes. Both EsV-1 and PBCV-1 contain approximately 2-kb inverted repeats adjacent to the terminal ends (Strasser et al. 1991; Delaroque et al. 2001). In addition to the terminal repeats, tandem repeats are located throughout the EsV-1 genome and comprise approximately 12% of the total genome size (Delaroque et al. 2001). A similar proportion of the *Feldmannia* sp. virus (FsV) genome also consists of repetitive DNA (Lee et al. 1995). Intriguingly, EhV-86 has three repeat families (none of which are located at the ends of the genome); one family is postulated to act as an origin of replication (adding credence to the circular mode of replication model), another family is postulated to contain immediate early promoter elements and the last family has a large repetitive proline rich domain that may bind calcium (Allen et al. 2006d). The repetitive regions in these *Phycodnaviridae* genomes, while hindering sequencing projects, may play a role in recombination between viruses that allows genetic information to be exchanged with themselves and with their hosts (Allen et al. 2006b, 2006c).

Limited sequence information is currently available for the phycodnavirus hosts, although this situation is changing. The *Chlorella* NC64A and *E. huxleyi* genomes are currently being sequenced by the Joint Genome Institute and drafts of the two genomes will soon be available. There is evidence that both *E. huxleyi* and EhV-86 share a putative phosphate permease gene, the large degree of difference between the genes suggesting this transfer was not recent; however, the best and only match of the virus was from the host (Wilson et al. 2005a). Host *E. huxleyi* also appears to contain numerous G+C-rich repetitive elements (B. Read, personal communication) similar to the repetitive elements in EhV-86 (Allen et al. 2006d); therefore the virus may have acquired these elements from its host.

The origin of the chlorella virus genes is a mystery. Microarrays that contain all 366 PBCV-1 genes are available and only two of the PBCV-1 genes hybridize with host chlorella DNA (G. Yanai-Balser et al., unpublished data). Furthermore, the G+C content of PBCV-1 and *Chlorella* NC64A differ significantly, 40% versus 67% (Van Etten et al. 1985). Therefore, it seems unlikely that PBCV-1 recently acquired its genes from this host.

Some of the phycodnavirus genomes have methylated bases. For example, genomes from 37 chlorella viruses contain 5-methylcytosine (5mC) in amounts varying from 0.12% to 47.5% of the total cytosine. In addition, 24 of the 37 viral DNAs contain N6-methyladenine (6 mA) in amounts varying from 1.5% to 37% of the total adenine (Van Etten et al. 1991). The methylated bases occur in specific DNA sequences; thus it is not surprising that the chlorella viruses encode multiple 5mC and 6 mA DNA methyltransferases. However, it was a surprise to discover that approximately 25% of the virus-encoded DNA methyltransferases have a companion DNA site-specific (restriction) endonuclease, some which have unique specificities (e.g., Xia et al. 1986, 1987; Nelson et al. 1998; Van Etten et al. 1991). Some of the virus encoded site-specific endonucleases are unusual in that they only cleave one strand of dsDNA, for example virus NYs-1 nickase cleaves /CC sites (Xia et al. 1988; Chan et al. 2004) and NY-2A nickase cleaves R/AG sites (Zhang et al. 1998; Chan et al. 2006). Thus, the virus-infected chlorellae are the first nonprokaryotic source of DNA site-specific endonucleases.

The phaeovirus EsV-1 genome also has low levels of methylated bases (1% and 3% of the total cytosines and adenines, respectively, are methylated) (Lanka et al. 1993; Delaroque et al. 2001). However, the EsV-1 DNA methyltransferases do not have companion site-specific endonucleases.

## Coding Potential

### *tRNAs, Introns, and Inteins*

In general, CDSs tend to be evenly distributed on both strands with minimal intergenic spaces. Exceptions to this rule include a 1788-nucleotide sequence in the middle of the PBCV-1 genome that contains a polycistronic gene encoding 11 tRNA genes (co-transcribed as a large precursor and then processed to mature tRNAs) (Dunigan et al. 2006). All the sequenced chloroviruses have a similar pattern, although the nature of the tRNAs and their genomic locations vary among the viruses (ATCV-1, 11 tRNAs, two duplicated; MT325, ten tRNAs, one duplicated;

FR483, nine tRNAs, one duplicated; NY-2A, seven tRNAs, one duplicated; AR158, six tRNAs, one duplicated) (Fitzgerald et al. 2007a, 2007b, 2007c). The coccolithovirus, EhV-86, encodes five tRNAs, four of which are clustered together (Wilson et al. 2005a). A second EhV strain (EhV-99B1) contains a sixth tRNA and also encodes a putative isoleucyl tRNA synthetase to complement the Ilu-tRNA (A. Lanzen, personal communication). In contrast, the phaeovirus EsV-1 does not have any tRNA encoding genes (Delaroque et al. 2001).

Codon usage in viral proteins is often correlated with the abundance and nature of virus-encoded tRNAs (Kropinski and Sibbald 1999). However, it is interesting that there is such a large variation in the number and type of tRNAs contained within these phycodnavirus genomes. Not only does the number of tRNAs in closely related viruses (often infecting the same host) differ, but the nature of the tRNAs also vary. For example, only three tRNAs are conserved among the six chloroviruses that have been sequenced (Fitzgerald et al. 2007c).

In addition to a 13-nt intron in a PBCV-1 Tyr-tRNA gene, PBCV-1 contains two other types of introns: a self-splicing intron in a transcription factor TFIIS-like gene and a spliceosomal-processed intron in its DNA polymerase gene. Chloroviruses MT325 and FR483 lack both of these introns (Fitzgerald et al. 2007a). A spliceosomal-processed intron was also identified in a pyrimidine dimer-specific glycosylase gene in 15 chloroviruses including NY-2A and AR158 (Sun et al. 2000; Fitzgerald et al. 2007b). An intron is also present at an identical location in the ATCV-1 pyrimidine dimer-specific glycosylase gene, but this intron has little sequence identity to those from viruses NY-2A and AR158 (Fitzgerald et al. 2007c). Like PBCV, NY-2A also contains an intron in its DNA polymerase and TFIIS-like genes; in addition, NY-2A also contains another self-splicing intron in a gene of unknown function. The coccolithovirus EhV-86 contains two self-splicing introns; one in the DNA polymerase gene and one in a nucleic acid independent nucleoside triphosphatase gene (Wilson et al. 2005a).

In addition to introns, some phycodnavirus proteins contain inteins. Inteins are the protein equivalent of introns: following translation they are autocatalytically spliced to create a mature protein product. Recently, two inteins were identified in the NY-2A genome (the first to be identified in the chloroviruses): one 337 amino acid intein is in a ribonucleotide reductase large subunit protein and the other is in a putative helicase protein (384 amino acids) (Fitzgerald et al. 2007b). However, inteins may be more common in a class of phycodnaviruses that infect the raphidophytes (Nagasaki et al. 2005). The raphidoviruses contain a 232 amino acid intein in their DNA polymerase protein, which is distantly related to an intein in the Mimivirus. This intein was present in all ten raphidoviruses examined.

## Characterized Gene Products

Many PBCV-1-encoded enzymes are either the smallest or among the smallest proteins of their family. The small sizes and the finding that many virus-encoded proteins are user-friendly have resulted in the biochemical and structural characterization of several PBCV-1 enzymes. Examples include:

1. The smallest eukaryotic ATP-dependent DNA ligase (Ho et al. 1997), which is the subject of intensive mechanistic and structural studies (Odell et al. 2003 and references cited therein).
2. The smallest type II DNA topoisomerase (Lavrukhin et al. 2000). This enzyme cleaves dsDNAs 30–50 times faster than the human type II DNA topoisomerase (Fortune et al. 2001); consequently, the enzyme from the chlorella viruses is used as a model to study the topoisomerase II DNA cleavage process (e.g., McClendon et al. 2006).
3. An RNA guanylyltransferase (Ho et al. 1996) that was the first enzyme of its type to have its crystal structure solved (Hakansson et al. 1997; Hakansson and Wigley 1998).
4. A small prolyl-4-hydroxylase that converts Pro-containing peptides into hydroxyl-Pro-containing peptides in a sequence-specific fashion (Eriksson et al. 1999).
5. An unusual thymidylate synthase, called Thy X, which usually only exists in some bacteria (Graziani et al. 2004). Unlike the traditional Thy A synthase, Thy X uses FAD for reduction (Myllykallio et al. 2002).
6. A dCMP deaminase that is unusual because the enzyme can also deaminate dCTP (Zhang et al. 2007). Cellular organisms use two enzymes that belong to different protein families to carry out these two reactions.
7. The smallest protein (94 amino acids) to form a functional $K^+$ channel (Plugge et al. 2000; Kang et al. 2004). The small size of the protein lends itself to molecular dynamics simulation studies to understand how ion channels function (Tayefeh et al. 2007). An even smaller $K^+$ ion channel protein (83 amino acids) was recently identified in virus ATCV-1 (Fitzgerald et al. 2007c) that forms a functional channel in *Xenopus* oocytes (S. Gazzarrini et al., unpublished data). These minimalist proteins may represent precursors of contemporary proteins, but it is also possible that they are products of evolutionary optimization during viral evolution.

Recently, a serine-palmitoyltransferase (SPT) and an elongase from EhV-86 have been functionally characterized (Han et al. 2006).

## Unique Genes

### Novel Virus Genes

The identification of genes such as DNA polymerase, RNA polymerase, DNA ligase, helicases and topoisomerases is an excellent starting point for virus characterization. It is the genes that differ among viruses that are the source of functional diversity and they are responsible for the individual niches that each virus occupies. As noted in this chapter, members of the *Phycodnaviridae* infect a range of hosts; hence they must be highly adapted to their individual niches. In addition to containing many of the genes commonly found in other DNA viruses, sequencing phycodnavirus genomes have revealed genes not previously found in viruses. These gene products may provide clues as to their niche adaptation.

An example of this is the putative sphingolipid biosynthesis pathway present in EhV-86 (Wilson et al. 2005a). Sphingolipid synthesis leads to the formation of ceramide, which can affect apoptosis; thus this is an intriguing choice of a pathway for a virus to manipulate. Sphingolipids are enriched in lipid rafts and may play a crucial role during viral release (Han et al. 2006). EhV-86 contains a cluster of genes encoding sphingolipid metabolic enzymes similar to those in other organisms, from yeast to mammals (Wilson et al. 2005a). A particularly surprising example is the viral gene, *ehv050*, which encodes both subunits of the eukaryotic serine palmitoyltransferases (SPT) (Han et al. 2006). SPT is the enzyme required for the first and rate-limiting step in the pathway and catalyses the condensation of serine with palmitoyl-CoA (Hanada and Nishijima 2003). The novel fusion protein encoded by EhV-86 is not only active, it has a preference for myristoyl-CoA (Han et al. 2006). The relevance of this is unclear, but it is likely to be fundamental to the infection strategy of the virus.

The chloroviruses are also unusual because they often encode enzymes involved in sugar metabolism. Three PBCV-1 encoded enzymes glutamine: fructose-6-phosphate amindotransferase, UDP-glucose dehydrogenase (UDP-GlcDH), and hyaluronan synthase (HAS) are involved in the synthesis of hyaluronan (hyaluronic acid), a linear polysaccharide composed of alternating $\beta$-1,4-glucuronic acid and $\beta$-1,3-N-acetylglucosamine residues (DeAngelis et al. 1997; Landstein et al. 1998). All three genes are transcribed early in PBCV-1 infection and hyaluronan accumulates on the external surface of the infected chlorella cells (Graves et al. 1999).

However, not all chloroviruses have a *has* gene. Surprisingly, many chloroviruses that lack a *has* gene have a gene encoding a functional chitin synthase (CHS). Furthermore, cells infected with these viruses produce chitin fibers on their external surface (Kawasaki et al. 2002). Chitin, an insoluble linear homopolymer of $\beta$-1,2-linked N-acetyl-glucosamine residues, is a common component of insect exoskeletons, shells of crustaceans, and fungal cell walls (Muzzarelli et al. 1986). Some chloroviruses contain both *has* and *chs* genes and form both hyaluronan and chitin on the surface of their infected cells (Kawasaki et al. 2002; Yamada and Kawasaki 2005). Finally, a few chloroviruses appear to lack both genes because no extracellular polysaccharides are formed on the host surface of cells infected with these viruses (Graves et al. 1999). The fact that many chloroviruses encode enzymes involved in these energy-demanding extracellular polysaccharide biosynthetic pathways suggests that the polysaccharides are important in the virus life cycles. However, at present this function is unknown.

Two PBCV-1-encoded enzymes, GDP-D-mannose dehydratase (GMD) and fucose synthase, comprise a highly conserved, three-step pathway that converts GDP-D-mannose to GDP-L-fucose (Tonetti et al. 2003). However, the PBCV-1 GMD not only has the predicted dehydratase activity, it has a reductase activity that produces GDP-D-rhamnose. Both fucose and rhamnose are present in the glycans attached to

PBCV-1 major capsid protein. Interestingly, another chlorella virus, ATCV-1, encodes a GMD that only has the dehydratase activity (Fruscione et al. 2008).

The chlorella viruses degrade the host cell wall during infection and also during virus release. Therefore, it is not surprising that the viruses encode several enzymes that may be involved in these processes (see Yamata et al. 2006 for details). For example, PBCV-1 encodes two chitinases, a chitosanase, a β-1–3 glucanase and an enzyme that cleaves polymers of either β-or α-1,4-linked glucuronic acids (Sugimoto et al. 2004). Enzymes packaged in the virion and involved in virus infection are typically expressed late in the infection cycle and one of the chitinases and the chitosanase meet this requirement and are packaged in the PBCV-1 virion. However, the composition of the host chlorella cell walls and the enzymatic digestion of the walls during infection are still under active investigation.

PBCV-1 encodes four enzymes involved in polyamine biosynthesis: ornithine decarboxylase (ODC), homospermidine synthase, agmatine iminohydrolase, and N-carbamoylputrescine amidohydrolase (Baumann et al. 2007; Kaiser et al. 1999; Morehead et al. 2002). ODC catalyzes the decarboxylation of ornithine to putrescine, which is the first and the rate-limiting enzymatic step in the polyamine biosynthetic pathway. The PBCV-1-encoded ODC is the smallest characterized ODC (Morehead et al. 2002). The PBCV-1 enzyme is also interesting because it decarboxylates arginine better than ornithine (Shah et al. 2004). These four enzymes form a complete polyamine biosynthetic pathway that allows the formation of homospermidine through the precursors arginine, agmatine, N-carbamoylputrescine and putrescine (Baumann et al. 2007). However, the relevance of this pathway to virus replication is unclear. Although polyamines are present in large amounts in the capsids of some viruses, that is not true for PBCV-1 (Kaiser et al. 1999).

As more phycodnavirus sequences become available, we predict that the number of unique and interesting gene pathways found in these viruses will increase. Each of the viruses sequenced to date appears to have its own quirks; this pattern is likely to be repeated as more diverse phycodnaviruses are studied. For example, EsV-1 contains at least six putative hybrid histidine kinases (two-component systems which form part of a stimulus-responsive transduction pathway) that are widespread in archaea and bacteria, but have not previously been found in viruses (Delaroque et al. 2000, 2001). The relevance of these genes to EsV-1 infection is unknown, but is likely to be fundamental to successful infection.

## Database ORFan Genes (Unknown Genes)

The wealth of genetic potential contained within the phycodnaviruses is perhaps best represented not by the unique proteins encoded by the viruses, but by the genes encoding proteins that do not have any matches in GenBank. For example, 86% of the EhV-86 gene products lack GeneBank homologs (Wilson et al. 2005a). These genes/CDSs/ORFs are commonly referred to as database ORFans (Fischer and Eisenberg 1999). Indeed, this is exemplified by a 100-kb region in EhV-86, which contains approximately 150 genes, none of which have a known function (Allen et al. 2006c). Half of these genes are associated with a unique promoter element that drives their expression, but only during the earliest stages of infection (Allen et al. 2006a, 2006d). Thus these genes encode proteins that undoubtedly play a crucial and integral role during virus infection.

## Core Genes and *Phycodnaviridae* Evolution

### *Inter-genus Variation*

There is accumulating evidence that the large dsDNA viruses referred to as nuclear-cytoplasmic large DNA viruses (NCLDV) (*Phycodnaviridae, Poxviridae, Mimiviridae, Asfariviridae,* and *Iridoviridae*) shared a common ancestor (Iyer et al. 2001, 2006; Raoult et al. 2004). Genome-wide bioinformatic surveys have suggested an ancestral virus genome composed of at least 50 genes that have subsequently been retained or discarded in a distinctive evolutionary traceable pattern in these highly divergent virus families (Iyer et al. 2001). Nine genes are present in all NCLDV members (known as group I genes), and 22 additional genes are present in at least three of the five families (group II and III genes). Lineage-specific gene loss and gain within NCLDV families obviously has led to the highly diverse properties of present day viruses.

High evolutionary rates, horizontal gene transfers and nonorthologous gene displacements make accurate phylogenetic resolution difficult over great periods of time, especially when using single-gene phylogenetic trees (Filee et al. 2003). Consequently, studies on whole genomes are becoming more popular when constructing evolutionary relationships between virus families (Iyer et al. 2001; Allen et al. 2006c). In a comparison of PBCV-1, EsV-1 and EhV-86, only 14 genes are common in all three viruses, indicating a large amount of inter-genus genomic variation between *Phycodnaviridae* genera (Allen et al. 2006c). A phylogenetic tree was constructed by concatenating six of the common gene products; the results indicate that the phycodnaviruses form a distinct clade from the other large DNA viruses, yet the tree fails to provide adequate resolution of the genera within the *Phycodnaviridae* (Fig. 7). However, the presence of genes encoding six RNA polymerase subunits in the EhV-86 genome (many of which are found in the other NCLDV virus families) clearly separates EhV-86 from PBCV-1 and EsV-1. This finding suggests that the ancestral *Phycodnaviridae* lineage diverged with one branch giving rise to EhV-86 and the second branch giving rise to the PBCV-1 and EsV-1 lineages. The change in lifestyle represented by this loss of RNA polymerase function (i.e., from nuclear independence to nuclear dependent transcription) probably contributes to the high diversity among present day genera in the *Phycodnaviridae*. Since ancestral NCLDVs contained RNA polymerase function, it is likely that of all the phycodnaviruses sequenced so far, EhV-86 represents the virus with the lifestyle most similar to the ancestral virus (Allen et al. 2006c).

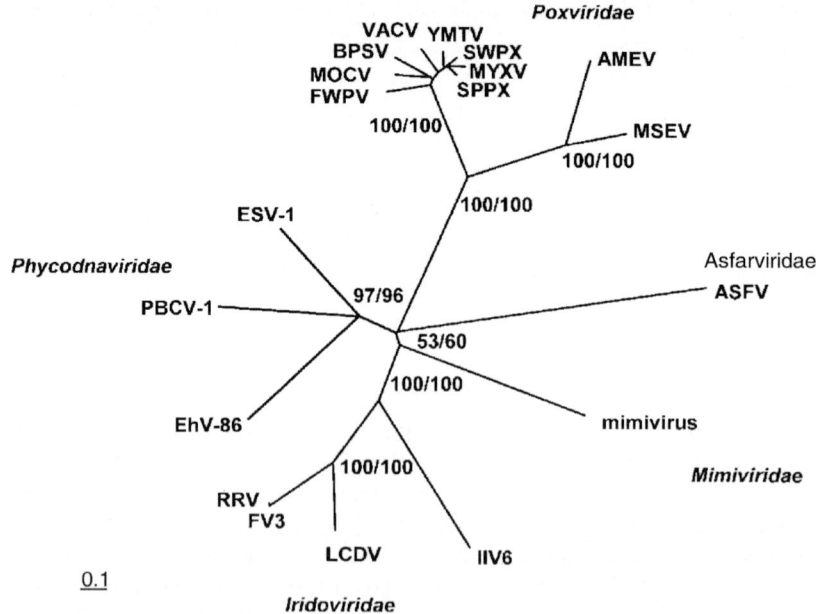

**Fig. 7** Phylogenetic inference tree based on a distance matrix algorithm between concatenated conserved domains from A18-like helicase, D6R-like helicase, A32-like ATPase, DNA polymerase, thio-oxidoreductase from members of the NCLDV group (Neighbor, in PHYLIP version 3.6b). Numbers at nodes indicate bootstrap values retrieved from 100 replicates for both the neighbor-joining and parsimony analyses. The bar depicts one base substitution per ten amino acids. (Diagram adapted from Allen et al. 2006d). Viruses included are African swine fever virus (*ASFV*), *Amsacta moorei* entomopoxvirus (*AMEV*), *Melanoplus sanguinipes* entomopoxvirus (*MSEV*), bovine papular stomatitis virus (*BPSV*), fowlpox virus (*FWPV*), sheeppox virus (*SPPX*), swinepox virus (*SWPX*), vaccinia virus (*VACV*), *Molluscum contagiosum* virus (*MOCV*), myxoma virus (*MYXV*), Yaba monkey tumor virus (*YMTV*), *Paramecium bursaria chlorella* virus 1 (*PBCV-1*), *Ectocarpus siliculosus* virus 1 (*ESV-1*), *Emiliania huxleyi* virus 86 (*EhV-86*), frog virus 3 (*FV3*), invertebrate iridescent virus 6 (*IIV6*), Regina ranavirus (*RRV*), lymphocystis disease virus 1 (*LCDV*) and mimivirus

As more phycodnavirus genomic sequences become available, we may be able to elucidate the complex history of phycodnavirus evolution.

## Intra-genus Variation

With little sequence and genomic data available from the majority of the phycodnaviruses identified so far, it is difficult to assess intra-genus variation. Currently, sequencing efforts have focused on members in just three families: the chloroviruses, phaeoviruses and coccolithoviruses. Therefore, we are just beginning to appreciate the genetic diversity that exists among these genera. Initial results indicate that the huge degree of inter-genus variation can, at least, be matched by a large intra-genus variation.

## Chloroviruses

The six sequenced chlorella viruses provide the best opportunity to study intra-genus variation within the *Phycodnaviridae*. The chlorella viruses are grouped by the host they infect: viruses PBCV-1, NY-2A and AR158 infect *Chlorella* NC64A (Fitzgerald et al. 2007b), MT325 and FR483 infect *Chlorella* Pbi (Fitzgerald et al. 2007a), and ATCV-1 infects *Chlorella* SAG 3.83 (Fitzgerald et al. 2007c). Approximately 80% of the genes are found in all six chloroviruses, suggesting that the majority of the genes are essential for successful viral infection of chlorella. As expected, viruses infecting the same host are most similar; the average protein amino acid identity between PBCV-1 and NY-2A or AR158 homologs is 73%, with 87% identity between NY-2A and AR158 homologs. PBCV-1 and MT325 or FR483 homologs have 50% amino acid identity; 86% identity exists between MT325 and FR483. PBCV-1 and ATCV-1 have 49% amino acid identity between homologs.

The amino acid differences between homologous proteins can be exploited to aid in understanding how certain proteins function. For example, electrophysiological experiments have been conducted with the 94 amino acid PBCV-1 encoded potassium ion channel protein (called Kcv) in *Xenopus* oocytes. Kcv-like proteins were cloned and sequenced from 40 additional chlorella viruses; 16 amino acid substitutions occurred in 94 of the amino acids, producing six new Kcv-like proteins that formed functional potassium selective channels in *Xenopus* oocytes. However, the biophysical properties of some of the new Kcv channels differed from PBCV-1 Kcv. The amino acid changes, together with the different electrophysiological properties, observed in the six Kcv-like channels were used to guide site-directed amino acid substitutions, either singularly or in combination, to identify key residues that conferred specific properties to Kcv (Kang et al. 2004; Gazzarrini et al. 2004).

Each of the six sequenced chloroviruses contains genes that encode unique proteins. For example, ATCV-1 is unique among these viruses in containing genes encoding dTDP-D-glucose 4,6 dehydratase, ribonucleotide-triphosphate reductase, and mucin-desulphatating sulphatase; MT325 encodes a functional aquaglyceroporin (Gazzarrini et al. 2006), FR483 an alkyl sulphatase and a functional potassium ion transporter (M. Kang et al., unpublished data); NY-2A an ubiquitin; AR158 a calcium transporting ATPase; and a functional Cu/Zn superoxide dismutase (M. Kang et al., unpublished data) is unique to PBCV-1.

Using the gene order of PBCV-1 as the model, there is a high degree of gene co-linearity between the three viruses, PBCV-1, NY-2A and AR158 that infect *Chlorella* NC64A. Unlike the co-linearity between the three NC64A viruses, PBCV-1 has only slight co-linearity with the two viruses, MT325 and FR483 that infect *Chlorella* Pbi and virus ATCV-1 that infects *Chlorella* SAG 3.83 (L.A. Fitzgerald et al., unpublished data).

Not all PBCV-1 genes are required for virus replication in the laboratory. For example, four spontaneously derived PBCV-1 mutants were isolated that contain 27- to 37-kb deletions at the left end of the 330-kb genome (Landstein et al. 1995). Two of these mutants have deletions beginning at nucleotide coordinates 4.9 or 16 kb and ending at 42 kb. In total, these two deleted regions, which probably resulted from recombination, encode 28 putative proteins. The other two mutants, which probably arose from nonhomologous recombination, lack the entire left terminal 37-kb of the PBCV-1 genome, including the 2.2-kb terminal inverted-repeat region. The deleted left terminus was replaced by the transposition of an inverted 7.7- or 18.5-kb copy from the right end of the PBCV-1 genome. These regions encode 26 single-copy ORFs, of which 23 are common to those deleted in the first two mutant viruses. Taken together, approximately 40 kb of single-copy DNA encoding 31 ORFs at the left end of the genome, or 12% of the PBCV-1 genome, is unnecessary for PBCV-1 replication in the laboratory. However, replication of the PBCV-1 deletion mutants is attenuated; i.e., their burst sizes are about half of wild-type virus.

The deletion mutants also indicate that the size of the inverted terminal repeats in PBCV-1 can vary. This conclusion is consistent with the finding that the size and sequence of the inverted repeat region is not conserved among chlorella viruses (Strasser et al. 1991; Yamada and Higashiyama 1993). This lack of conservation is somewhat surprising because one predicts that the DNA termini might be essential for either virus DNA replication and/or DNA packaging. These results also indicate that the virus DNA packaging machinery tolerates significant differences in genome size; for example the largest deletion in PBCV-1 creates a genome of approximately 302 kb, whereas chlorella virus NY-2A has a genome of roughly 370 kb. Similar large deletions were reported in the left terminus of the chlorella virus CVK1 genome (Songsri et al. 1995).

Large insertions also occur in the chlorella viruses. Comparison of PBCV-1 and CVK2 genomes revealed that an approximately 15 kb region in the PBCV-1 left terminal region is absent in the CVK2 genome. However, CVK2 contains a 22.2-kb insert in this region that contains five gene copies of a homolog of PBCV-1 glycoprotein Vp260; this 22.2-kb sequence is absent in the PBCV-1 genome (Chuchird et al. 2002; Nishida et al. 1999).

The sizes and locations of the deletions and transpositions in the chlorella viruses resemble poxviruses (Turner and Moyer 1990) and ASFV (Blasco et al. 1989) deletion mutants. Like PBCV-1, poxviruses and ASFV genomes have inverted terminal repeats and covalently closed hairpin ends. Models to explain the generation of deletions and deletion/transpositions in the poxvirus genomes (Shchelkunov and Totmenin 1995; Turner and Moyer 1990) may be relevant to the chlorella viruses.

Several other observations reflect the diversity of chlorella virus genomes.

1. The NC64A virus NY-2A genome is roughly 40 kb larger than the PBCV-1 330-kb genome and contains approximately 30 more genes (Fitzgerald et al. 2007b).
2. Southern hybridization and DNA sequence analyses indicate that not all PBCV-1 genes exist in all of the NC64A virus isolates. Also, an extra ORF is often inserted between co-linear genes. The insertion of extra genetic elements,

referred to as morons (for more DNA), between adjacent genes also occurs in related lambda phages (Hendrix et al. 2000).

3. Sequence analyses of a gene (*pdg*) encoding a UV-specific DNA repair enzyme from 42 NC64A viruses revealed that 15 of them contain a 98-nucleotide spliceosomal-processed intron that is 100% conserved; four other viruses contain an identically positioned 81-nucleotide intron that is nearly 100% conserved (Sun et al. 2000). In contrast, the nucleotides in the *pdg* coding regions (exons) from the intron-containing viruses are 84%–100% identical. The 100% identity of the 98-nucleotide intron sequence in 15 viruses and the near 100% identity of an 81-nucleotide intron sequence in another four viruses imply that either the intron was acquired recently or that there is strong selective pressure to maintain the DNA sequence of the intron once it is in the *pdg* gene. However, the abilities of intron-containing and intron-lacking viruses to repair UV-damaged DNA in the dark were indistinguishable (Sun et al. 2000). These findings contradict the dogma that intron sequences are more variable than exon sequences.

4. Yamada and his colleagues (Nishida et al. 1998; Yamada et al. 1994) reported that 8% of the NC64A viruses isolated in Japan contain a self-splicing group I intron. This intron is inserted in the gene encoding either a transcriptional elongation factor TFIIS (~60% of the viruses) or an unidentified ORF encoding a 14.2-kDa polypeptide (~40% of the viruses); however, in a few viruses the intron is in the major capsid protein gene. Yamada et al. (1994) suggested that the self-splicing intron might function as a mobile genetic element; e.g., transferring genes between algae and protists.

To summarize, considerable variation occurs in the chlorovirus genomes and the total number of genes in the chlorovirus gene pool exceeds that of a single isolate. The different sizes of the chlorovirus genomes as well as the large deletions and insertions, suggest that dynamic and frequent rearrangements of virus genomes occur in natural environments. The fact that the left end of the chlorella virus genome is tolerant to deletions/insertions/rearrangements suggests that a recombinational hotspot in this region allows viruses to exchange genes among themselves and possibly with their host(s).

## Coccolithoviruses

In addition to the fully sequenced English Channel coccolithovirus, EhV-86, there is good sequence coverage (>80%) of a second Norwegian virus, EhV-163 (Allen et al. 2006b). Both strains are capable of infecting many, but not all, of the same *E. huxleyi* strains and they have nearly identical sized genomes (Allen et al. 2007). The sequence of a third strain, another Norwegian strain closely related to EhV-163, is also near completion. Sequencing the coccolithoviruses is very difficult because of the highly repetitive nature of the genomes. Direct comparison of EhV-86 and EhV-163 genomic sequences reveal that of the 202 CDSs for which there is full sequence in both genomes, only 20 are completely identical at the nucleotide level and an additional 17 at the protein level (Allen et al. 2006b). Although these

37 conserved CDSs are distributed throughout the genome, some of them appear to be clustered together in four regions. Many of the remaining fully sequenced EhV-163 CDSs have 95%–99% identity with their EhV-86 counterparts. One CDS, *ehv142*, is highly variable (86.9% at the nucleotide level, 79.1% at the amino acid level), the function of this gene product is unknown, but it may be a KELCH-like protein. KELCH-like proteins in the poxviruses contribute to virulence, host range and reproduction; therefore this EhV-86 protein may contribute to the variation in host range displayed by these viruses (Tulman et al. 2002; Pires de Miranda et al. 2003; Kochneva et al. 2005). Since the EhV-163 genome sequence is unfinished, it is not possible to draw any firm conclusions as to which genes are absent from EhV-163, but present in EhV-86. However, a genomic rearrangement occurs in EhV-163. A putative EhV-86 phosphate permease (*ehv117*) has been partially deleted and replaced with a putative endonuclease (*ehv117A*) (Allen et al. 2006b). The significance of this replacement is unknown, but it may represent an overall like-for-like replacement because both gene products have the potential to increase the phosphate available to the virus, though by very different mechanisms. One mechanism is by increasing phosphate uptake and the other by degrading host nucleic acid.

A genome-wide microarray-based survey, used to assess the genomic variation without sequencing, has revealed a large amount of variation in genomic content among the coccolithoviruses (Allen and Wilson 2006; Allen et al. 2007). In a screen of a dozen virus isolates, 71 out of 425 CDSs tested were either absent or highly variable in at least one isolate. All core NCLDV genes identified in EhV-86 are present in all isolates tested to date. The advantage of this genomic snapshot analysis can be seen when considering the most phylogenetically and geographically distinct isolates, EhV-163 and EhV-V2. EhV-163 and EhV-V2 have at least 27 and 28 missing or highly variable CDSs, respectively, in comparison with EhV-86 (26 common to both strains). The largest variation occurred between isolates obtained from the same geographic location, but separated by time of sampling (Schroeder et al. 2002), suggesting that the coccolithoviruses are excellent examples of the classic evolutionary rule of thumb that variation within populations is often greater than variation between populations. Despite the large degree of variation in genome content, the overall genome size is similar between coccolithoviruses, suggesting that physical genome size, in addition to genome content, may be crucial to the virus (Allen et al. 2007). Whether this is due to genomic packaging, replication or some other unknown process is not known.

## Phaeoviruses

A comparison of EsV-1 (231 genes, 335 kbp) and FirrV-1 (156 genes, 231 kbp) genomes revealed that 93 FirrV-1 genes are related to EsV-genes (Delaroque et al. 2003). The most highly conserved genes (>50% amino acid identity) predictably include the core NCLDV genes. The majority of the virus-encoded proteins have 31%–50% amino acid identity. However, the gene order is remarkably different between the two viral genomes, suggesting a high rate of recombination and divergence since their last common ancestor. Whereas the chloroviruses and coccolithoviruses (discussed in Sects. 7.2.1 and 7.2.2) only infect one algal species, phaeoviruses often infect two distinct, but taxonomically similar, filamentous brown algae, *Ectocarpus* and *Feldmannia* (Delaroque et al. 2003). Despite the closely related host organisms and similar life styles, their genomes are remarkably different. Partial genomic sequence is also available for another *Feldmannia* virus, FsV, and the gene order is similar to that of FiirV-1. Of particular interest are large sections of the EsVgenome that have no FiirV-1 homologs; presumably these sections of the genome have either been lost in the FiirV-1 lineage or acquired by the EsV-1 lineage (Delaroque et al. 2001, 2003). It is possible that the genes obtained in this region are essential for successful infection of *Ectocarpus* hosts, but are not needed for *Feldmannia* infection. However, these viruses have almost indistinguishable lysogenic replication cycles and the extra genomic regions in EsV-1 may be predominantly redundant (Delaroque et al. 2003).

## Summary

The *Phycodnaviridae* is an ancient, genetically diverse, yet morphologically similar family of large icosahedral viruses that infect eukaryotic algae; they have dsDNA genomes ranging from 180 to 560 kb. Currently, the *Phycodnaviridae* consist of six genera named after the hosts they infect (*Chlorovirus*, *Coccolithovirus*, *Prasinovirus*, *Prymnesiovirus*, *Phaeovirus* and *Raphidovirus*). Their hosts, eukaryotic algae, are an incredibly diverse group of organisms that represent at least five distinct evolutionary lineages. They are ubiquitous in marine, freshwater and terrestrial habitats, and the number of algal species may be as high as several million. Their role in global carbon and nutrient cycling make $O_2$-evolving eukaryotic algae arguably the most important microorganisms for maintaining the status quo of life on the planet. It is likely that viruses infect all eukaryotic algae and it is reasonable to assume that many of these viruses will be assigned to the *Phycodnaviridae*. Therefore, it is perhaps surprising that only a few hundred (at most) phycodnaviruses have been isolated, less than a dozen have been sequenced and only one (PBCV-1) has been studied in any depth. Thus it is obvious that a huge gap exists in our knowledge and understanding of the *Phycodnaviridae*.

The *Phycodnaviridae* clearly group into a single family. Their common evolutionary origin is reflected not only in their appearance (despite their range in sizes, these viruses are structurally and morphologically similar) but also in their genomic content. As with the other NCLDV viruses (*Poxviridae, Iridoviridae, Mimiviridae,* and *Asfarviridae*), their common evolutionary origin is revealed by the so-called core genes, which are responsible for essential viral functions such as DNA replication, RNA transcription and virion assembly; these genes have been conserved since their ancient last common ancestor. Yet the enormous divergence that has occurred since that ancestral virus graced the world has created a truly fascinating group of viruses that will enthrall researchers for many years to come. We have hardly scratched the surface in studies on the *Phycodnaviridae*, yet it is already obvious that the highly diverse nature of the host algae is also reflected in the diversity of the phycodnaviruses. As the host organisms have diverged to occupy their own specific niches, these viruses have responded to the challenge and evolved to survive in the wide range of environments they now find themselves. This is reflected in the huge degree of variation among the viruses. So far, each phycodnavirus genera appear to be unique in lifestyle, with each virus having its own peculiarities and novelties. Nearly everything that can vary, does vary within the *Phycodnaviridae*, from genome size, structure (circular or linear) and content, to propagation strategy and life style. Variety is the spice of life, which makes the *Phycodnaviridae* the spiciest of the viruses. As researchers recognize the scientific and commercial worth of studying these viruses, we predict they will not be regarded as lesser known for very long.

**Acknowledgements** Research in the Wilson laboratory has been supported by Natural Environment Research Council (NERC) programs Marine and Freshwater Microbial Biodiversity (grant NER/T/S/2000/00640); Environmental Genomics (grant NE/A509332/1); NERC responsive mode grants NE/D001455/1 and NER/A/S/2003/00296 and from Marine Genomics Europe, through framework programme FP6 of the European Commission and National Science Foundation grant EF-0723730. Research in the Van Etten laboratory has been supported in part by Public Health Service grant GM32441 from the National Institute of General Medical Sciences, the National Science Foundation grant EF-0333197, and NIH grant P20-RR1565 from the COBRE program of the National Center for Research Resources.

# References

Agarkova IV, Dunigan DD, Van Etten JL (2006) Virion-associated restriction endonucleases of chloroviruses. J Virol 80:8114–8123

Allen MJ, Wilson WH (2006) The coccolithovirus microarray: an array of uses. Brief Funct Genomic Proteomic 5:273–279

Allen MJ, Forster T, Schroeder DC, Hall M, Roy D, Ghazal P, Wilson WH (2006a) Locus-specific gene expression pattern suggests a unique propagation strategy for a giant algal virus. J Virol 80:7699–7705

Allen MJ, Schroeder DC, Donkin A, Crawfurd KJ, Wilson WH (2006b) Genome comparison of two Coccolithoviruses. Virology J 3:15

Allen MJ, Schroeder DC, Holden MTG, Wilson WH (2006c) Evolutionary history of the Coccolithoviridae. Mol Biol Evol 23:86–92

Allen MJ, Schroeder DC, Wilson WH (2006d) Preliminary characterization of repeat families in the genome of EhV-86, a giant algal virus that infects the marine microalga *Emiliania huxleyi*. Arch Virology 151:525–535

Allen MJ, Martinez-Martinez J, Schroeder DC, Somerfield PJ, Wilson WH (2007) Use of microarrays to assess viral diversity: from genotype to phenotype. Environ Microbiol 9:971–982

Baldauf SL (2003) The deep roots of eukaryotes. Science 300:1703–1706

Bamford DH, Burnett RM, Stuart DI (2002) Evolution of viral structure. Theor Popul Biol 61:461–470

Baumann S, Sander A, Gurnon JR, Yanai-Balser GM, Van Etten JL, Piotrowski M (2007) Chlorella viruses contain genes encoding a complete polyamine biosynthetic pathway. Virology 360:209–217

Benson SD, Bamford JK, Bamford DH, Burnett RM (2004) Does common architecture reveal a viral lineage spanning all three domains of life? Mol Cell 16:673–685

Bergh O, Borsheim KY, Bratbak G, Heldal M (1989) High abundance of viruses found in aquatic environments. Nature 340:467–468

Blasco R, De La Dega I, Almazan F, Aguero M, Vinuela E (1989) Genetic variation of African swine fever virus: variable regions near the ends of the viral DNA. Virology 173:251–257

Bown PR (1998) Calcareous nannofossil biostratigraphy. Chapman and Hall, London

Bratbak G, Egge JK, Heldal M (1993) Viral mortality of the marine alga *Emiliania huxleyi* (Haptophyceae) and termination of algal blooms. Mar EcolProgr 93:39–48

Bratbak G, Wilson W, Heldal M (1996) Viral control of *Emiliania huxleyi* blooms? J Marine Syst 9:75–81

Bräutigam M, Klein M, Knippers R, Müller DG (1995) Inheritance and meiotic elimination of a virus genome in the host *Ectocarpus siliculosus* (Phaeophyceae). J Phycol 31:823–827

Brown CW, Yoder JA (1994) Coccolithophorid blooms in the global ocean. J Geophys Res Oceans 99:7467–7482

Brussaard CPD (2004) Viral control of phytoplankton populations—a review. J Eukary Microbiol 51:125–138

Brussaard CPD, Kempers RS, Kop AJ, Riegman R, Heldal M (1996) Virus-like particles in a summer bloom of *Emiliania huxleyi* in the North Sea. Aquatic Microbial Ecol 10:105–113

Brussaard CPD, Noordeloos AAM, Sandaa RA, Heldal M, Bratbak G (2004a) Discovery of a dsRNA virus infecting the marine photosynthetic protist *Micromonas pusilla*. Virology 319:280–291

Brussaard CPD, Short SM, Frederickson CM, Suttle CA (2004b) Isolation and phylogenetic analysis of novel viruses infecting the phytoplankton *Phaeocystis globosa* (Prymnesiophyceae). Appl Environ Microbiol 70:3700–3705

Brussaard CPD, Kuipers B, Veldhuis MJW (2005a) A mesocosm study of *Phaeocystis globosa* population dynamics: I. Regulatory role of viruses in bloom control. Harmful Algae 4:859–874

Brussaard CPD, Mari X, Bleijswijk JDLV, Veldhuis MJW (2005b) A mesocosm study of *Phaeocystis globosa* (Prymnesiophyceae) population dynamics: II. Significance for the microbial community. Harmful Algae 4:875–893

Bubeck JA, Pfitzner AJ (2005) Isolation and characterization of a new type of chlorovirus that infects an endosymbiotic chlorella strain of the heliozoon *Acanthocystis turfacea*. J Gen Virol 86:2871–2877

Burkill PH, Archer SD, Robinson C (2002) Dimethyl sulphide biogeochemistry within a coccolithophore bloom (DISCO): an overview. Deep Sea Research Part II: Topical Studies in Oceanography 49:2863–2885

Castberg T, Thyrhaug R, Larsen A, Sandaa R-A, Heldal M, Van Etten JL, Bratbak G (2002) Isolation and characterization of a virus that infects *Emiliania huxleyi* (Haptophyta). J Phycol 38:767–774

Chan SH, Zhu Z, Van Etten JL, Xu SY (2004) Cloning of CviPII nicking and modification system from chlorella virus NYs-1 and application of Nt.CviPII in random DNA amplification. Nucleic Acids Res 32:6187–6199

Chan SH, Zhu Z, Dunigan DD, Van Etten JL, Xu SY (2006) Cloning of Nt.CviQII nicking endonuclease and its cognate methyltransferase: M.CviQII methylates AG sequences. Prot Express Purifi 49:138–150

Charlson RJ, Lovelock JE, Andreae MO, Warren SG (1987) Oceanic phytoplankton, atmospheric sulfur, cloud albedo and climate. Nature 326:655–661

Chen F, Suttle CA (1995) Amplification of DNA-polymerase gene fragments from viruses infecting microalgae. Appl Environ Microbiol 61:1274–1278

Chen F, Suttle CA, Short SM (1996) Genetic diversity in marine algal virus communities as revealed by sequence analysis of DNA polymerase genes. Appl Environ Microbiol 62:2869–2874

Chrétiennot-Dinet MJ, Courties C, Vaquer A, Neveux J, Claustre H, Lautier J, Machado MC (1995) A new marine picoeucaryote: *Ostreococcus tauri* gen. et sp. nov. (Chlorophyta, Prasinophyceae). Phycologia 34:285–292

Chuchird N, Nishida K, Kawasaki T, Fujie M, Usami S, Yamada T (2002) A variable region on the chlorovirus CVK2 genome contains five copies of the gene for Vp260, a viral glycoprotein. Virology 295:289–298

Clokie MRJ, Mann NH (2006) Marine cyanophages and light. Environ Microbiol 8:2074–2082

Cottrell MT, Suttle CA (1991) Wide-spread occurrence and clonal variation in viruses which cause lysis of a cosmopolitan, eukaryotic marine phytoplankter, *Micromonas pusilla*. Mar Ecol Progr 78:1–9

Cottrell MT, Suttle CA (1995a) Dynamics of a lytic virus infecting the photosynthetic marine ico-flagellate *Micromonas pusilla*. Limnol Oceanography 40:730–739

Cottrell MT, Suttle CA (1995b) Genetic diversity of algal viruses which lyse the photosynthetic picoflagellate *Micromonas pusilla* (Prasinophyceae). Appl Environ Microbiol 61:3088–3091

Crick FH, Watson JD (1956) Structure of small viruses. Nature 177:473–475

Dahl E, Bagoien E, Edvardsen B, Stenseth NC (2005) The dynamics of *Chrysochromulina* species in the Skagerrak in relation to environmental conditions. J Sea Res 54:15–24

DeAngelis PL, Jing W, Graves MV, Burbank DE, Van Etten JL (1997) Hyaluronan synthase of chlorella virus PBCV-1. Science 278:1800–1803

Delaroque N, Maier I, Knippers R, Muller DG (1999) Persistent virus integration into the genome of its algal host, *Ectocarpus siliculosus* (Phaeophyceae). J Gen Virol 80:1367–1370

Delaroque N, Boland W, Muller DG, Knippers R (2003) Comparisons of two large phaeoviral genomes and evolutionary implications. J Mol Evol 57:613–622

Delaroque N, Muller DG, Bothe G, Pohl T, Knippers R, Boland W (2001) The complete DNA sequence of the *Ectocarpus siliculosus* virus EsV-1 genome. Virology 287:112–132

Derelle E, Ferraz C, Rombauts S, Rouzé P, Worden AZ, Robbens S, Partensky F, Degroeve S, Echeynié S, Cooke R, Saeys Y, Wuyts J, Jabbari K, Bowler C, Panaud O, Piégu B, Ball SG, Ral JP, Bouget FY, Piganeau G, De Baets B, Picard A, Delseny M, Demaille J, Van de Peer Y, Moreau H (2006) Genome analysis of the smallest free-living eukaryote *Ostreococcus tauri* unveils many unique features. Proc Nat Acad Sci U S A 103:11647–11652

Diez B, Pedros-Alio C, Massana R (2001) Study of genetic diversity of eukaryotic picoplankton in different oceanic regions by small-subunit rRNA gene cloning and sequencing. Appl Environ Microbiol 67:2932–2941

Dundas I, Johannessen OM, Berge G, Heimdal B (1989) Toxic algal bloom in Scandinavian waters, May–June 1988. Oceanography 2:9–14

Dunigan DD, Fitzgerald LA, Van Etten JL (2006) Phycodnaviruses: a peek at genetic diversity. Virus Res 117:119–132

Eikrem W, Throndsen J (1990) The ultrastructure of *Bathycoccus* gen-nov and *Bathycoccus prasinos* sp-nov, a nonmotile picoplanktonic alga (Chlorophyta, Prasinophyceae) from the Mediterranean and Atlantic. Phycologia 29:344–350

Elderfield H (2002) Climate change: carbonate mysteries. Science 296:1618–1621

Eriksson M, Myllyharju J, Tu H, Hellman M, Kivirikko KI (1999) Evidence for 4-hydoxyproline in viral proteins: characterization of a viral prolyl 4 hydroxylase and its peptide substrates. J Biol Chem 274:22131–22134

Estep KW, MacIntyre F (1989) Taxonomy, life cycle, distribution and dasmotrophy of *Chrysochromulina*: a theory accounting for scales, haptonema, muciferous bodies and toxicity. Mar Ecol Progr 57:11–21

Evans C, Archer SD, Jacquet S, Wilson WH (2003) Direct estimates of the contribution of viral lysis and microzooplankton grazing to the decline of a *Micromonas spp.* population. Aquatic Microbial Ecol 30:207–219

Evans C, Kadner SV, Darroch LJ, Wilson WH, Liss PS, Malin G (2007) The relative significance of viral lysis and microzooplankton grazing as pathways of dimethylsulfoniopropionate (DMSP) cleavage: an *Emiliania huxleyi* culture study. Limnol Oceanogr 52:1036–1045

Figueroa RI, Rengefors K (2006) Life cycle and sexuality of the freshwater raphidophyte *Gonyostomum semen* (Raphidophyceae). J Phycol 42:859–871

Filee J, Forterre P, Laurent J (2003) The role played by viruses in the evolution of their hosts: a view based on informational protein phylogenies. Res Microbiol 154:237–243

Fischer D, Eisenberg D (1999) Finding families for genomic ORFans. Bioinformatics 15:759–762

Fitzgerald LA, Graves MV, Li X, Feldblyum T, Hartigan J, Van Etten JL (2007a) Sequence and annotation of the 314-kb MT325 and the 321-kb FR483 viruses that infect *Chlorella* Pbi. Virology 358:459–471

Fitzgerald LA, Graves MV, Li X, Feldblyum T, Nierman WC, Van Etten JL (2007b) Sequence and annotation of the 369-kb NY-2A and the 345-kb AR158 viruses that infect *Chlorella* NC64A. Virology 358:472–484

Fitzgerald LA, Graves MV, Li X, Hartigan J, Pfitzner AJ, Hoffart E, Van Etten JL (2007c) Sequence and annotation of the 288-kb ATCV-1 virus that infects an endosymbiotic chlorella strain of the heliozoon *Acanthocystis turfacea*. Virology 362:350–361

Fortune JM, Lavrukhin OV, Gurnon JR, Van Etten JL, Lloyd RS, Osheroff N (2001) Topoisomerase II from chlorella virus PBCV-1 has an exceptionally high DNA cleavage activity. J Biol Chem 276:24401–24408

Frohns F, Käsmann A, Kramer D, Schäfer B, Mehmel M, Kang M, Van Etten JL, Gazzarrini S, Moroni A, Thiel G (2006) Potassium ion channels of chlorella viruses cause rapid depolarization of host cells during infection. J Virol 80:2437–2444

Fruscione F, Sturla L, Duncan, G, Van Etten JL, Valbuzzi, P, De Flora A, Di Zanni E, Tonetti M (2008) Differential role of NADP⁺ and NADPH in the activity and structure of GDP-D-mannose 4,6-dehydratase from two chlorella viruses. J. Biol Chem 283:184–193

Fuller NJ, Campbell C, Allen DJ, Pitt FD, Zwirglmaierl K, Le Gall F, Vaulot D, Scanlan DJ (2006) Analysis of photosynthetic picoeukaryote diversity at open ocean sites in the Arabian Sea using a PCR biased towards marine algal plastids. Aquatic Microbial Ecol 43:79–93

Gazzarrini S, Kang M, Van Etten JL, Tayefeh S, Kast SM, DiFrancesco D, Thiel G, Moroni A (2004) Long-distance interactions within the potassium channel pore are revealed by molecular diversity of viral proteins. J Biol Chem 279:28443–28449

Gazzarrini S, Kang M, Epimashko S, Van Etten JL, Dainty J, Theil G, Moroni A (2006) Chlorella virus MT325 encodes water and potassium channels that interact synergistically. Proc Natl Acad Sci U S A 103:5355–5360

Graves MV, Burbank DE, Roth R, Heuser J, DeAngelis PL, Van Etten JL (1999) Hyaluronan synthesis in virus PBCV-1 infected chlorella-like green algae. Virology 257:15–23

Graves MV, Bernadt CT, Cerny R, Van Etten JL (2001) Molecular and genetic evidence for a virus-encoded glycosyltransferase involved in protein glycosylation. Virology 285:332–345

Graziani S, Xia Y, Gurnon JR, Van Etten JL, Leduc D, Skouloubris S, Myllykallio, Hiebl U (2004) Functional analysis of FAD-dependent thymidylate synthase ThyX from *Paramecium bursaria* chlorella virus-1. J Biol Chem 279:54340–54347

Green JC, Pienaar RN (1977) The taxonomy of the order Isochrysidales (Prymnesiophyceae) with special reference to the genera *Isochrysis* Parke, *Dicrateria* Parke and *Imantonia* Reynolds. J Mar Biol Assoc UK 57:7–17

Hakansson K, Wigley DB (1998) Structure of a complex between a cap analogue and mRNA guanylyl transferase demonstrates the structural chemistry of RNA capping. Proc Natl Acad Sci USA 95:1505–1510

Hakansson K, Doherty AJ, Shuman S, Wigley DB (1997) X-ray crystallography reveals a large conformational change during guanyl transfer by mRNA capping enzymes. Cell 89:543–553

Han G, Gable K, Yan L, Allen MJ, Wilson WH, Moitra P, Harmon JM, Dunn TM (2006) Expression of a novel marine viral single-chain serine palmitoyltransferase and construction of yeast and mammalian single-chain chimera. J Biol Chem 281:39935–39942

Hanada K, Nishijima M (2003) Purification of mammalian serine palmitoyltransferase, a heterosubunit enzyme for sphingolipid biosynthesis, by affinity-peptide chromatography. Methods Mol Biol 228:163–174

Hendrix RW, Lawrence JG, Hatfull GF, Casjens S (2000) The origins and ongoing evolution of viruses. Trends Microbiol 8:504–508

Henry EC, Meints RH (1992) A persistent virus-infection in *Feldmannia* (*Phaeophyceae*). J Phycol 28:517–526

Hiroishi S, Okada H, Imai I, Yoshida T (2005) High toxicity of the novel bloom-forming species *Chattonella ovata* (*Raphidophyceae*) to cultured fish. Harmful Algae 4:783–787

Ho CK, Van Etten JL, Shuman S (1996) Expression and characterization of an RNA capping enzyme encoded by chlorella virus PBCV-1. J Virology 70:6658–6664

Ho CK, Van Etten JL, Shuman S (1997) Characterization of an ATP dependent DNA ligase encoded by chlorella virus PBCV-1. J Virology 71:1931–1937

Holligan PM, Viollier M, Harbour DS, Camus P, Champagnephilippe M (1983) Satellite and ship studies of coccolithophore production along a continental-shelf edge. Nature 304:339–342

Holligan PM, Fernandez E, Aiken J, Balch WM, Boyd P, Burkill PH, Finch M, Groom SB, Malin G, Muller K, Purdie DA, Robinson C, Trees CC, Turner SM, Vanderwal P (1993) A biogeochemical study of the Coccolithophore, *Emiliania huxleyi,* in the North-Atlantic. Global Biogeochem Cycles 7:879–900

Honjo T (1993) Overview on bloom dynamics and physiological ecology of *Heterosigma akashiwo*. In: Smayda TJ, Shimizu Y (eds) Toxic phytoplankton blooms in the sea. Elsevier, New York, pp 33–41

Iyer LM, Aravind L, Koonin EV (2001) Common origin of four diverse families of large eukaryotic DNA viruses. J Virol 75:11720–11734

Iyer LM, Balaji S, Koonin EV, Aravind L (2006) Evolutionary genomics of nucleo-cytoplasmic large DNA viruses. Virus Res 117:156–184

Jacobsen A, Bratbak G, Heldal M (1996) Isolation and characterization of a virus infecting *Phaeocystis pouchetii* (Prymnesiophyceae). J Phycol 32:923–927

Jacquet S, Heldal M, Iglesias-Rodriguez D, Larsen A, Wilson W, Bratbak G (2002) Flow cytometric analysis of an *Emiliana huxleyi* bloom terminated by viral infection. Aquatic Microbial Ecol 27:111–124

Kai AKL, Cheung YK, Yeung RKK, Wong JTY (2006) Development of single-cell PCR methods for the Raphidophyceae. Harmful Algae 5:649–657

Kaiser A, Vollmert M, Tholl D, Graves MV, Xing W, Lisec AD, Gurnon JR, Nickerson KW, Van Etten JL (1999) Chlorella virus PBCV-1 encodes a functional Homospermidine synthase. Virology 263:254–262

Kang M, Moroni A, Gazzarrini S, DiFrancesco D, Thiel G, Severino M, Van Etten JL (2004) Small potassium ion channel proteins encoded by chlorella viruses. Proc Natl Acad Sci U S A 101:5318–5324

Kang M, Dunigan DD, Van Etten JL (2005) Chlorovirus: a genus of Phycodnaviridae that infects certain chlorella-like green algae. Mol Plant Pathol 6:213–224

Kawakami H, Kawakami N (1978) Behavior of a virus in a symbiotic system, *Paramecium bursaria*-zoochlorella. J Protozool 25:217–225

Kawasaki T, Tanaka M, Fujie M, Usami S, Sakai K, Yamada T (2002) Chitin synthesis in chlorovirus CVK2-infected chlorella cells. Virology 302:123–131

Khan S, Arakawa O, Onoue Y (1995) Effects of physiological factors on morphology and motility of *Chattonella antiqua* (Raphidophyceae). Botanica Marina 38:347–353

Khan S, Arakawa O, Onoue Y (1997) Neurotoxins in a toxic red tide of *Heterosigma akashiwo* (*Raphidophyceae*) in Kagoshima Bay, Japan. Aquaculture Res 28:9–14

Kochneva G, Kolosova I, Maksyutova T, Ryabchikova E, Shchelkunov S (2005) Effects of deletions of kelch-like genes on cowpox virus biological properties. Arch Virol 150:1857–1870

Kropinski AM, Sibbald MJ (1999) Transfer RNA genes and their significance to codon usage in the *Pseudomonas aeruginosa* lamboid bacteriophage D3. Can J Microbiol 45:791–796

Lancelot C, Billen G, Sournia A, Weisse T, Colijn F, Veldhuis MJW, Davies A, Wassmann P (1987) Phaeocystis blooms and nutrient enrichment in the continental coastal zones of the North Sea. Ambio 16:38–46

Landstein D, Burbank D, Nietfeldt JW, Van Etten JL (1995) Large deletions in antigenic variants of the chlorella virus PBCV-1. Virology 214:413–420

Landstein D, Graves MV, Burbank DE, DeAngelis P, Van Etten JL (1998) Chlorella virus PBCV-1 encodes functional glutamine: fructose-6-phosphate amidotransferase and UDP-glucose dehydrogenase enzymes. Virology 250:388–396

Lanka STJ, Klein M, Ramsperger U, Müller DG, Knippers R (1993) Genome structure of a virus infecting the marine brown alga *Ectocarpus siliculosus*. Virology 193:802–811

Lavrukhin OV, Fortune JM, Wood TG, Burbank DE, Van Etten JL, Osheroff N, Lloyd RS (2000) Topoisomerase II from chlorella virus PBCV-1. Characterization of the smallest known type II topoisomerase. J Biol Chem 275:6915–6921

Leadbeater BSC (1972) Identification, by means of electron microscopy, of flagellate nanoplankton from the coast of Norway. Sarsia 49:107–124

Lee AM, Ivey RG, Henry EC, Meints RH (1995) Characterization of a repetitive DNA element in a brown algal virus. Virology 212:474–480

Li WKW (1994) Primary production of Prochlorophytes, Cyanobacteria, and eukaryotic ultraphytoplankton—measurements from flow cytometric sorting. Limnol Oceanogr 39:169–175

Li Y, Lu Z, Sun L, Ropp S, Kutish GF, Rock DL, Van Etten JL (1997) Analysis of 74 kb of DNA located at the right end of the 330-kb chlorella virus PBCV-1 genome. Virology 237:360–377

Liss PS, Malin G, Turner SM, Holligan PM (1994) Dimethyl sulfide and *Phaeocystis* – a review. J Marine Sys 5:41–53

Maier I, Müller DG, Katsaros C (2002) Entry of the DNA virus, *Ectocarpus fasciculatus* virus type 1 (Phycodnaviridae), into host cell cytosol and nucleus. Phycol Res 50:227–231

Malin G (1997) Biological oceanography—sulphur, climate and the microbial maze. Nature 387:857–859

Malin G, Wilson WH, Bratbak G, Liss PS, Mann NH (1998) Elevated production of dimethylsulfide resulting from viral infection of cultures of *Phaeocystis pouchetii*. Limnol Oceanogr 43:1389–1393

Manton I, Leadbeater BSC (1974) Fine structural observations on six species of *Chrysochromulina* from wild Danish marine nanoplankton, including a description of *C. campanulifera* sp. nov. and a preliminary summary of the nanoplankton as a whole. Det Kongelige Danske Videnskabernes Selskab Biologiske Skrifter 20:1–26

Manzur KL, Farooq A, Zeng L, Plotnikova O, Koch AW, Sachchidanand, Zhou MM (2003) A dimeric viral SET domain methyltransferase specific to Lys[27] of histone H3. Nature Struct Biol 10:187–196

Markine-Goriaynoff N, Gillet L, Van Etten JL, Korres H, Verma N, Vanderplasschen A (2004) Glycosyltransferases encoded by viruses. J Gen Virol 85:2741–2754

Mayer JA, Taylor FJR (1979) A virus which lyses the marine nanoflagellate *Micromonas pusilla*. Nature 281:299–301

McClendon AK, Dickey JS, Osheroff N (2006) Ability of viral topoisomerase II to discern the handedness of supercoiled DNA: bimodal recognition of DNA geometry by type II enzymes. Biochemistry 45:11674–11680

McHugh DJ (1991) Worldwide distribution of commercial resources of seaweeds including Geldium. Hydrobiologia 221:19–29

Mehmel M, Rothermel M, Meckel T, Van Etten JL, Moroni A, Thiel G (2003) Possible function for virus encoded K⁺ channel Kcv in the replication of chlorella virus PBCV-1. FEBS Lett 552:7–11

Meints RH, Van Etten JL, Kuczmarski D, Lee K, Ang B (1981) Viral-infection of the symbiotic chlorella-like alga present in *Hydra viridis*. Virology 113:698–703

Meints RH, Lee K, Burbank DE, Van Etten JL (1984) Infection of a chlorella-like alga with the virus, PBCV-1: Ultrastructural studies. Virology 138:341–346

Meints RH, Lee K, Van Etten JL (1986) Assembly site of the virus PBCV-1 in a chlorella-like green alga: ultrastructural studies. Virology 154:240–245

Meints RH, Burbank DE, Van Etten JL, Lamport DTA (1988) Properties of the *Chlorella* receptor for the virus PBCV-1. Virology 164:15–21

Moon-van der Staay SY, De Wachter R, Vaulot D (2001) Oceanic 18S rDNA sequences from picoplankton reveal unsuspected eukaryotic diversity. Nature 409:607–610

Morehead TA, Gurnon JR, Adams B, Nickerson KW, Fitzgerald LA, Van Etten JL (2002) Ornithine decarboxylase encoded by chlorella virus PBCV-1. Virology 301:165–175

Müller DG (1991) Mendelian segregation of a virus genome during host meiosis in the marine brown alga *Ectocarpus siliculosus*. J Plant Physiol 137:739–743

Müller DG (1996) Host-virus interactions in marine brown algae. Hydrobiologia 327:21–28

Müller DG, Kawai H, Stache B, Lanka S (1990) A virus-infection in the marine brown alga *Ectocarpus siliculosus* (*Phaeophyceae*). Bot Acta 103:72–82

Müller DG, Stache B (1992) Worldwide occurrence of virus-infections in filamentous marine brown-algae. Helgolander Meeresunters 46:1–8

Müller DG, Kapp M, Knippers R (1998) Viruses in marine brown algae. In: Advances in virus research, Vol. 50. Academic, San Diego, pp 49–67

Muzzarelli R, Jeuniaux C, Gooday GW (eds) (1986) Chitin in nature and technology. Plenum Press, New York

Myllykallio H, Lipowski G, Leduc D, Filee J, Forterre P, Liebl U (2002) An alternative flavin-dependent mechanism for thymidylate synthesis. Science 297:105–107

Nagasaki K, Yamaguchi M (1997) Isolation of a virus infectious to the harmful bloom causing microalga *Heterosigma akashiwo* (*Raphidophyceae*). Aquatic Microbial Ecol 13:135–140

Nagasaki K, Yamaguchi M (1998a) Effect of temperature on the algicidal activity and the stability of HaV (*Heterosigma akashiwo* virus). Aquatic Microbial Ecol 15:211–216

Nagasaki K, Yamaguchi M (1998b) Intra-species host specificity of HaV (*Heterosigma akashiwo* virus) clones. Aquatic Microbial Ecol 14:109–112

Nagasaki K, Tarutani K, Yamaguchi M (1999a) Cluster analysis on algicidal activity of HaV clones and virus sensitivity of *Heterosigma akashiwo* (*Raphidophyceae*). J Plankton Res 21:2219–2226

Nagasaki K, Tarutani K, Yamaguchi M (1999b) Growth characteristics of *Heterosigma akashiwo* virus and its possible use as a microbiological agent for red tide control. Appl Environ Microbiol 65:898–902

Nagasaki K, Tomaru Y, Katanozaka N, Shirai Y, Nishida K, Itakura S, Yamaguchi M (2004) Isolation and characterization of a novel single-stranded RNA virus infecting the bloom-forming diatom *Rhizosolenia setigera*. Appl Environ Microbiol 70:704–711

Nagasaki K, Shirai Y, Tomaru Y, Nishida K, Pietrokovski S (2005) Algal viruses with distinct intraspecies host specificities include identical intein elements. Appl Environ Microbiol 71:3599–3607

Nandhagopal N, Simpson AA, Gurnon JR, Yan X, Baker TS, Graves MV, Van Etten JL, Rossmann MG (2002) The structure and evolution of the major capsid protein of a large, lipid-containing DNA virus. Proc Natl Acad Sci U S A 99:14758–14763

Nelson M, Burbank DE, Van Etten JL (1998) Chlorella viruses encode multiple DNA methyltransferases. Biol Chem 379:423–428

Neupartl M, Meyer C, Woll I, Frohns F, Kang M, Van Etten JL, Kramer D, Hertel B, Moroni A, Thiel G (2008) Chlorella viruses evoke a rapid release of K$^+$ from host cells during early phase of infection. Virology 372:340–348

Nishida K, Suzuki S, Kimura Y, Nomura N, Fujie M, Yamada T (1998) Group I introns found in chlorella viruses: Biological implications. Virology 242:319–326

Nishida K, Kimura Y, Kawasaki T, Fujie M, Yamada T (1999) Genetic variation of chlorella viruses: variable regions localized on the CVK2 genomic DNA. Virology 255:376–384

Not F, Latasa M, Marie D, Cariou T, Vaulot D, Simon N (2004) A single species, *Micromonas pusilla* (Prasinophyceae), dominates the eukaryotic picoplankton in the western English channel. Appl Environ Microbiol 70:4064–4072

Not F, Massana R, Latasa M, Marie D, Colson C, Eikrem W, Pedros-Alio C, Vaulot D, Simon N (2005) Late summer community composition and abundance of photosynthetic picoeukaryotes in Norwegian and Barents Seas. Limnol Oceanogr 50:1677–1686

Odell M, Malinina L, Sriskanda V, Teplova M, Shuman S (2003) Analysis of the DNA joining repertoire of chlorella virus DNA ligase and a new crystal structure of the ligase-adenylate intermediate. Nucleic Acids Res 31:5090–5100

O'Kelly CJ, Sieracki ME, Thier EC, Hobson IC (2003) A transient bloom of *Ostreococcus* (Chlorophyta, Prasinophyceae) in the West Neck Bay, Long Island, New York. J Phycol 39:850–854

Onimatsu H, Suganuma K, Uenoyama S, Yamada T (2006) C-terminal repetitive motifs in Vp130 present at the unique vertex of the chlorovirus capsid are essential for binding to the host chlorella cell wall. Virology 353:432–442

Palenik B, Grimwood J, Aerts A, Rouzé P, Salamov A, Putnam N, Dupont C, Jorgensen R, Derelle E, Rombauts S, Zhou K, Otillar R, Merchant SS, Podell S, Gaasterland T, Napoli C, Gendler K, Manuell A, Tai V, Vallon O, Piganeau G, Jancek S, Heijde M, Jabbari K, Bowler C, Lohr M, Robbens S, Werner G, Dubchak I, Pazour GJ, Ren Q, Paulsen I, Delwiche C, Schmutz J, Rokhsar D, Van de Peer Y, Moreau H, Grigoriev IV (2007) The tiny eukaryote *Ostreococcus* provides genomic insights into the paradox of plankton speciation. Proc Natl Acad Sci U S A 104:7705–7710

Parke M, Manton I, Clarke B (1959) Studies on marine flagellates. V. Morphology and microanatomy of *Cyrysochromulina strobilus* sp. nov. J Marine Biol Assoc UK 38:169–188

Peters AF, Marie D, Scornet D, Kloareg B, Cock JM (2004) Proposal of *Ectocarpus siliculosus* (Ectocarpales, Phaeophyceae) as a model organism for brown algal genetics and genomics. J Phycol 40:1079–1088

Pires de Miranda M, Reading PC, Tscharke DC, Murphy BJ, Smith GL (2003) The vaccinia virus kelch-like protein C2L affects calcium-independent adhesion to the extracellular matrix and inflammation in a murine intradermal model. J Gen Virol 84:2459–2471

Plugge B, Gazzarrini S, Nelson M, Cerana R, Van Etten JL, Derst C, DiFrancesco D, Moroni A, Thiel G (2000) A potassium channel protein encoded by chlorella virus PBCV-1. Science 287:1641–1644

Que Q, Li Y, Wang IN, Lane LC, Chaney WG, Van Etten JL (1994) Protein glycosylation and myristylation in chlorella virus PBCV-1 and it antigenic variants. Virology 203:320–327

Raoult D, Audic S, Robert C, Abergel C, Renesto P, Ogata H, La Scola B, Suzan M, Claverie JM (2004) The 1.2 megabase genome sequence of mimivirus. Science 306:1344–1350

Reisser W (ed) (1992) Algae and symbioses. Biopress, Bristol, UK

Rodriguez-Trelles F, Tarrio R, Ayala FJ (2006) Origins and evolution of spliceosomal introns. Annu Rev Genet 40:47–76

Rohozinski J, Girton LE, Van Etten JL (1989) Chlorella viruses contain linear nonpermuted double strand DNA genomes with covalently closed hairpin ends. Virology 168:363–369

Sahlsten E (1998) Seasonal abundance in Skagerrak-Kattegat coastal waters and host specificity of viruses infecting the marine photosynthetic flagellate *Micromonas pusilla*. Aquatic Microbial Ecol 16:103–108

Sahlsten E, Karlson B (1998) Vertical distribution of virus-like particles (VLP) and viruses infecting *Micromonas pusilla* during late summer in the southeastern Skagerrak, North Atlantic. J Plankton Res 20:2207–2212

Sandaa RA, Heldal M, Castberg T, Thyrhaug R, Bratbak G (2001) Isolation and characterization of two viruses with large genome sizes infecting *Chrysochromulina ericina* (Prymnesiophyceae) and *Pyramimonas orientalis* (Prasinophyceae). Virology 290:272–280

Schroeder DC, Oke J, Malin G, Wilson WH (2002) Coccolithovirus (Phycodnaviridae): characterisation of a new large dsDNA algal virus that infects *Emiliania huxleyi*. Arch Virology 147:1685–1698

Schroeder DC, Oke J, Hall M, Malin G, Wilson WH (2003) Virus succession observed during an *Emiliania huxleyi* bloom. Appl Environ Microbiol 69:2484–2490

Schuster AM, Girton L, Burbank DE, Van Etten JL (1986) Infection of a chlorella-like alga with the virus PBCV-1: transcriptional studies. Virology 148:181–189

Shah R, Coleman CS, Mir K, Baldwin J, Van Etten JL, Grishin NV, Pegg AE, Stanley BA, Phillips MA (2004) *Paramecium bursaria* chlorella virus-1 encodes an unusual arginine decarboxylase that is a close homolog of eukaryotic ornithine decarboxylases. J Biol Chem 279:35760–35767

Shchelkunov SN, Totmenin AV (1995) Two types of deletions in orthopoxvirus genomes. Virus Genes 9:231–245

Shihra I, Krauss RW (1965) Chlorella physiology and taxonomy of forty-one isolates. College Park, MD, University of Maryland Press

Short SM, Suttle CA (2002) Sequence analysis of marine virus communities reveals that groups of related algal viruses are widely distributed in nature. Appl Environ Microbiol 68:1290–1296

Simpson AA, Nandhagopal N, Van Etten JL, Rossmann MG (2003) Structural analyses of Phycodnaviridae and Iridoviridae. Acta Crystallogr D Biol Crystallogr 59:2053–2059

Songsri P, Hamazaki T, Ishikawa Y, Yamada T (1995) Large deletions in the genome of chlorella virus CVK1. Virology 214:405–412

Stoltz DB (1971) The structure of icosahedral cytoplasmic deoxyriboviruses. J Ultrastruct Res 37:219–239

Strasser P, Zhang YP, Rohozinski J, Van Etten JL (1991) The termini of the chlorella virus PBCV-1 genome are identical 2.2-kbp inverted repeats. Virology 180:763–769

Sugimoto I, Onimatsu H, Fujie M, Usami S, Yamada T (2004) vAL-1, a novel polysaccharide lyase encoded by chlorovirus CVK2. FEBS Lett 559:51–56

Sun L, Li Y, McCullough AK, Wood TG, Lloyd RS, Adams B, Gurnon JR, Van Etten JL (2000) Intron conservation in a UV-specific DNA repair gene encoded by chlorella viruses. J Mol Evol 50:82–92

Suttle CA (2000a) Cyanophages and their role in the ecology of cyanobacteria. In: Whitton BA, Potts M (eds) The ecology of cyanobacteria: their diversity in time and space. Kluwer, Boston, pp 563–589

Suttle CA (2000b) Ecological, evolutionary and geochemical consequences of viral infection of cyanobacteria and eukaryotic algae. In: Hurst CJ (ed) Viral ecology. Academic, pp 247–296

Suttle CA, Chan AM (1995) Viruses Infecting the marine Prymnesiophyte *Chrysochromulina spp* – isolation, preliminary characterization and natural-abundance. Mar Ecolo Progr 118:275–282

Tai V, Lawrence JE, Lang AS, Chan AM, Culley AI, Suttle CA (2003) Characterization of HaRNAV, a single-stranded RNA virus causing lysis of *Heterosigma akashiwo* (Raphidophyceae). J Phycol 39:343–352

Tarutani K, Nagasaki K, Yamaguchi M (2000) Viral impacts on total abundance and clonal composition of the harmful bloom-forming phytoplankton *Heterosigma akashiwo*. Appl Environ Microbiol 66:4916–4920

Tarutani K, Nagasaki K, Yamaguchi M (2006) Virus adsorption process determines virus susceptibility in *Heterosigma akashiwo* (Raphidophyceae). Aquat Microbial Ecol 42:209–213

Tayefeh S, Kloss T, Thiel G, Hertel B, Moroni A, Kast SM (2007) Molecular dynamics simulation of the cytosolic mouth in Kcv-type potassium channels. Biochemistry 46:4826–4839

Tett P (1990) The photic zone. In: Herring PJ, Campbell AK, Whitfield M, Maddock L (eds) Light and life in the sea. Cambridge University Press, Cambridge, pp 59–87

Throndsen J (1996) Note on the taxonomy of *Heterosigma akashiwo* (Raphidophyceae). Phycologia 35:367

Tomaru Y, Tarutani K, Yamaguchi M, Nagasaki K (2004) Quantitative and qualitative impacts of viral infection on a *Heterosigma akashiwo* (Raphidophyceae) bloom in Hiroshima Bay, Japan. Aquatic Microbial Ecol 34:227–238

Tonetti M, Zanardi D, Gurnon JR, Fruscione F, Armirotti A, Damonte G, Sturla L, De Flora A, Van Etten JL (2003) *Paramecium bursaria* chlorella virus 1 encodes two enzymes involved in the biosynthesis of GDP-L-fucose and GDP-D-rhamnose. J Biol Chem 278:21559–21565

Tulman ER, Afonso CL, Lu Z, Zsak L, Sur JH, Sandybaev NT, Kerembekova UZ, Zaitsev VL, Kutish GF, Rock DL (2002) The genomes of sheeppox and goatpox viruses. J Virol 76:6054–6061

Turner PC, Moyer RW (1990) The molecular pathogenesis of poxviruses. Curr Top Microbiol Immunol 163:125–151

Van Etten JL (2003) Unusual life style of giant chlorella viruses. Annu Rev Genet 37:153–195

Van Etten JL, Meints RH (1999) Giant viruses infecting algae. Annu Rev Microbiol 53:447–494

Van Etten JL, Meints RH, Burbank DE, Kuczmarski D, Cuppels DA, Lane LC (1981) Isolation and characterization of a virus from the intracellular green-alga symbiotic with *Hydra viridis*. Virology 113:704–711

Van Etten JL, Meints RH, Kuczmarski D, Burbank DE, Lee K (1982) Viruses of symbiotic chlorella-like algae isolated from *Paramecium bursaria* and *Hydra viridis*. Proc Natl Acad Sci U S A 79:3867–3871

Van Etten JL, Burbank DE, Kuczmarski D, Meints RH (1983a) Virus-infection of culturable chlorella-like algae and development of a plaque-assay. Science 219:994–996

Van Etten JL, Burbank DE, Xia Y, Meints RH (1983b) Growth cycle of a virus, PBCV-1, that infects chlorella-like algae. Virology 126:117–125

Van Etten JL, Burbank DE, Joshi J, Meints RH (1984) DNA synthesis in a chlorella-like alga following infection with the virus PBCV-1. Virology 134:443–449

Van Etten JL, Schuster AM, Girton L, Burbank DE, Swinton D, Hattman S (1985) DNA methylation of viruses infecting a eukaryotic chlorella-like alga. Nucleic Acids Res 13:3471–3478

Van Etten JL, Burbank DE, Meints RH (1986) Replication of the algal virus PBCV-1 in UV-irradiated chlorella. Intervirology 26:115–120

Van Etten JL, Lane LC, Meints RH (1991) Viruses and virus-like particles of eukaryotic algae. Microbiol Rev 55:586–620

Van Etten JL, Graves MV, Muller DG, Boland W, Delaroque N (2002) Phycodnaviridae—large DNA algal viruses. Arch Virol 147:1479–1516

Veldhuis MJW, Wassmann P (2005) Bloom dynamics and biological control of a high biomass HAB species in European coastal waters: a Phaeocystis case study. Harmful Algae 4:805–809

Verity P, Brussaard C, Nejstgaard J, van Leeuwe M, Lancelot C, Medlin L (2007) Current understanding of *Phaeocystis* ecology and biogeochemistry, and perspectives for future research. Biogeochemistry 83:311–330

Villarreal LP, DeFilippis VR (2000) A hypothesis for DNA viruses as the origin of eukaryotic replication proteins. J Virol 74:7079–7084

Wang IN, Li Y, Que Q, Bhattacharya M, Lane LC, Chaney WG, Van Etten JL (1993) Evidence for virus-encoded glycosylation specificity. Proc Natl Acad Sci U S A 90:3840–3844

Westbroek P, Brown CW, Vanbleijswijk J, Brownlee C, Brummer GJ, Conte M, Egge J, Fernandez E, Jordan R, Knappertsbusch M, Stefels J, Veldhuis M, Vanderwal P, Young J (1993) A model system approach to biological climate forcing—the example of *Emiliania huxleyi*. Glob Planet Change 8:27–46

Wilson WH, Tarran GA, Schroeder D, Cox M, Oke J, Malin G (2002) Isolation of viruses responsible for the demise of an *Emiliania huxleyi* bloom in the English Channel. J Marine Biol Assoc UK 82:369–377

Wilson WH, Schroeder DC, Allen MJ, Holden MT, Parkhill J, Barrell BG, Churcher C, Hamlin N, Mungall K, Norbertczak H, Quail MA, Price C, Rabbinowitsch E, Walker D, Craigon M, Roy D, Ghazal P (2005a) Complete genome sequence and lytic phase transcription profile of a Coccolithovirus. Science 309:1090–1092

Wilson WH, Van Etten JL, Schroeder DS, Nagasaki K, Brussaard C, Delaroque N, Bratbak C, Suttle C (2005b) Family: Phycodnaviridae. In: Fauquet CM, Mayo MA, Maniloff J, Dusselberger U, Ball LA (eds) Virus taxonomy, VIIIth ICTV report. Elsevier/Academic, London, pp 163–175

Wolf S, Maier I, Katsaros C, Müller DG (1998) Virus assembly in *Hincksia hincksiae* (Ecocarpales, Phaeophyceae). An electron and fluorescence microscopic study. Protoplasma 203:153–167

Wilson WH, Schroeder DC, Ho J, Canty M (2006) Phylogenetic analysis of PgV-102P, a new virus from the English Channel that infects *Phaeocystis globosa*. J Marine Biol Assoc UK 86:485–490

Wolf S, Müller DG, Maier I (2000) Assembly of a large icosahedral DNA virus, MclaV-1, in the marine alga *Myriotrichia clavaeformis* (Dictyosiphonales, Phaeophyceae). Eur J Phycol 35:163–171

Worden AZ (2006) Picoeukaryote diversity in coastal waters of the Pacific Ocean. Aquatic Microbial Ecol 43:165–175

Worden AZ, Nolan JK, Palenik B (2004) Assessing the dynamics and ecology of marine picophytoplankton: the importance of the eukaryotic component. Limnol Oceanogr 49:168–179

Wrigley NG (1969) An electron microscope study of the structure of Serocesthis iridescent virus. J Gen Virol 5:123–134

Xia Y, Burbank DE, Uher L, Rabussay D, Van Etten JL (1986) Restriction endonuclease activity induced by PBCV-1 virus infection of a chlorella-like green alga. Mol Cell Biol 6:1430–1439

Xia Y, Burbank DE, Uher L, Rabussay D, Van Etten JL (1987) IL-3A virus infection of chlorella-like green alga induces a DNA restriction endonuclease with novel sequence specificity. Nucleic Acids Res 15:6075–6090

Xia Y, Morgan R, Schildkraut I, Van Etten JL (1988) A site-specific single strand endonuclease activity induced by NYs-1 virus infection of a chlorella like green alga. Nucleic Acids Res 16:9477–9487

Yamada T, Higashiyama T (1993) Characterization of the terminal inverted repeats and their neighboring tandem repeats in the chlorella CVK1 virus genome. Mol Gen Genet 241:554–563

Yamada T, Kawasaki T (2005) Microbial synthesis of hyaluronan and chitin: New approaches. J Biosci Bioeng 99:521–528

Yamada T, Tamura K, Aimi T, Songsri P (1994) Self-splicing group I introns in eukaryotic viruses. Nucleic Acids Res 22:2532–2537

Yamada T, Onimatsu H, Van Etten JL (2006) Chlorella viruses. In: Advances in virus research. Academic, pp 293–336

Yan X, Olson NH, Van Etten JL, Bergoin M, Rossmann MG, Baker TS (2000) Structure and assembly of large lipid-containing dsDNA viruses. Nat Struct Biol 7:101–103

Yan X, Chipman PR, Castberg T, Bratbak G, Baker TS (2005) The marine algal virus PpV01 has an icosahedral capsid with T=219 quasisymmetry. J Virol 79:9236–9243

Zeidner G, Preston CM, Delong EF, Massana R, Post AF, Scanlan DJ, Béjà O (2003) Molecular diversity among marine picophytoplankton as revealed by psbA analyses. Environ Microbiol 5:212–216

Zhang Y, Nelson M, Nietfeldt J, Xia Y, Burbank DE, Ropp S, Van Etten JL (1998) Chlorella virus NY-2A encodes at least twelve DNA endonuclease/methyltransferase genes. Virology 240:366–375

Zhang Y, Maley F, Maley GF, Duncan G, Dunigan DD, Van Etten JL (2007) Chloroviruses encode a bifunctional dCMP-dCTP deaminase that produces two key intermediates in dTTP formation. J Virol 81:7662–7671

Zhang YP, Strasser P, Grabherr R, Van Etten JL (1994) Hairpin loop structure at the termini of the chlorella virus PBCV-1 genome. Virology 202:1079–1082

Zingone A, Sarno D, Forlani G (1999) Seasonal dynamics in the abundance of *Micromonas pusilla* (Prasinophyceae) and its viruses in the Gulf of Naples (Mediterranean Sea). J Plankton Res 21:2143–2159

Zingone A, Natale F, Biffali E, Borra M, Forlani G, Sarno D (2006) Diversity in morphology, infectivity, molecular characteristics and induced host resistance between two viruses infecting *Micromonas pusilla*. Aquatic Microbial Ecol 45:1–14

Ziveri P, Broerse ATC, van Hinte JE, Westbroek P, Honjo S (2000) The fate of coccoliths at 48 degrees N 21 degrees W, northeastern Atlantic. Deep-Sea Research. Part II. Topical Studies in Oceanography 47:1853–1875

# African Swine Fever Virus

**E.R. Tulman, G.A. Delhon, B.K. Ku, D.L. Rock(⊠)**

**Contents**

**Abstract** African swine fever virus (ASFV) is a large, intracytoplasmically-replicating DNA arbovirus and the sole member of the family *Asfarviridae*. It is the etiologic agent of a highly lethal hemorrhagic disease of domestic swine and therefore extensively studied to elucidate the structures, genes, and mechanisms affecting viral replication in the host, virus–host interactions, and viral virulence. Increasingly apparent is the complexity with which ASFV replicates and interacts with the host cell during infection. ASFV encodes novel genes

D.L. Rock
Department of Pathobiology, College of Veterinary Medicine,
University of Illinois, Urbana, IL 61802, USA
dlrock@uiuc.edu

James L. Van Etten (ed.) *Lesser Known Large dsDNA Viruses.*
Current Topics in Microbiology and Immunology 328.
© Springer-Verlag Berlin Heidelberg 2009

involved in host immune response modulation, viral virulence for domestic swine, and in the ability of ASFV to replicate and spread in its tick vector. The unique nature of ASFV has contributed to a broader understanding of DNA virus/host interactions.

## Introduction

African swine fever virus (ASFV) is a large, enveloped virus containing a double stranded (ds) DNA genome of approximately 190 kilobase pairs (kbp). ASFV shares aspects of genome structure and replication strategy with other large dsDNA viruses, most notably with poxviruses. ASFV and poxviruses replicate in the cytoplasm of the infected cell, primarily in discrete perinuclear assembly sites referred to as virus factories. They also exhibit temporal regulation of gene expression and have similar genome structures, including terminal inverted repeats, terminal crosslinks, a central conserved region and variable regions at each end of the genome.

Although initially classified as an iridovirus based largely on virion morphology, increasing knowledge of ASFV molecular biology led to its reclassification as the sole member of a new DNA virus family, *Asfarviridae* (*Asfar*, African swine fever and related viruses), which shares general features with other families of large, nucleocytoplasmically replicating DNA viruses, including the *Poxviridae*, *Iridoviridae*, and *Phycodnaviridae* (Dixon et al. 2000). Cladistic analysis of gene complements supports monophyly of ASFV, poxviruses, iridoviruses, phycodnaviruses, and Mimiviruses (Iyer et al. 2001, 2006). These viruses either replicate exclusively in the cytoplasm, or start their life cycle in the nucleus but complete it in the cytoplasm of the infected cell, and some are relatively independent of the host cell transcriptional machinery for replication.

ASFV is the only known DNA arbovirus. In sub-Saharan Africa, ASFV is maintained in a sylvatic cycle between wild swine (warthogs and bushpigs) and argasid ticks of the genus *Ornithodoros*. Unlike domestic swine, wild swine infected with ASFV are generally asymptomatic, with low viremia titers. Most adult warthogs in ASFV enzootic areas are seropositive and persistently infected. ASFV persistently infects ticks of the *Ornithodoros* spp. from which ASFV can be isolated years postinfection (p.i.). ASFV infection of domestic swine results in several disease forms, ranging from highly lethal to subclinical depending on contributing viral and host factors. The ability of ASFV to replicate and efficiently induce marked cytopathology in macrophages in vivo appears to be a critical factor in ASFV virulence in domestic swine. Currently, there is no vaccine available for ASF and the disease is controlled by animal quarantine and slaughter.

Progress has been made in defining the genetic basis of ASFV virulence and host range; however, our overall understanding of the complex mechanisms underlying viral-swine/tick–host interactions and how they impact infection outcomes remains rudimentary.

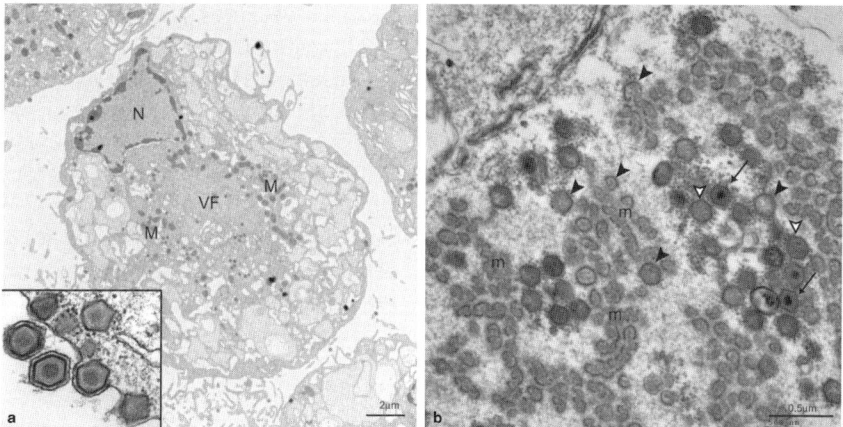

**Fig. 1a, b** Electron micrographs of ASFV-infected swine macrophages. Primary porcine macrophage cell cultures were infected with a pathogenic African ASFV isolate (Pr4) and examined 16 h postinfection (p.i.). **a** ASFV-infected macrophage with a cytoplasmic virus factory (*VF*) adjacent to the nucleus (*N*) and recruiting mitochondria (*M*). Insert: mature virions budding from cell membrane. **b** Higher magnification of a virus factory with virions at different stages of morphogenesis, including membrane precursor material (*m*), intermediate partially formed capsids (*filled arrowheads*), empty capsids (*open arrowheads*), and virions containing nucleoid (*arrows*). (Electron micrographs kindly provided by Dr. Thomas Burrage, Plum Island Animal Disease Center, ARS USDA)

## ASFV Virion

The ASFV virion is comprised of more than 50 polypeptides and has a complex but regular structure by electron microscopy, icosahedral in symmetry and containing several concentric layers for an overall diameter of approximately 200 nm (Breese and DeBoer 1966; Carrascosa et al. 1984, 1985; Estevez et al. 1986; Schloer 1985) (Fig. 1A). The 80-nm virion core is composed of an electron-dense nucleoprotein structure, or nucleoid, enclosed by a thick protein layer, referred to as the core shell or matrix, which contains several viral proteins estimated to comprise about one-third of the virion protein mass (Andres et al. 1997, 2002). Surrounding the core are two lipid bilayers, originally called the inner membrane and likely derived from collapsed cisternae of host endoplasmic reticulum (Andres et al. 1997, 1998; Rouiller et al. 1998).

External to the inner membrane is the capsid, composed of the structural protein p72 (also referred to as p73), which comprises approximately one-third the protein content of the virion, and providing the icosahedral structure to the virion (Andres et al. 1997; Carrascosa et al. 1986; Garcia-Escudero et al. 1998; Tabares et al. 1980a). ASFV capsomeres are arranged in a hexagonal lattice and individually consist of a hexagonal arrangement 13 nm in diameter with a central hole (Carrascosa et al. 1984). The triangulation number of the ASFV capsid has been estimated using capsid dimensions and intercapsid distances (7.4–8.1 nm) to be T=189 to T=217, suggesting that 1892 to 2172 capsomers comprise the ASFV capsid (Carrascosa et al. 1984). Covering the capsid is a loose external membrane obtained by virion budding through the plasma membrane and removable with osmotic shock or detergent; this membrane is not required for virus infection (Andres et al. 2001b; Breese and DeBoer 1966; Carrascosa et al. 1984; Moura Nunes et al. 1975).

ASFV virions are closely associated with cellular membranes, which previously were a source of cellular or immature viral component contamination in semi-purified virion preparations (Carrascosa et al. 1985; Estevez et al. 1986). Isolation of ASFV particles free from cellular membrane contaminants is greatly improved using Percoll density gradients, which yield virions with a sedimentation coefficient of 3,500±300S, DNA-protein ratio of 0.18±0.02, and a specific infectivity of $2.7\times10^7$ PFU/mg of protein (Carrascosa et al. 1985). As many as 15 protein components of the virion migrate in SDS-PAGE at the same molecular weight as cellular proteins and/or share immunoreactivity with cellular proteins, including actin, α-tubulin, and β-tubulin (Carrascosa et al. 1985; Estevez et al. 1986). An early immunoelectron microscopy study using monoclonal antibodies localized several viral proteins within the virion, including those in external (p14, p24), intermediate (p12, p72, p17, p37), and nucleoid (p150) layers (Carrascosa et al. 1986). Similar to what has been found in poxvirus virions, ASFV virions contain enzymatic activities that contribute to early events in, and activities critical for, viral replication in the cell cytoplasm, including RNA polymerase, nucleoside triphosphate phosphohydrolase, topoisomerase, mRNA capping, and protein kinase activity (Kuznar et al. 1980, 1981; Polatnick 1974; Salas et al. 1981 1983). A lipid-modified form of ubiquitin has also been found incorporated into mature ASFV virions (Webb et al. 1999).

# The ASFV Genome

The ASFV genome is a dsDNA molecule whose general properties have been reviewed (Viñuela 1985). Initial studies of the genome from Vero cell-adapted ASFV derived from the virulent Badajoz 1971 (BA71) Spanish outbreak isolate (BA71V) indicated a size of approximately 170 kbp (Almendral et al. 1984; Enjuanes et al. 1976). Sedimentation rates and/or reannealing kinetics following genomic and restriction fragment denaturation, and the formation of S1 nuclease-resistant heteroduplexes by terminally located restriction fragments, indicated the presence of terminal-genomic covalent crosslinks (Almendral et al. 1984; Ley et al. 1984; Ortin et al. 1979). BA71V contained terminal hairpin loops of 37 A+T-rich, incompletely base-paired nucleotides that were inverted and complementary relative to the opposite terminus and which could be isolated as dimeric forms from infected cells, suggestive of head–head and tail–tail concatemeric replicative intermediates (Gonzalez et al. 1986). Additional inverted sequence repetition extended 2.1 kbp from the genomic termini, as seen through electron microscope examination of heteroduplexes formed upon reannealing, and by southern cross-hybridization, of denatured terminal restriction fragments (Almendral et al. 1984; Sogo et al. 1984). These terminal inverted repeat (TIR) sequences were subsequently shown to largely comprise tandem repeats of 27–35 base pairs (bp) and to be identical between genomic termini (de la Vega et al. 1994; Meireles and Costa 1994). Also present near ASFV terminal hairpins were sequences resembling those found to mediate resolution of concatemeric replicative intermediates in poxviruses (de la Vega et al. 1994; Meireles and Costa 1994). Overall, these properties reflected size and structural similarities between the ASFV genome and those of viruses within the family *Poxviridae* (Viñuela 1985).

The Vero cell-adapted ASFV strain BA71V became a prototype for molecular and genomic studies, although it was recognized that cell-culture-passaged viruses had genomic differences, expressed as restriction fragment differences, relative to field isolates of ASFV, some of which have genome lengths approaching 190 kbp (Blasco et al. 1989a; Santurde et al. 1988; Tabares et al. 1987; Wesley and Pan 1982). Comparative restriction mapping studies revealed notable aspects of ASFV genomic variability. Diverse African ASFV isolates have less conserved restriction profiles than European and American isolates, indicating greater genomic heterogeneity among viruses from geographic regions supporting the natural transmission cycle and indicating a common origin for non-African isolates (Blasco et al. 1989a; Dixon and Wilkinson 1988; Wesley and Tuthil 1984). Genomic homogeneity among African ASFV isolates associated with disease outbreaks in domestic swine relative to isolates isolated from ticks has also been noted (Dixon and Wilkinson 1988; Sumption et al. 1990). Subsequent molecular phylogenetic studies utilizing part of the p72 gene support some of these findings, including relative homogeneity among West African, European, and American isolates, homogeneity among certain African lineages associated with outbreaks in domestic swine, and relative heterogeneity among isolates from southern and East Africa (Bastos et al. 2003; Lubisi et al. 2003; Wambura et al. 2006).

The ASFV genome contains discrete regions and elements of variability discernible through restriction mapping. ASFV isolates and cell culture-adapted viruses contain a conserved, centrally located 125-kbp genomic core in which major insertion-deletion events are rare, leaving larger-scale variability confined to the left 38- to 47-kbp and right 13- to 16-kbp terminal regions of the ASFV genome (Blasco et al. 1989a; Sumption et al. 1990; Tabares et al. 1987) (Fig. 2). Terminal region variability predominates in repetitive sequence 7–27 kbp from the left genomic terminus and accounts for 9.5 kbp deleted in BA71V relative to parental BA71 (Aguero et al. 1990; Blasco et al. 1989a, 1989b; Sumption et al. 1990). Sequence analysis of these larger variable regions reveal multigene families (MGFs), arrays of often tandemly arranged genes that share sequence similarity and are of different copy number and complement among ASFV isolates (Almendral et al. 1990; Gonzalez et al. 1990). Small-scale variability has been noted in the ASFV genome, including central variable and minisatellite-like regions containing tandem repeats of different lengths and ten variable genomic regions discernible between closely related African isolates (Almazan et al. 1995; Dixon et al. 1990; Irusta et al. 1996; Sumption et al.

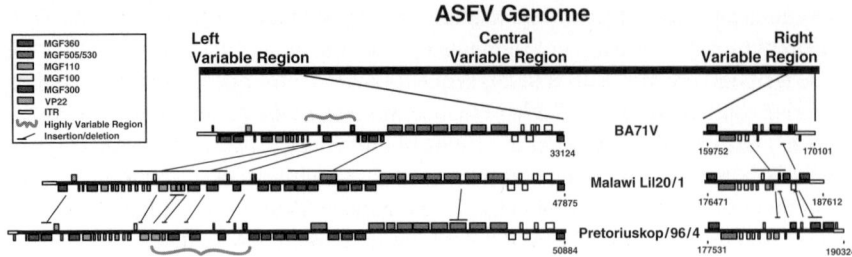

**Fig. 2** Variable genomic regions of ASFV. Major features of the ASFV genome (*top*) include a central conserved region, central variable region, left and right terminal genomic regions, and terminal inverted repeat sequences. Terminal genomic regions contain tandem arrays of multigene families (*MGF*). MGF copy numbers are variable between viral strains, which here include the prototypical, but cell culture-adapted, strain BA71V and virulent field isolates (*bottom*, GenBank Accession numbers ASU18466, AY261361, and AY261363). Gene insertion and deletion occurs in multiple MGF regions (indicated here by *lines* marking regions absent at homologous loci), and often in the region containing MGF110 and in a region identified as highly variable among field isolates (indicated by *bracket*). Given the function of certain MGF genes in viral virulence and host range, these terminal genomic differences likely contribute to phenotypic differences between viral isolates. (Figure modified from Balinsky and Afonso, unpublished data)

1990). These variable regions, in particular the central variable region, have been used as epidemiological markers to discriminate between similar genotypes (Bastos et al. 2004; Boshoff et al. 2007; Lubisi et al. 2007; Nix et al. 2006 ; Phologane et al. 2005 ). Length heterogeneity has been found among clones of a single isolate, both in TIR repeats and in terminal unique sequences (Aguero et al. 1990; Blasco et al. 1989a, 1989b; de la Vega et al. 1994; Santurde et al. 1988).

The ASFV genome contains several distinct MGFs, originally characterized as genes present in repetitive sequence in terminal genomic regions and named to reflect the average lengths of the predicted gene product (i.e., MGF110 and MGF360 genes encode proteins of approximately 110 and 360 amino acids, respectively; Almendral et al. 1990; Gonzalez et al. 1990) (Fig. 2). MGF110 and MGF360 genes are present in the highly variable left-terminal genomic region, including genes absent in BA71V and other European isolates, and in the right terminal genomic region (Almendral et al. 1990; de la Vega et al. 1990; Gonzalez et al. 1990; Vydelingum et al. 1993). Multiple MGF110 gene deletions have been found in variants of a single ASFV isolate, including viruses that appeared to lack all copies of MGF110 but remained virulent in domestic swine (Aguero et al. 1990). Other ASFV MGF include MGF300, MGF505, and MGF100 (J.M. Rodriguez et al. 1994; Vydelingum et al. 1993; Yozawa et al. 1994). MGF300 is located adjacent to highly variable noncoding repeat sequences in the left terminal region (Almazan et al. 1995; Yozawa et al. 1994). MGF505 is referred to as MGF530 in strain Malawi Lil20/1 and, like MGF300, encodes products sharing similarity to those of MGF360 (J.M. Rodriguez et al. 1994; Yozawa et al. 1994). Proposed mechanisms of MGF variability include in-frame recombination of homologous genes, duplications, deletions, and terminal genomic sequence transpositions (de la Vega et al. 1990; Vydelingum et al. 1993).

Initial ASFV genomic sequence analysis was limited to BA71V, 55 kbp from the right end of the pathogenic African isolate Malawi Lil20/1, and partial sequences from a number of pathogenic and nonpathogenic isolates (Dixon 1994; Yanez 1995). While structurally resembling poxviral genomes, inter- and intragenic tandem repetition occurs with greater frequency in the ASFV genome (Dixon 1994; Yanez 1995). The BA71V genome contains 151 genes predicted as major open reading frames (ORFs). Although two small ORFs are present in the TIR, all others are present in the unique coding region and distributed on both strands, with a preference for transcriptional orientation toward the genomic terminus, as often observed in poxvirus genomes, apparent only in the left-terminal 17 kbp of the BA71V genome (Dixon 1994; Yanez 1995). Analysis of the complete BA71V genome sequence confirmed the predominance of MGF in ASFV terminal genomic regions, including five MGF110, two MGF100, 12 MGF360, eight MGF505/530 and four MGF300 genes. Lil20/1 contained an additional two MGF100, one MGF110 and one p22 structural membrane protein gene present only in the left terminus of BA71V (Camacho and Viñuela 1991; Vydelingum et al. 1993). Subsequent genomic sequencing of seven pathogenic African field isolates has confirmed that while MGF genes contribute to most of the ASFV variability in gene complement, particularly among MGF110 genes which are present in six to 14 copies, ASFV field isolates retained more MGF genes than BA71V, and that these sequences are largely responsible for the larger genome lengths of approximately 185–194 kbp (D.L. Rock, unpublished data; GenBank accession numbers AY261360–AY261366) (Fig. 2).

The ASFV genome encodes a complement of proteins indicative of its complex virion structure and intracytoplasmic replication cycle, with most encoded within the central genomic core identified as relatively conserved among different virus isolates (Dixon 1994; Yanez 1995). These include membrane and other structural proteins known to be present in the virus particle, and those that more recently have been shown to affect different stages of virion morphogenesis in the infected cell (Afonso et al. 1992; Alcami et al. 1992, 1993; Brookes et al. 1998b; Camacho and Viñuela 1991; Lopez-Otin et al. 1988, 1990; Munoz et al. 1993; F. Rodriguez et al. 1994; Simon-Mateo et al. 1995; Sun et al. 1995, 1996). Other ASFV proteins share sequence similarity to cellular proteins or enzymes, including those involved in aspects of nucleotide metabolism, DNA replication and repair, transcription, and protein modification, and those that likely account for enzymatic activities present in ASFV virions or induced in infected cells (Baylis et al. 1992, 1993a; Blasco 1990; Boursnell et al. 1991; Freije et al. 1993; Hammond et al. 1992; Lu et al. 1993; Martin Hernandez and Tabares 1991; Martins et al. 1994; Rodriguez et al. 1993b; Yanez 1993; Yanez et al. 1993a, 1993b, 1993c). Several of these proteins appear to be distantly related to homologs identified in poxviruses (Baylis et al. 1993b; Blasco 1990; Boursnell et al. 1991; Freije et al. 1993; Martin Hernandez and Tabares 1991; Roberts et al. 1993; Yanez et al. 1993b). Indeed, cladistic analysis of conserved gene complements, including genes assumed to be distantly related protein orthologs, have led to the conclusion that ASFV not only shares a common origin with poxviruses but also with other large, nuclear-cytoplasmically replicating DNA viruses, including members of the families *Iridoviridae*, *Phycodnaviridae* and *Mimiveridae* (Iyer et al. 2001; Iyer et al. 2006).

Despite these similarities, the ASFV genome remains divergent relative to those of other viral families, with the few homologs in other viruses distantly related, a lack of gene co-linearity, and relatively unique overall gene complement (Yanez 1995). Additional enzymatic components encoded in the ASFV genome include homologs of cellular ubiquitin conjugating enzyme, trans-prenyltransferase, NifS-like protein, and components of a base-excision repair pathway (Hingamp et al. 1992; Rodriguez et al. 1992). ASFV also encodes proteins predicted to mediate virus–host interaction, virulence, and mechanisms that enhance the ability of the virus to successfully replicate within the host, including homologs of cellular inhibitor of apoptosis (IAP), Bcl-2, IKB, myeloid differentiation primary response antigen MyD116, lectin-like, and CD2 proteins (Borca et al. 1994b; Neilan et al. 1993a; Rodriguez et al. 1993a; Sussman et al. 1992). Notably, several of these putative virulence/host range proteins, along with certain MGF proteins, the central variable region protein 9-RL (pB602L as annotated in BA71V), and the variable tandem repeat-containing structural protein p54 (pE183L) (Irusta et al. 1996; F. Rodriguez et al. 1994; Sun et al. 1995), are among the most variable among multiple field isolates (D.L. Rock, unpublished data). While these proteins have been characterized through sequence similarity and/or biologically, the many ASFV proteins currently lacking predicted functions may yet prove to mediate novel functions involving virus–host interactions.

## ASFV-Cell Interactions

### *Virus Entry*

Several viral proteins bind to the cell surface, including the conserved late protein p12 (pO61R), the variable structural protein p54, and the early membrane protein p30 (also referred to as p32 and pCP204L) (Carrascosa et al. 1991; Gomez-Puertas et al. 1998). p12 is a predicted transmembrane protein that is localized external to the capsid layer of the virion and that binds to susceptible cells, including swine macrophages (Angulo et al. 1993; Carrascosa et al. 1993 1991). Although purified recombinant p12 blocked specific binding of virus particles to susceptible cells, anti-p12 antibodies failed to neutralize virus infectivity in vitro or protect swine against ASFV infection (Angulo et al. 1993; Carrascosa et al. 1995). Neutralizing antibodies specific for p54 or for the major capsid protein p72 (pB646L) inhibited virus attachment to swine macrophages, while neutralizing antibodies specific for p30 blocked infection even after virus attachment to these cells (Gomez-Puertas et al. 1996). In addition, p54 and p30 bound to macrophages, with p54 blocking specific binding of virus particles and p30 blocking virus internalization (Gomez-Puertas et al. 1998).

These results suggest a role for p72 and p54 in virus attachment and for p30 in virus internalization. Notably, neutralizing antibodies specific for these proteins are not sufficient for mediating protection in swine (Neilan et al. 2004).

Although the identity of the cell receptor(s) that mediates ASFV entry is unknown, saturable binding sites for ASFV have been demonstrated on the surface of susceptible cells, including swine macrophages (Alcami et al. 1989a, 1989b, 1990). In a study on cell susceptibility to virus infection, it was found that the most restricted cell lines showed reduced efficiency of virus binding, virus internalization, and fewer p12 binding sites (Carrascosa et al. 1999). Correlation of CD163 expression on the surface of swine monocyte/macrophages with susceptibility to ASFV infection, and the ability to inhibit ASFV infection of and binding to macrophages with anti-CD163 antibody, suggests some role for CD163 in macrophage infection (Sánchez-Torres et al. 2003).

Internalization of viral particles is a temperature-, energy-, and acid-dependent process with morphological features suggestive of receptor-mediated endocytosis, with ASFV-cell fusion likely occurring not at the plasma membrane but rather between the viral envelope and endosome membrane (Alcami et al. 1989a; Geraldes and Valdeira 1985; Valdeira et al. 1998; Valdeira and Geraldes 1985). It has been hypothesized that cytosolic viral cores are transported via retrograde transport on microtubules to viral assembly sites near the microtubule organizing center (MTOC) (Heath et al. 2001). Indeed, ASFV interacts with microtubules in vitro, the p54 attachment protein interacts with the microtubular motor complex through binding of the motor protein dynein light-chain, and pharmacologic disruption of microtubules or expression of the dominant negative dynein motor protein p50/dynamitin inhibits DNA synthesis, late protein accumulation, and viral replication, indicating that microtubule-mediated transport may be important for early events in ASFV infection (Alonso et al. 2001; Carvalho et al. 1988; de Matos and Carvalho 1993; Heath et al. 2001). Fluorescently tagged ASFV and time-lapse confocal microscopy indicates that, at early times p.i., microtubule-dependent intracellular movement of individual virions in perinuclear areas is intermittent and without apparent direction, averaging 0.2–0.5 μm/s (Hernaez et al. 2006).

## Viral Gene Expression

### Viral Transcription

The ASFV virion contains DNA-dependent RNA polymerase activity and the ability to synthesize and posttranscriptionally modify viral transcripts in vitro when permeabilized and incubated with ribonucleotides, suggesting that ASFV initiates viral gene transcription immediately upon infection and independent of cellular enzymes (Kuznar et al. 1980; Salas et al. 1981). Indeed, inhibition of cellular RNA polymerase II indicates it is not essential for ASFV replication (Salas et al. 1988a).

In addition, the ASFV genome encodes multiple subunits of a viral RNA polymerase (Lu et al. 1993; Yanez et al. 1993a; Yanez 1995). ASFV infection of cells in the presence of protein or DNA synthesis inhibitors results in viral transcripts that resemble those derived from ASFV virions in vitro, mapping to similar regions in the viral genome and yielding translation products of similar size distribution (Salas 1986). These viral transcripts and proteins synthesized in the presence of the DNA synthesis inhibitor cytosine arabinoside are considered early ASFV genes, with genes expressed exclusively after viral DNA replication classified as late genes (Carvalho and Rodrigues-Pousada 1986; Estevez et al. 1986; Salas 1986; Santaren and Viñuela 1986). Thus, ASFV gene expression, like that of members of the *Poxviridae*, consists of early transcription of specific viral genes and a subsequent late transcriptional phase in which de novo protein synthesis and the onset of viral DNA replication is required (Moss et al. 1991).

Early ASFV transcripts synthesized in vitro contain methylated cap structures and are polyadenylated, and they lack internal methylations (Salas et al. 1981). The majority (~80%) of these in vitro ASFV cap structures are $m^7G(5')ppp(5')A^m$, indicating preference for capped mRNA to be initiated with ATP and presence of RNA triphosphatase, guanylyltransferase, (guanine-N7) methyltransferase, and (nucleoside-2'-O-)methyltransferase activity in the ASFV virion (Salas et al. 1981). A small proportion of ASFV cap structures are methylated only at the penultimate nucleoside, suggesting that 2'-hydroxyl methylation of the ASFV transcript's initial 5' nucleoside does not require a methylated terminal guanosine as does the vaccinia virus 2'-O-methyltransferase (Barbosa and Moss 1978; Salas et al. 1981). The average length of 3' poly(A) chains found on ASFV transcripts in vitro are 33 nucleotides (Salas et al. 1981). ASFV encodes a guanylyltransferase (pNP868R) similar to the heterodimeric vaccinia virus capping enzyme large subunit that performs both triphosphatase and guanylyltransferase activities; however, it is not known if ASFV proteins have the two additional capping activities and poly(A) polymerase functions (Pena et al. 1993; Shuman 1989; Yanez 1995). An ASFV-encoded Nudix (nucleoside diphosphate X) hydrolase (pD250R), while sharing sequence similarity with the vaccinia virus D10 protein that affects transcript stability through removal of mRNA cap structures, exhibits higher hydrolytic activity toward diphosphoinositol polyphosphates and dinucleoside polyphosphates than toward cap analogs in vitro (Cartwright et al. 2002; Parrish et al. 2007).

Detailed mapping and kinetic analysis have resulted in the grouping of ASFV genes into four transcriptional classes. ASFV early gene expression in infected cells is detectable using a nuclease protection assay as early as 2 h p.i., with a plateau in synthesis at 4–8 h p.i. (Almazan et al. 1992). Early ASFV transcripts contain short 5' untranslated leader sequences and have multiple, discrete 3' ends that map to runs of seven or more intergenic thymidylate residues in the coding strand downstream of the translational stop codon, sequences potentially acting as signals for mRNA termination (Almazan et al. 1992; Rodriguez et al. 1993b). Late gene expression is completely dependent on viral DNA replication and reaches a maximum at 12–16 h p.i., with data suggesting that late transcripts also contain

5' untranslated sequences and terminate at polythymidylate tracts (Almazan et al. 1993; Cistue and Tabares 1992; Galindo et al. 2000b).

While the majority of ASFV early genes are expressed throughout infection at both early and late times but strongest prior to DNA replication, a subclass of early genes are expressed only transiently prior to DNA synthesis (Almazan et al. 1992; Carvalho and Rodrigues-Pousada 1986; Estevez et al. 1986; Salas 1986; Santaren and Viñuela 1986). These transcripts have been classified as immediate early and have maximum expression at 3 h p.i., undergoing a subsequent decrease in expression until ultimately ceasing prior to DNA replication (Almazan et al. 1992). The decrease in immediate early gene expression is also dependent on protein expression, as immediate early genes exhibit early transcriptional kinetics in cells infected in the presence of the protein synthesis inhibitor cycloheximide (Almazan et al. 1992).

An intermediate ASFV transcriptional class with distinct transcriptional kinetics has also been characterized (J.M. Rodriguez et al. 1996). While dependent on DNA replication similar to late transcripts, intermediate transcripts utilize distinct transcriptional start sites and display distinct kinetics, first detected 4–6 h p.i., maximal expression occurring 6–8 h p.i., and decreasing levels during maximal late gene expression. Similar intermediate genes in vaccinia virus are transcribed from replicated viral DNA and they encode transcription factors required for late gene expression (Keck et al. 1990).

Cis- and trans-acting transcriptional elements specific for different temporal classes remain to be completely defined. The late gene encoding the p72 capsid protein utilizes a promoter region within −36 and +5 of the transcriptional start, including regions of greater importance both around position −13 and a conserved TATA sequence at the transcriptional start site (Garcia-Escudero and Viñuela 2000). Similar TATA sequences arc required for efficient transcription of at least three additional late genes (Garcia-Escudero and Viñuela 2000). Two characterized intermediate genes share highly conserved sequence at positions −25 to −15 and −9 to +9 relative to the translational start codon, but the significance of this sequence to intermediate transcriptional regulation is unknown (J.M. Rodriguez et al. 1996). Sequences 139 bases upstream of one of these genes (*I243L* or *K9L*) were required for efficient expression of a reporter gene transfected into ASFV-infected cells (Yates et al. 1995). Notably, ASFV promoter regions can be recognized, and ASFV genes expressed, in the context of vaccinia virus infection, further suggesting functional similarities between ASFV and poxvirus replicative machinery and transcriptional regulation despite a lack of obvious sequence similarity in the promoter sequences (Hammond and Dixon 1991).

ASFV trans-acting factors remain largely unidentified, with few obvious homologs to vaccinia virus-encoded transcription factors. One ASFV gene expressed at early, intermediate and late times encodes a protein (pI243L or K9L) similar to eukaryotic transcription factor SII family proteins and to vaccinia virus E4L, a multifunctional protein that acts as both a subunit of the viral RNA polymerase and as a transcription factor for vaccinia virus intermediate gene expression (J.M. Rodriguez et al. 1996; Rosales et al. 1994). A second ASFV gene (*Q706L*) has been speculated to encode a divergent homolog of the poxvirus early transcription factor small subunit, but since it is a helicase-like protein, it also shares sequence similarity with the pox-viral early transcription termination factor (Baxter et al. 1996; Baylis et al. 1993b; Freije et al. 1993; Roberts et al. 1993; Yanez et al. 1993b).

## Protein Synthesis and Modification

ASFV protein expression begins at early times postinfection and follows transcriptional kinetics, yielding nonstructural and structural proteins that may undergo various posttranslational modifications. Viral early protein expression is detected as soon as 2 h p.i., and late protein expression begins around the onset of DNA replication. Approximately 100 virus-induced proteins of 10–220 kDa can be detected in infected cells, including about twice as many late proteins as early proteins (Escribano and Tabares 1987; Estevez et al. 1986; Rodriguez et al. 2001; Santaren and Viñuela 1986; Tabares et al. 1980b, 1983; Urzainqui et al. 1987). More than 50 structural proteins of 11.5–150 kDa can be detected in purified virions by 2D gel electrophoresis, with one- to two-thirds as many detected in 1D gels (Carrascosa et al. 1985; Estevez et al. 1986; Schloer 1985; Tabares et al. 1980a). While viral proteins are often localized to viral factories, other patterns of expression have been observed, including nuclear, diffuse cytoplasmic, and cellular membrane staining using monoclonal antibodies or the presence of viral proteins in nuclear cell fractions (Sanz et al. 1985; Tabares et al. 1980b, 1983). The observation that p72 mRNA is evenly distributed throughout the cytoplasm while p72 is localized predominantly to viral factories suggests that proteins translated in the cytoplasm are targeted to viral factories (Oura et al. 1998b). Radiolabeling and immunoprecipitation experiments indicate that at least 15 virally induced proteins are phosphorylated and five viral proteins may be glycosylated in infected cells, with additional nonprotein and cellular components also glycosylated (del Val et al. 1986; del Val and Viñuela 1987; Tabares et al. 1983; Urzainqui et al. 1987). ASFV encodes a serine/threonine-like protein kinase (pR298L) that is incorporated into the virion and active toward serine residues in vitro and it is likely responsible for phosphorylating ASFV proteins in vivo (Baylis et al. 1993a; Salas et al. 1988b). At least one ASFV protein undergoes posttranslational N-terminal acylation (Alfonso et al. 2007).

ASFV encodes two large polyproteins, pp220 (pCP2475L) and pp62 (pCP530R), which undergo posttranslational proteolytic processing by a viral protease during virion morphogenesis to yield mature structural proteins. The first structural polyprotein described in a DNA virus, pp220, is a 2,475 amino acid myristoylated polyprotein precursor that undergoes temporally ordered processing to yield discrete intermediates of 90 and 55 kDa and mature structural proteins p150, p37, p34, and p14 (Lopez-Otin et al. 1989; Lopez-Otin et al. 1988; Simon-Mateo et al. 1993). pp62 is processed in an ordered fashion to yield an intermediate of 46 kDa and mature structural proteins p35 and p15 (Simon-Mateo et al. 1997). Both viral polyproteins are expressed at late times postinfection and yield processed products 1–3 h after synthesis, and their products comprise the major components of the virion core shell (Andres et al. 1997, 2002).

Processing of both ASFV polyproteins occurs after the second Gly in a Gly-Gly-X amino acid motif (Lopez-Otin et al. 1988, 1989 ; Simon-Mateo et al. 1993). This cleavage site is utilized by other proteolytically processed viral and cellular proteins, including adenoviral structural proteins and cellular polyubiquitin, and it is similar to the Ala-Gly-X cleaved in poxviral structural protein precursors (Lopez-Otin et al. 1989; Simon-Mateo et al. 1993). The pp220 precursor contains a number of proteolytic target motifs not processed in mature structural proteins, but which may be processed to yield minor cleavage products excluded from mature virions (Heath et al. 2003).

An ASFV protein (pS273R) sharing sequence similarity with the core domain of cellular SUMO-1-specific, vaccinia virus I7, and adenovirus E3 proteases is an active cysteine protease capable of processing pp220 and pp62 at Gly-Gly-X sites, and it is incorporated into the virion core shell (Andres et al. 2001a; Li and Hochstrasser 1999; Rubio et al. 2003). Processing of pp62 by ASFV protease in vitro indicates that both cleavage sites are independently processed with the same efficiency, suggesting that the preferential use of the first site observed during infection in cells may rely on an additional mechanism by which proteolysis is temporally regulated (Rubio et al. 2003). While correct proteolytic processing of structural proteins is required for poxviral virion assembly and maturation, the use of ordered polyprotein processing by ASFV may be a strategy for regulating gene expression that is novel in a large DNA virus (Simon-Mateo et al. 1993).

ASFV may have additional translational and posttranslational affects on protein targeting, stability, and maturation because it encodes proteins affecting or potentially affecting protein targeting, disulfide bond formation, ubiquitination, and chaperone functions. The ASFV j4R protein (pH339R) interacts with the alpha chain of the nascent polypeptide-associated complex (NAC), a cellular ribosome-binding heteroduplex that contacts nascent polypeptides and is thought to affect co-translational processes such as protein folding and targeting (Goatley et al. 2002; Wang et al. 1995). An ASFV-encoded flavin adenine dinucleotide (FAD)-dependent sulfhydryl oxidase (pB119L) shares sequence similarity with cellular Erv1p/Alrp-family FAD-dependent sulfhydryl oxidases, is able to catalyze formation of disulfide bonds comparable to Erv1p in vitro, and is likely part of a viral redox pathway for intracytoplasmic disulfide bond formation similar to that conserved among poxviruses (Lewis et al. 2000; Rodriguez et al. 2006; Senkevich et al. 2002).

ASFV encodes the only known viral homolog (pI215L) of cellular ubiquitin conjugating enzymes (UBCs), proteins catalyzing an intermediate step in the process by which ubiquitin protein is conjugated to other proteins to target them for degradation by the proteosome or for other regulatory functions (Hingamp et al. 1992). The ASFV UBC, or vUBC1, is active in vitro, forming thiolester bonds with activated ubiquitin which is then transferred to other substrates (including vUBC1), suggesting that ASFV could manipulate ubiquitin-mediated host cell responses or modify viral or host proteins (Hingamp et al. 1992). Indeed, vUBC1 interacts with a cellular ARID family DNA-binding domain protein and is specifically required for in vitro ubiquitination of p15, a cleavage product of pp62 and one of several virion proteins that react with antiubiquitin antibody, suggesting that vUBC1 and ubiquitin modification of viral proteins may be important for virion formation (Bulimo et al. 2000;

Hingamp et al. 1995). Ubiquitination has not yet been shown to affect structural protein stability or targeting during ASFV assembly. Posttranslational stability of viral structural proteins is affected by the 80-kDa 9-RL/pB602L capsid-associated protein (CAP80). CAP80, encoded in the ASFV central variable genomic region, provides chaperone functions for capsid protein p72, preventing its aggregation and allowing normal folding kinetics in vitro (Cobbold et al. 2001).

## Viral DNA Replication

ASFV DNA replication occurs maximally 5–10 h p.i., including in infected monkey cell cultures treated to inhibit cellular DNA replication (Moulton and Coggins 1968b; Pan et al. 1980; Tabares and Sanchez Botija 1979). DNA synthesis in infected swine peripheral blood monocyte cultures begins 3–4 h p.i. and is maximal at 5 h p.i. (Pan et al. 1980). The ASFV genome encodes enzymes that are likely involved in intracytoplasmic DNA replication, including DNA topoisomerase, helicase, polymerase, ligase, and binding proteins (Esteves et al. 1987; Yanez 1995), and the phosphonoacetic acid-sensitive DNA polymerase activity induced upon infection is required for viral replication (Moreno et al. 1978; Polatnick and Hess 1972).

ASFV DNA replication has two phases, an early phase in the cell nucleus and a later prominent phase within the virus factories (Brookes et al. 1998; Garcia-Beato et al. 1992; Moulton and Coggins 1968b; Oura et al. 1998b; Pan et al. 1980; Rojo et al. 1999; Tabares and Sanchez Botija 1979; Vigario et al. 1967). Notably and consistent with a role for the nucleus in DNA replication, ASFV growth was markedly inhibited in enucleated Vero cells (Ortin and Viñuela 1977). Cytoplasmic extracts of ASFV-infected Vero cells at 8 h p.i. supported viral DNA synthesis in vitro, however, indicating that DNA synthesis is independent of the nucleus at later times postinfection (Caeiro et al. 1990). In situ hybridization and electron microscopic autoradiography indicated that ASFV DNA was present predominantly in the nuclei at 6 h p.i. and exclusively in the cytoplasm after 12 h p.i. (Garcia-Beato et al. 1992; Rojo et al. 1999). Sedimentation analysis indicated that, at 4.5 h p.i. in swine macrophages, newly synthesized DNA in the nucleus sediments at a lower rate than that present in the cytoplasm (Rojo et al. 1999). Similar to iridoviruses, ASFV genomic replication appears to initiate with a nuclear phase in which smaller-sized, replicative intermediates are produced and subsequently move to the cytoplasm, where larger intermediates are formed and mature in cytoplasmic factories (Garcia-Beato et al. 1992; Goorha 1982; Rojo et al. 1999; Tabares and Sanchez Botija 1979).

Notably, two viral structural proteins, which are components of the virion core shell, possess nucleocytoplasmic transport activities (Eulalio et al. 2004). One of these proteins (p37) is present in distinct nuclear regions at early times postinfection, is present exclusively in the cytoplasm late in infection, and is actively imported and exported as a bone fide nucleocytoplasmic shuttling protein (Eulalio et al. 2004, 2007). In yeast and transfection-based assays, p37 nuclear export is mediated by both nuclear export factor CRM1-dependent and independent pathways and involves three distinct domains on p37; however, in ASFV-infected cells, inhibition of CRM1-dependent export has no effect on viral replication kinetics or on p37 distribution (Eulalio et al. 2004, 2006, 2007). Similarly, a potential DNA repair enzyme encoded by ASFV is localized to the nucleus at early times postinfection (Redrejo-Rodriguez et al. 2006). While the function of these proteins and the cell nucleus in an early phase of ASFV DNA replication remains to be determined, co-localization of p37 with viral DNA suggests a role in nuclear transport of viral DNA during the replication cycle (Eulalio et al. 2007).

Also unresolved is the precise molecular mechanism by which ASFV genome replication occurs. Dimeric forms of terminal genomic sequences isolated from infected cells are suggestive of head–head and tail–tail concatemeric genomic replicative intermediates (Caeiro et al. 1990; Gonzalez et al. 1986; Rojo et al. 1999). Dimeric genomic forms are present in infected macrophages at late times postinfection, although other multimeric intermediates have not been identified (Rojo et al. 1999). In addition, pulse-chase radiolabeling and sedimentation analysis indicate that the introduction of terminal cross-links is a sequential process. Similar features have been observed during poxvirus genome replication, where concatemeric intermediates are resolved into genome-length units and have led to the suggestion that ASFV genomic replication may proceed by a de novo start model as proposed for poxviruses (Baroudy et al. 1982; Rojo et al. 1999).

In situ hybridization and ultrastructural analysis at late times postinfection indicate that viral DNA in the virus factory condenses into a pronucleoid structure that is inserted into icosahedral particles during virion maturation (Brookes et al. 1998). Interaction of viral DNA with virion precursors are conceivably mediated by structural DNA-binding proteins expressed by ASFV, including a protein (pE120R or p14.5) localized to viral factories and a protein (pA104R or 5-AR) with sequence similarity to bacterial histone-like proteins that is present in the virion nucleoid (Borca et al. 1996; Esteves et al. 1987; Martinez-Pomares et al. 1997; Neilan et al. 1993b).

ASFV encodes proteins likely mediating functions that indirectly enhance or ensure the fidelity of viral DNA replication, including nucleotide metabolism and DNA repair functions. ASFV-induced increases in enzyme activity and/or the presence of active or virally encoded homologs indicate that ASFV thymidine kinase, thymidylate kinase, ribonucleotide reductase, and dUTPase enzymes boost available nucleotide precursor pools in infected cells or, in the case of dUTPase, minimize misincorporation of genotoxic deoxyuridine into the viral DNA (Boursnell et al. 1991; Cunha and Costa 1992; Martin Hernandez and Tabares 1991; Oliveros et al. 1999; Polatnick and Hess 1970; Yanez et al. 1993c).

ASFV encodes several proteins that are potential components of a base excision DNA repair (BER) pathway for removal of misincorporated nucleotides. BER usually requires a DNA glycosylase for base removal, an apurinic/apyrimidinic (AP) endonuclease that nicks the damaged strand upstream of the AP site, a phosphodiesterase, a DNA repair polymerase to fill the AP site, and a DNA ligase to repair the phosphodiester backbone. ASFV encodes a homolog of the X family polymerases (pol X or pO174L), an AP endonuclease (pE296R), and a DNA ligase (pNP419L); however, no identifiable DNA glycosylase gene occurs in the ASFV genome (Yanez 1995). ASFV pol X is the smallest known nucleotide polymerase, sharing sequence similarity with the eukaryotic repair protein polymerase β (pol β) and certain functional and mechanistic similarities with pol β and other cellular polymerases (Bakhtina et al. 2007; Jezewska et al. 2007; Kumar et al. 2007; Oliveros et al. 1997; Yanez 1995). In vitro, pol X binds intermediates in the BER process, catalyzes template-dependent single nucleotide gap repair, and has AP lyase activity (Garcia-Escudero et al. 2003; Oliveros et al. 1997). The ASFV AP endonuclease has AP-site-specific endonucleolytic, 3'-phosphodiesterase, and 3'-5' proofreading exonuclease activities, further indicating that ASFV may use a reparative BER-like pathway to maintain genome integrity (Lamarche and Tsai 2006; Redrejo-Rodriguez et al. 2006).

Notably, ASFV pol X contains novel structural and functional features. These include (lack of) amino-terminal pol β-like lyase and DNA binding domains, a unique overall shape and a unique secondary structure in the catalytic subdomain, the lowest reported fidelity of characterized polymerases including an inability to discriminate between G:C and G:G base pairs, and a unique DNA binding mechanism that may affect active-site organization and contribute to low fidelity (Beard and Wilson 2001; Jezewska et al. 2006; Maciejewski et al. 2001; Sampoli Benitez et al. 2006; Showalter et al. 2001; Showalter and Tsai 2001). These characteristics have led to the suggestion that pol X may act as a mutase or have a role in a mutagenic repair system responsible for generating ASFV genetic heterogeneity. Similarly, the ASFV DNA ligase has been characterized as error-prone, sealing 3'-OH mismatched nicks with high efficiency, further supporting the concept that low fidelity may be of significance for ASFV BER-like components (Lamarche et al. 2005, 2006; Showalter et al. 2001). Conversely, a higher steady state G:C fidelity reported for pol X and the activities of the viral AP endonuclease may contribute to the overall fidelity of an ASFV BER pathway (Garcia-Escudero et al. 2003; Lamarche and Tsai 2006).

## Virion Morphogenesis

Virion morphogenesis has been examined ultrastructurally, and the functions of specific viral proteins during this process have been identified either using immunoelectron microscopy and viral mutants lacking or inducibly expressing specific viral genes (Andres et al. 1997; Arzuza et al. 1992; Breese and DeBoer 1966; Brookes et al. 1996; Garcia-Escudero et al. 1998; Moura Nunes et al. 1975; Rouiller et al. 1998). ASFV replication primarily occurs in virus factories, elements of which are first observed by 6–8 h p.i. Viral factories contain accumulations of amorphic and circular membranous material and increasing numbers of empty immature capsids and mature viral particles 12–24 h p.i. Immature virions include one- to six-sided intermediate forms; mature virions contain electron-dense, DNA-containing nucleoid (Fig. 1B).

During the earliest stages of morphogenesis, the major late structural protein p72, despite a lack of apparent signal or transmembrane sequences, is recruited from the cytoplasm and becomes associated with endoplasmic reticulum (ER) membranes as revealed by cell fractionation experiments (Cobbold et al. 1996). Laminar membranous structures take on a polyhedral shape though progressive formation of the capsid on the convex surface and core shell on the concave surface of the membrane (Andres et al. 1997). These virion precursor membranes appear as two bilayers contiguous with cellular ER membranes. The presence of cellular ER-associated proteins and a viral ER luminal protein (pXP124L) in virus factories and purified virions suggest that collapsed cisternae from host ER are the likely source of internal virion membranes (Andres et al. 1998; Rouiller et al. 1998). However, antibodies against ER markers generally do not react with viral precursor membranes, indicating exclusion of some cellular proteins during condensation of ER cisternae (Andres et al. 1998). Notably, a profound redistribution of resident ER proteins occurs in cells expressing pXP124L, a MGF110 family protein (Netherton et al. 2004b). pXP124L also contains a unique KDEL-like ER retention motif, which affects its predominantly pre-Golgi/ER-Golgi intermediate compartment (ERGIC) subcellular localization and is necessary for redistribution of ER luminal proteins to abnormal, perinuclear ERGIC (Netherton et al. 2004b). This finding led to the suggestion that MGF110 proteins might play a role in preparing ER membranes for virus envelopment. The p54 attachment and dynein-binding protein also is critical for early events involving recruitment of the ER into viral precursor membranes, as it is localized to both of these structures and its suppression arrests morphogenesis prior to formation of envelope precursors (Brookes et al. 1998; F. Rodriguez et al. 1996, 2004).

Suppression of p72 arrests morphogenesis with partially and fully collapsed ER cisternae enwrapping core material, indicating that it is essential for capsid formation (Andres et al. 1998; Garcia-Escudero et al. 1998). The requirement of p72 expression for development of polyhedral membrane forms, the similar kinetics of intermediate complex formation and p72 envelopment, and association of ER-bound p72 with 50,000-kDa intermediate assembly complexes all indicate that p72/capsid-associated assembly and ER envelopment are functionally linked (Cobbold and Wileman 1998). Newly synthesized p72 associates with, is stabilized by, and conformationally matures while bound to the ASFV CAP80 chaperone. CAP80 is a late nonstructural protein that dissociates from mature p72 prior to capsid assembly. However, CAP80 is essential for virion morphogenesis, as its repression results in morphologic arrest similar to that induced with p72 mutants (Cobbold et al. 2001; Epifano et al. 2006a). While capsid formation and p72 has been observed on the outer, convex membrane face of immature particles (Andres et al. 1997, 1998), p72 may also be present on the inner side of the inner viral envelope, assembling on both cytoplasmic faces of the cisternae (Cobbold et al. 1996; Cobbold and Wileman 1998; Rouiller et al. 1998).

Concomitant with formation of the capsid, the core shell protein layer forms on the inner face of the membrane, comprised largely of proteolytic processed products of pp220 and pp62 polyproteins (Andres et al. 1997, 2002). pp220 is essential for core assembly, because its suppression results in icosahedral particles that lack core structures, mature core proteins, and viral DNA (Andres et al. 2002). Polyprotein processing occurs concomitant with virus assembly and is critical for core morphogenesis because expression of the ASFV proteinase (pS273R) is essential for proper core morphogenesis and virus infectivity (Alejo et al. 2003; Andres et al. 2002). Association with cellular membranes may be important for polyprotein processing because only correctly processed products associated with membranes are incorporated into mature virions (Heath et al. 2003). The late structural protein p49 (pB438L) is an integral membrane protein localized to virus factories and required for morphogenesis, since its repression inhibits pp220 and pp62 processing and results in the formation of noninfectious, nonicosahedral virions that still contain p72 and pE120R capsid proteins (Epifano et al. 2006b; Galindo et al. 2000b). Similarly, other temporal aspects of the process also affect virion maturation, because p54, p72, and CAP80 expression is required for pp220 and pp62 processing and pp220 processing is required for pp62 processing in vivo, indicating coordinated regulation of polyprotein processing in infected cells (Andres et al. 2002; Epifano et al. 2006a; Rodriguez et al. 2004).

In addition to the viral proteinase, other ASFV-encoded enzymes are, or potentially are, involved in morphogenesis. ASFV sulfhydryl oxidase mutants form aberrant virions that contain acentric nucleoids (Lewis et al. 2000). This morphogenic effect potentially involves the role of sulfhydryl oxidase in a viral cytoplasmic redox pathway for disulfide bond formation, as the ASFV structural protein pE248R is likely a final substrate in the pathway and is similar to vaccinia virus L1R, a structural protein essential for morphogenesis (Ravanello and Hruby 1994; Rodriguez et al. 2006). Consistent with the importance of cytoplasmic redox potential in viral morphogenesis, oxidation of the infected cell cytosol inhibits normal p72 folding and recruitment to the ER and affects capsid assembly, with resistance of mature virions to oxidation conceivably dependent on the viral sulfhydryl oxidase (Cobbold et al. 2007).

Localization of the ASFV Nudix hydrolase to the ER and its effect on diphosphoinositol polyphosphates has led to the suggestion that it may also involve manipulation of cellular secretory pathway components during viral morphogenesis (Cartwright et al. 2002). The ASFV trans-prenyltransferase (pB318L), homolog of cellular enzymes catalyzing synthesis of prenyl diphosphate intermediates during isoprenoid synthesis, is an active trans-geranylgeranyl-diphosphate synthase and a nonstructural, integral membrane protein that also localizes to ER and ER-derived viral membrane precursors, conceivably providing substrates for prenylation of cellular or viral proteins during viral morphogenesis (Alejo et al. 1997, 1999). An actual role for these latter two enzymes in viral morphogenesis remains to be demonstrated.

Although not required for recruitment of p72 to the ER membrane, ATP and intracellular calcium gradients are important for later morphogenic stages, including capsid assembly on, and envelopment by, the ER cisternae (Cobbold et al. 2000). Inhibition of glycosylation in infected cells also affects ASFV morphogenesis, reducing virus replication and development of mature intracellular virions (del Val and Viñuela 1987).

## Cellular Changes Induced by ASFV Infection

ASFV infection induces a general downregulation or shutoff of host protein synthesis starting at early times postinfection, becoming more pronounced at later times postinfection (Estevez et al. 1986; Rodriguez et al. 2001; Tabares et al. 1980a). This inhibition has been shown, using high-resolution two-dimensional gel electrophoresis, to affect up to 77% of the proteins in Vero cells infected with BA71V and 92% of the proteins in swine macrophages infected with a pathogenic ASFV isolate (Rodriguez et al. 2001). High-resolution proteomics has also revealed upregulation of a small subset of cellular proteins in response to ASFV infection (Alfonso et al. 2004; Rodriguez et al. 2001). Mass spectrometry allowed identification of 12 proteins upregulated in ASFV-infected Vero cells, including redox-related proteins, nucleotide diphosphate kinases, heat shock proteins, apolipoproteins and members of the Ran-Gppnhp-Ranbd1 complex (Alfonso et al. 2004). The significance of these changes in host cell protein expression for virus replication, cytopathology, and pathogenesis remains largely unknown. Conversely, directed and global gene expression studies have not demonstrated an equivalent global shutoff of host cell transcription at early times postinfection, but these studies did indicate early transcriptional upregulation of select proinflammatory cytokines, cell signaling molecules, and molecules normally secreted or expressed on the cell surface (Afonso et al. 2004; Gil et al. 2003; Gomez del Moral et al. 1999; Zhang et al. 2006).

Formation of viral factories involves dramatic changes in the cytoplasm of infected cells, including rearrangement of organelles, membranes and cytoskeleton (Breese and DeBoer 1966; Carvalho et al. 1988; Moura Nunes et al. 1975). ASFV infection induces loss of the trans-Golgi network (TGN) late secretory pathway compartment (McCrossan et al. 2001). Mitochondria migrate toward, and accumulate near, viral factories in a microtubule-dependent fashion, assuming a morphology consistent with increased respiration and concurrent with induction of mitochondrial stress-response proteins including Hsp60 (Rojo et al. 1998) (Fig. 1A). The cellular microtubule network becomes disorganized after the onset of viral DNA replication and formation of virus factories, likely resulting from redistribution of centrosome proteins and functional disruption of centrosomes, which lose the ability to nucleate microtubules (Jouvenet and Wileman 2005). Microtubule-dependent disruption of the TGN is linked to the ability of ASFV to slow protein traffic to the plasma membrane, a potential mechanism for evading the immune system (Netherton et al. 2006). ASFV infection also increased the formation of a stable, acylated subpopulation of microtubules in infected cells (Jouvenet et al. 2004).

In addition to being modified during ASFV infection, microtubules serve critical roles in the development and stability of viral factories. Consistent with a role for intact microtubules and retrograde microtubular transport in establishing factory sites, their inhibition at early times postinfection allowed ASFV early protein synthesis but restricted factory formation, DNA synthesis, late gene expression, and viral replication (Alonso et al. 2001; Carvalho et al. 1988; Heath et al. 2001). Disruption of the microtubule network also disrupts ASFV factories, suggesting that microtubules contribute to the factory structure (Carvalho et al. 1988; Heath et al. 2001). Viral factories resemble aggresomes, perinuclear inclusions containing accumulations of cellular protein aggregates (Heath et al. 2001). Similarities include the recruitment of chaperones and mitochondria, microtubule-dependent formation near the MTOC, rearrangement of intermediate filaments, and collapse of vimentin into distinct cage structures (Heath et al. 2001). Engagement of vimentin into viral factories occurs early after infection when a microtubule-dependent concentration of vimentin is rearranged into an aster near the MTOC (Stefanovic et al. 2005). The aster is subsequently redistributed to the factory margins and converted to a cage structure in a manner dependent on viral DNA replication (Stefanovic et al. 2005). ASFV DNA replication is also required for vimentin phosphorylation and activation of cellular calcium calmodulin kinase II (CaMKII), and inhibition of CaMKII prevents vimentin cage formation and virus DNA replication (Stefanovic et al. 2005). The role of the vimentin cage during ASFV replication is unknown, but it has been suggested to prevent diffusion of viral components in the cytoplasm or constitute a physical scaffold facilitating formation of the viral factory (Heath et al. 2001; Stefanovic et al. 2005).

## Virion Egress

Although intracellular mature virions are infectious (Andres et al. 2001b; Moura Nunes et al. 1975), they are transported to the plasma membrane where they are released by budding to yield extracellular enveloped virions (Breese and DeBoer 1966) (Fig. 1C). The ability of late virion structural protein p14.5 to bind DNA and interact with p72 suggests a role in encapsidation of the ASFV genome (Martinez-Pomares et al. 1997); however, suppression of p14.5 indicates an additional function for this protein which involves movement of intracellular virions from factories to the plasma membrane (Andres et al. 2001b). Similar to the microtubule-mediated mechanism by which ASFVs migrate to factory sites early in infection, ASFVs align along microtubules late in infection, and anterograde transport of mature virus away from factories toward the plasma membrane is dependent on the motor protein conventional kinesin, which is recruited to the viral factory and to cytoplasmic virions (Jouvenet et al. 2004). ASFV virions at the cell surface can also induce actin nucleation, similar to vaccinia viruses, which utilize actin tails to facilitate cell–cell viral spread; ASFV virions appear in the tip of filopodia-like projections (Jouvenet et al. 2006).

## Viral Host Range

While appropriate cellular receptors are very important in conferring susceptibility to ASFV, arrest of viral replication at steps after virus internalization suggests that additional virus–host interactions play a role in determining the host range (Carrascosa et al. 1999). It is increasingly apparent that the terminal genomic regions and MGF genes play a significant role in ASFV host range. MGF530/MGF360 genes perform a macrophage host range function that involves promoting infected cell survival (Zsak et al. 2001). Deletion of one to three MGF360/MGF530 genes from a pathogenic ASFV isolate had no effect on its ability to replicate in macrophages. However, a larger deletion of six MGF360 genes and two MGF530 genes significantly reduced viral replication in macrophages, and deletion of two additional MGF360 genes completely eliminated virus replication. These experiments established that while MGF530/MGF360 genes are essential for macrophage host range, either gene dosage or gene complementation is important for efficient replication. Notably, MGF360 genes are also important host range determinants in *Ornithodoros* ticks (see ASFV-Tick Host Interactions below).

Implicated in macrophage host range are ASFV proteins involved in nucleotide and nucleic acid metabolism and which, similar to those in other large DNA viruses, may provide the deoxynucleotide pools favorable for efficient virus replication in specific cell types, including the highly differentiated, nondividing macrophage, which does not synthesize thymidylate de novo. Deletion of the dUTPase (*E165R*) and thymidine kinase (*K196R*) genes from ASFV reduces its ability to replicate in macrophages without affecting its replication in monkey cell lines (Moore et al. 1998; Oliveros et al. 1999). Deletion of thymidine kinase from a pathogenic ASFV attenuated the virus for swine, again correlating macrophage host range with virulence in swine (Moore et al. 1998). The requirement for the ASFV AP endonuclease in macrophage host range suggests that viral DNA repair is critical for virus viability in the oxidizing intracellular environment of the macrophage (Lamarche and Tsai 2006; Redrejo-Rodriguez et al. 2006).

## Viral Modulation of Apoptosis

Apoptosis, or programmed cell death, appears to play a prominent role during ASFV infection. Infection of pigs with ASFV results in lymphocyte, macrophage, and megakaryocyte apoptosis (Carrasco et al. 1996; Gomez-Villamandos et al. 1995; Oura et al. 1998b; Ramiro-Ibañez et al. 1996; Salguero et al. 2004). Apoptosis of ASFV-infected macrophages has also been observed in vitro at late times postinfection (Neilan et al. 1997a; Ramiro-Ibañez et al. 1996). Lymphocyte apoptosis is significant in lymph nodes, spleen, and thymus, and is likely the primary cause of the lymphoid cell depletion and immunodeficiency that characterize ASF. Unlike macrophages, lymphocytes are not permissive for ASFV infection, suggesting that indirect mechanisms, possibly involving cytokines secreted by ASFV-infected macrophages, are responsible for lymphocyte apoptosis (Oura et al. 1998c). Higher levels of TNF-α expression were observed in ASFV-infected macrophages, and the ability of supernatants from these infected macrophages to induce apoptosis in uninfected lymphocytes was inhibited with anti-TNF-α antibody (Gomez del Moral et al. 1999). Additionally, increased levels of TNF-α and other cytokines were detected in lymphoid tissues at times and locations coincident with lymphocyte apoptosis (Fernández de Marco et al. 2007; Salguero et al. 2005).

Apoptosis and activation of caspase-9 and -3, cysteine proteinases central to the apoptotic regulatory cascade, were observed in Vero cells early after inoculation with infectious but not UV-inactivated ASFV (Carrascosa et al. 2002; Hernaez et al. 2004). Caspase activation did not require viral DNA replication and protein synthesis, suggesting that induction of apoptosis in these cells is a very early event, perhaps involving ASFV uncoating (Carrascosa et al. 2002; Hernaez et al. 2004). Interestingly, p54, the essential attachment, dynein-binding, and morphogenesis protein induced caspase-3 activation and apoptosis when transfected into cells (Hernaez et al. 2004). Caspase activation is dependent on the presence of the dynein-binding motif in p54.

Levels of host p53, a protein with central roles in cell survival and cell cycle regulation, and Bax, a pro-apoptotic protein, are elevated throughout the course of ASFV infection, suggesting that a p53-dependent apoptotic pathway may play a role in apoptosis of ASFV-infected cells (Granja et al. 2004a). As an endoplasmic reticulum (ER)-tropic virus that induces increased levels of Hsp60 in infected cells, ASFV was predicted to activate the pro-apoptotic transcription factor CHOP/GADD153 through ER and mitochondrial stress pathways similar to other ER-tropic viruses; however, ASFV infection failed to activate CHOP/GADD153 and inhibited its induction by other stimuli (Netherton et al. 2004).

Several ASFV proteins inhibit apoptosis. ASFV *A179L (5-HL)*, an early gene encoding a homolog of cellular Bcl-2, inhibited apoptosis when transfected into human myeloid leukemia cells, an effect that required an intact pA179L BH1 domain (Afonso et al. 1996; Revilla et al. 1997). ASFV lacking *A224L (4CL)*, a late gene encoding a protein of the inhibitor of apoptosis protein (IAP) family, induced greater levels of caspase-3 activity than wild-type virus (Nogal et al. 2001). pA224L inhibited caspase activity and cell death induced by various means when stably overexpressed in Vero cells, an effect likely resulting from the direct interaction of pA224L with the catalytic fragment of caspase-3 (Nogal et al. 2001). Caspase-3 activity is also affected by ASFV pEP153R, a C-type lectin-like protein expressed throughout the infectious cycle (Galindo et al. 2000a; Yanez 1995). Increased levels of caspase-3 and apoptosis occurred in cells infected with ASFV mutants lacking *EP153R* (Hurtado et al. 2004). Additionally, transfection of the *EP153R* gene into Vero and COS cells resulted in enhanced cell survival following ASFV infection (Hurtado et al. 2004). The role of MGF360 and MGF530 genes in macrophage host range also appears to involve critical survival functions that prevent early cell death (Zsak et al. 2001).

## Cellular Genes Required for ASFV Infection

Cellular genes important for ASFV replication have been identified using a novel phenotype-based screen in cultured cells (Chang et al. 2006). A random HeLa cell library with clones having single genes inactivated through antisense expression of expressed sequence tags (EST) was used to screen for a phenotype in which ASFV replication was inhibited. Six genes were identified in the screen to affect viral replication, and these included genes likely involved in the host cell immune response, signal transduction, mitochondrial stability, and functions related to actin cytoskeleton reorganization. Three of these genes specifically reduced viral replication more than 100-fold when reversibly inhibited with inducible antisense transcripts and included HLA-B-associated transcript 3 (BAT3), C1q and tumor necrosis factor-related protein 6 (C1qTNF), and TOM40 (translocase of outer mitochondrial membrane 40). Additional experiments suggested that BAT3-dependent inhibition of ASFV replication involved perturbation of apoptosis-related signaling, consistent with a critical role for apoptosis and apoptosis-inhibition during ASFV infection.

## ASFV–Host Interactions

In sub-Saharan Africa, ASFV is maintained in a sylvatic cycle between wild swine (warthogs and bushpigs) and argasid ticks of the genus *Ornithodoros* (Fig. 3) (Plowright et al. 1969a, 1969b; Thomson et al. 1983; Wilkinson 1989). Unlike domestic swine, wild swine infected with ASFV are generally asymptomatic with low viremia titers (Heuschele and Coggins 1969; Montgomery 1921; Plowright 1981; Thomson 1985). Most adult warthogs in ASFV enzootic areas are seropositive and are likely to be persistently infected. Like warthogs, bushpigs demonstrate subclinical infection and are more resistant to direct-contact transmission than are domestic species; however, the duration of ASFV viremia may be extended (Anderson et al. 1998). Although ASFV replication in blood leukocytes of domestic swine, warthogs, and bushpigs in vitro is similar, ASFV replication, spread, and induction of lymphocyte apoptosis in vivo is reduced in bushpigs when compared to domestic swine (Anderson et al. 1998; Oura et al. 1998a).

ASFV infection of domestic swine results in several disease forms, ranging from highly lethal to subclinical depending on contributing viral and host factors. In the acute form of the disease, the incubation time ranges from 5 to 15 days. Affected animals exhibit fever (41–42°C) and anorexia followed by congestion and cyanosis of the skin, increased respiratory and heart rates, nasal discharge, incoordination, vomiting and, finally, coma and death. Survival times for animals infected with African ASFV strains range from 2 to 9 days (Conceicao 1949; Creig and Plowright 1970; Haresnape et al. 1988; Mendes 1961; Thomson et al. 1979). Typical pathological findings in acute ASF include leukopenia (Detray and Scott 1957; Edwards et al. 1985; Wardley and Wilkinson 1977), B and T cell lymphopenia (Sánchez Vizcaino et al. 1981; Wardley and Wilkinson 1980), thrombocytopenia (Anderson et al. 1987; Edwards 1983; Edwards et al. 1985), lymphocyte and mononuclear cell apoptosis (Carrasco et al. 1996; Gomez-Villamandos et al. 1995; Oura et al. 1998c; Ramiro-Ibañez et al. 1996; Salguero et al. 2004), hemorrhage in lymph nodes, spleen, kidneys, and respiratory and gastrointestinal tracts, congestion of skin and serosae, and severe interlobular lung edema (DeKock et al. 1994; Detray 1963;

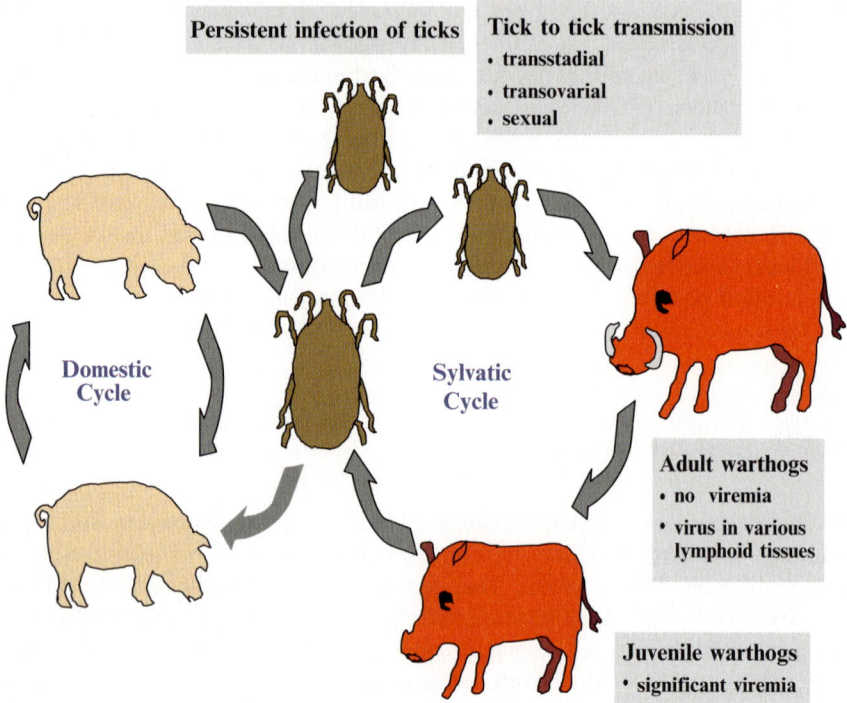

**Fig. 3** Transmission cycle of ASFV. In sub-Saharan Africa, ASFV is maintained in a sylvatic cycle between warthogs and argasid ticks of the genus *Ornithodoros* (*right*). In endemic areas, infected ticks and warthogs are the source of virus responsible for disease outbreaks in domestic swine (*left*). Once established, virus is efficiently contact-transmitted between domestic swine. Persistent infections in both vertebrate and invertebrate hosts likely contribute to perpetuation of ASFV in nature

Konno et al. 1972; Manso Ribeiro and Rosa Azevedo 1961; Maurer et al. 1958; Montgomery 1921; Nunes Petisca 1965; Steyn 1928, 1932). The extensive necrosis in affected tissues and severe hemostatic and hemodynamic changes are likely important factors leading to death. Acute ASF also induces significant changes in acute-phase proteins (Carpintero et al. 2007; Sanchez-Cordon et al. 2007). Subacute cases last 3–4 weeks and the most prominent signs include remittent fever, loss of condition, pneumonia, dyspnea, cardiac insufficiency and swelling of the joints. While hemorrhage of lymph nodes and other tissues may be found, it is not as prominent as in acute ASF (Moulton and Coggins 1968a).

The primary cell types infected by ASFV are those belonging to the mononuclear-phagocytic system, including fixed tissue macrophages and specific lineages of reticular cells (Colgrove et al. 1969; Konno et al. 1971a, 1971b; Mebus 1988; Moulton and Coggins 1968a). Affected tissues show extensive damage after infection with highly virulent viral strains. Moderately virulent ASFV strains also appear to infect these cell types, but the degree of tissue involvement and the resulting tissue damage are much less severe. The ability of ASFV to replicate and efficiently induce marked cytopathology in macrophages in vivo appears to be a critical factor in ASFV virulence.

Persistent infection with ASFV is reported to occur in warthogs and in domestic pigs surviving acute viral infection (DeKock et al. 1994; Detray 1957). Under experimental conditions, long-term persistent infection is the sequel to infection with ASFV in domestic pigs (Carrillo et al. 1994). In these animals, viral DNA was detected in the peripheral blood monocyte fraction more than 500 days p.i. by PCR; however, infectious virus could not be isolated from these samples.

Currently, there is no vaccine available for ASF and the disease is controlled by animal quarantine and slaughter. Attempts to vaccinate animals using infected cell extracts, supernatants of infected pig peripheral blood leukocytes, purified and inactivated virions, infected glutaraldehyde-fixed macrophages, or detergent-treated infected alveolar macrophages failed to induce protective immunity (Coggins 1974; Forman et al. 1982; Kihm et al. 1987; Mebus 1988). Homologous protective immunity does develop in pigs surviving viral infection. Pigs surviving acute infection with moderately virulent or attenuated variants of ASFV develop long-term resistance to homologous, but rarely to heterologous, virus challenge (Hamdy and Dardiri 1984; Ruiz-Gonzalvo et al. 1981). Pigs immunized with live attenuated ASF viruses containing engineered deletions of specific ASFV virulence/host range genes were protected when challenged with homologous parental virus (Lewis et al. 2000; Moore et al. 1998; Zsak et al. 1996, 1998).

Humoral and cellular immunity are significant components of the protective immune response to ASF. Antibodies to ASFV are sufficient to protect pigs from lethal ASFV infection (Hamdy and Dardiri 1984; Onisk et al. 1994; Ruiz-Gonzalvo et al. 1981). Although ASFV neutralizing antibodies directed against virion proteins p30, p54, and p72 have been described (Borca et al. 1994a; Gomez-Puertas et al. 1996; Zsak et al. 1993), they are not sufficient for antibody-mediated protection (Neilan et al. 2004). CD8[+] lymphocytes also appear to have a role in the protective immune response to ASFV infection (Oura et al. 2005).

## ASFV Virulence-Associated Genes

Several ASFV genes or gene regions are associated with viral pathogenesis and virulence in domestic swine but do not affect viral replication in macrophages in vitro. Two of these, *UK* (*DP96R*) and *23-NL* (*DP71L* or *l14L*), are located adjacent to each other in the right variable region of the genome. UK is an early protein that contains four to ten tandem repeats, is 92–156 amino acids in length depending on the viral isolate, and lacks similarity to other known proteins (Zsak et al. 1998). Deletion of the *UK* gene from pathogenic ASFV does not affect viral growth in macrophages in vitro, but it markedly attenuates the virus in swine (Zsak et al. 1998). Given its lack of similarity to known proteins, UK is a completely novel virulence factor. The *23-NL* gene encodes NL, a protein with similarity to cellular MyD116 and to the herpes simplex virus neurovirulence factor ICP34.5 (Sussman et al. 1992; Zsak et al. 1996). ICP34.5 functions in viral maturation and egress in a cell-type- and cell-state-dependent manner, and it prevents host protein shutoff by directing dephosphorylation of eIF-2α by protein phosphatase 1α (PP1) (He et al. 1998). NL may have a similar role in ASFV infection, since deletion of *23-NL* from the ASFV E70 strain reduces its virulence in swine without affecting viral replication in macrophages in vitro; deletion of *NL* from BA71V reduces PP1 activation in infected Vero cells, and NL specifically interacts with the catalytic subunit of PP1 in a yeast two-hybrid system (Rivera et al. 2007; Zsak et al. 1996). Long (184 amino acid) or short (70–72 amino acid) forms of NL are encoded by all known ASFV isolates where they localize to the cell nucleus (Goatley et al. 1999; Zsak et al. 1996). Interestingly, deletion of *23-NL* from other pathogenic ASFV isolates has little to no effect on virulence, indicating complementation of NL function or interaction of NL with other viral determinants in viral virulence (Afonso et al. 1998). MGF360 and MGF530 genes affect and may complement an NL-related virulence function (Neilan et al. 2002).


## ASFV Immune Evasion Genes


ASFV, similar to other large DNA viruses, affects and modulates host immune responses. ASFV-infected macrophages mediate changes in cellular immune function, and they likely play a role in the severe apoptosis observed in lymphoid tissue (Childerstone et al. 1998; Oura et al. 1998c; Ramiro-Ibañez et al. 1996; Takamatsu et al. 1999). ASFV inhibits phorbol myristic acid-induced expression of proinflammatory cytokines such as TNF-α, IFN-α, and IL-8 while inducing production of TGF-β from infected macrophages (Powell et al. 1996). Conversely, increased TNF-α expression has been reported after ASFV infection in vitro and in vivo and TNF-α may play a key role in ASFV pathogenesis, including changes in vascular permeability, coagulation, and induction of apoptosis in uninfected lymphocytes (Gomez del Moral et al. 1999; Salguero et al. 2002, 2005). Notably, ASFV strains with different virulence phenotypes differ in their ability to induce expression of proinflammatory cytokine or IFN-related genes in macrophages early in infection (Afonso et al. 2004; Gil et al. 2003; Zhang et al. 2006).

The ASFV ankyrin repeat-containing protein pA238L (5EL) is the only known viral homolog of cellular IκB proteins, the cytoplasmic inhibitors of the NFκB/Rel family of cellular transcription factors, and it is thought to be important in evading host immune responses (Miskin et al. 1998; Powell et al. 1996). Posttranslationally modified pA238L co-precipitates with the RelA subunit of NFκB and it is able to prevent binding of NFκB to DNA target sequences and inhibit NFκB-dependent gene expression (Miskin et al. 1998; Powell et al. 1996; Revilla et al. 1998; Tait et al. 2000). pA238L lacks the serine residues that target IκB-α for degradation in proteasomes and, unlike cellular IκB-α, is not degraded upon stimulation of NFκB following virus infection (Dixon et al. 2004; Tait et al. 2000). Interestingly, pA238L is actively imported to the nucleus and undergoes CRM1-dependent nuclear export, but it does not appear to prevent NFκB nuclear translocation as does IκB, indicating that functional inhibition of NFκB may rely on pA238L interference at DNA-binding sites (Silk et al. 2007). pA238L also inhibits cellular transcription independent of the NFκB pathway. The C-terminus of pA238L binds the catalytic subunit of calcineurin to inhibit its phosphatase activity and ultimately inhibit activation of the cellular NFAT transcription factor (Miskin et al. 1998, 2000). The dual activity of pA238L provides a novel mechanism for ASFV to modulate the response of host cells to infection, especially considering the role of NFκB and NFAT transcriptional pathways in inducing expression of a wide range of proinflammatory and antiviral mediators and cytokines. Consistent with this role, pA238L is able to regulate expression of cyclooxygenase-2 (COX-2), TNF-α, and inducible nitric-oxide synthase (iNOS). COX-2 downregulation occurs in an NFκB-independent, but NFAT-dependent, manner (Granja et al. 2004b). TNF-α expression is inhibited at cAMP-responsive element (CRE) and κ3 sites on the TNF-α promoter, likely through downregulation of NFκB, NFAT, and c-Jun transcription factor function and potentially by displacing coactivators CRE binding protein (CBP)/p300 (Granja et al. 2006a). Similarly, pA238L inhibits expression of iNOS, and ultimately production of nitric oxide, by a mechanism likely involving p300 transactivation, RelA transacylation, and the ability of RelA and p300 to interact and bind NFκB sequences in the iNOS promoter (Granja et al. 2006b). Interestingly, deletion of *A238L* from pathogenic ASFV does not affect viral growth in macrophages in vitro or viral pathogenesis and virulence in domestic swine (Neilan et al. 1997b).

Additional ASFV-encoded proteins modulate or interfere with host immune responses. The ASFV 8DR protein (pEP402R) is the only known viral homolog of cellular CD2, a T cell protein involved in co-regulation of cell activation (Borca et al. 1994b; Rodriguez et al. 1993a). 8DR is necessary and sufficient for mediating heme adsorption by ASFV-infected cells (Borca et al. 1994b; Rodriguez et al. 1993a). Deletion of the *8DR* gene from the ASFV genome led to decreased early virus replication and generalization of infection in swine, and 8DR suppressed cellular immune responses in vitro (Borca et al. 1998). The 8DR cytoplasmic tail interacts with cellular actin-binding adaptor SH3P7, a protein involved in signal transduction and vesicle transport, and co-localizes with SH3P7 near virus factories in ASFV-infected cells (Kay-Jackson et al. 2004). The significance of this interaction for virus replication is unknown. The ASFV pEP153R (8CR) protein is similar to cellular and poxviral proteins resembling C-type lectin-like proteins, including membrane-bound immunoactivation and immunoregulatory proteins CD69 and NKG2 (Neilan et al. 1999; Yanez 1995). It has been suggested that pEP153R stabilizes 8DR-mediated interactions with other ligands (Galindo et al. 2000a). A potential role for pEP153R in immunomodulation may be subtle, however, since pEP153R does not affect viral pathogenesis or virulence in domestic swine (Neilan et al. 1999). pA224L, in addition to caspase-3-dependent inhibition of apoptosis, can induce IκB kinase-dependent activation of NFκB, potentially affecting cytokine expression but also potentially inhibiting apoptosis through transcriptional upregulation of antiapoptotic genes (Rodriguez et al. 2002). Evidence also suggests that ASFV dramatically affects Th2/B cell responses, including upregulation of Th2 cytokines by a soluble virulence factor (p36) released from ASFV-infected monocytes and the nonspecific activation and apoptosis seen in B cell populations from ASFV-infected animals (Takamatsu et al. 1999; Vilanova et al. 1999).

ASFV multigene family 360 and 530 genes play a role in modulating host innate responses. Unlike wild type virus, infection of macrophages with Pr4Δ35, a mutant virus lacking MGF360/530 genes, resulted in increased mRNA levels for several type I interferon early-response genes (Afonso et al. 2004). Analysis of IFN-α mRNA and secreted IFN-α levels at 3, 8, and 24 h p.i. revealed undetectable IFN-α in mock and wild type-infected macrophages but significantly increased IFN-α levels at 24 h p.i. in Pr4Δ35-infected macrophages, indicating that MGF360/530 genes either directly or indirectly suppress a type I IFN response. This effect may account for the growth defect of Pr4Δ35 in macrophages and its attenuation in swine (Zsak et al. 2001, unpublished data).

## ASFV–Tick Host Interactions

ASFV persistently infects ticks of the *Ornithodoros* spp. from which ASFV can be isolated many years postinfection. Natural infection of *Ornithodoros* ticks with ASFV occurs at all developmental stages, with infection rates ranging from 0.3% to 1.7% (Pini 1977; Plowright et al. 1969a, 1969b, 1977; Thomson et al. 1983; Wilkinson et al. 1988). Ticks may become infected by feeding on viremic warthogs or by sexual, transstadial, and transovarial transmission (Endris and Hess 1994; Kleiboeker et al. 1999; Parker et al. 1969; Plowright et al. 1970a, 1994; Rennie et al. 2001). Virus titers in infected ticks collected from warthog burrows range between $10^4$ and $10^6$ $HAD_{50}$ (Pini 1977; Plowright 1977).

ASFV infection in ticks is characterized by the establishment of a long-term, persistent infection with relatively high levels of virus replication occurring in a number of tissues and organs (Basto et al. 2006; Greig 1972; Hess et al. 1989; Kleiboeker et al. 1998; Plowright et al. 1970b; Rennie et al. 2000). Infected ticks secrete virus in both saliva and coxal fluids (Plowright et al. 1969b, 1970a, 1970b, 1974). Differences in infection rates, infectious doses, and persistent infections have been observed when ticks were exposed to different ASFV isolates (Basto et al. 2006; Greig 1972; Plowright et al. 1970b).

Following experimental infection of *Ornithodoros porcinus porcinus*, initial ASFV replication occurred in phagocytic digestive cells of the midgut epithelium. Subsequent infection and replication of ASFV in undifferentiated midgut cells was observed at 15 days p.i.. Generalization of virus infection from midgut to other tissues, including the salivary and coxal glands, required 2–3 weeks and most likely involved virus movement across the basal lamina of the midgut into the hemocoel. Secondary sites of virus replication included hemocytes (type I and II), connective tissue, coxal gland, salivary gland, and reproductive tissue (Kleiboeker et al. 1998).

The importance of midgut replication for successful generalization of ASFV infection in the tick has been demonstrated by the failure of the pathogenic Malawi Lil20/1 virus, a strain that is not able to replicate in midgut epithelial cells, to infect ticks (Kleiboeker et al. 1999).

ASFV MGF360 genes encode significant tick host range determinants. An ASFV African field isolate lacking three MGF360 genes (Pr4Δ3-C2) had reduced replication in infected *O. porcinus porcinus* ticks, with virus titers reduced 100- to 1,000-fold relative to wild type virus. Pr4Δ3-C2 replication was not detected in the midgut, hemolymph, salivary gland, coxal gland or reproductive organs at 15 weeks p.i. (Burrage et al. 2004). The impaired virus replication of Pr4Δ3-C2 in the tick midgut likely accounts for lack of generalized infection in the tick necessary for natural transmission of virus from ticks to pigs.

## Final Statements

Considerable progress has been made in the understanding this complex virus and important pathogen in the past few years. Hopefully, these findings, which include ASFV replicative processes and aspects of ASFV-host interaction, will lead to novel approaches for ASF disease control and also contribute to a broader understanding of DNA virus–host interactions in general.

**Acknowledgements** We wish to acknowledge and thank all past members of the African Swine Fever Research Group, Plum Island Animal Disease Center, ARS USDA, for their efforts and research accomplishments on ASF.

## References

Afonso CL, Alcaraz C, Brun A, Sussman MD, Onisk DV, Escribano JM, Rock DL (1992) Characterization of p30, a highly antigenic membrane and secreted protein of African swine fever virus. Virology 189:368–373

Afonso CL, Neilan JG, Kutish GF, Rock DL (1996) An African swine fever virus bcl-2 homolog, 5-HL, suppresses apoptotic cell death. J Virol 70:4858–4863

Afonso CL, Zsak L, Carrillo C, Borca MV, Rock DL (1998) African swine fever virus NL gene is not required for virus virulence. J Gen Virol 79:2543–2547

Afonso CL, Piccone ME, Zaffuto KM, Neilan JG, Kutish GF, Lu Z, Balinsky CA, Gibb TR, Bean TJ, Zsak L, Rock DL (2004) African swine fever virus multigene family 360 and 530 genes affect host interferon response. J Virol 78:1858–1864

Aguero M, Blasco R, Wilkinson P, Viñuela E (1990) Analysis of naturally occurring deletion variants of African swine fever virus: multigene family 110 is not essential for infectivity or virulence in pigs. Virology 176:195–204

Alcami A, Carrascosa AL, Viñuela E (1989a) The entry of African swine fever virus into Vero cells. Virology 171:68–75

Alcami A, Carrascosa AL, Viñuela E (1989b) Saturable binding sites mediate the entry of African swine fever virus into Vero cells. Virology 168:393–398

Alcami A, Carrascosa AL, Viñuela E (1990) Interaction of African swine fever virus with macrophages. Virus Res 17:93–104

Alcami A, Angulo A, Lopez-Otin C, Munoz M, Freije JM, Carrascosa AL, Viñuela E (1992) Amino acid sequence and structural properties of protein p12, an African swine fever virus attachment protein. J Virol 66:3860–3868

Alcami A, Angulo A, Viñuela E (1993) Mapping and sequence of the gene encoding the African swine fever virion protein of M(r) 11500. J Gen Virol 74:2317–2324

Alejo A, Yanez RJ, Rodriguez JM, Viñuela E, Salas ML (1997) African swine fever virus transprenyltransferase. J Biol Chem 272:9417–9423

Alejo A, Andres G, Viñuela E, Salas ML (1999) The African swine fever virus prenyltransferase is an integral membrane trans-geranylgeranyl-diphosphate synthase. J Biol Chem 274:18033–18039

Alejo A, Andres G, Salas ML (2003) African swine fever virus proteinase is essential for core maturation and infectivity. J Virol 77:5571–5577

Alfonso P, Rivera J, Hernaez B, Alonso C, Escribano JM (2004) Identification of cellular proteins modified in response to African swine fever virus infection by proteomics. Proteomics 4:2037–2046

Alfonso P, Quetglas JI, Escribano JM, Alonso C (2007) Protein pE120R of African swine fever virus is post-translationally acetylated as revealed by post-source decay MALDI mass spectrometry. Virus Genes 35:81–85

Almazan F, Rodriguez JM, Andres G, Perez R, Viñuela E, Rodriguez JF (1992) Transcriptional analysis of multigene family 110 of African swine fever virus. J Virol 66:6655–6667

Almazan F, Rodriguez JM, Angulo A, Viñuela E, Rodriguez JF (1993) Transcriptional mapping of a late gene coding for the p12 attachment protein of African swine fever virus. J Virol 67:553–556

Almazan F, Murguia JR, Rodriguez JM, de la Vega I, Viñuela E (1995) A set of African swine fever virus tandem repeats shares similarities with SAR-like sequences. J Gen Virol 76:729–740

Almendral JM, Blasco R, Ley V, Beloso A, Talavera A, Viñuela E (1984) Restriction site map of African swine fever virus DNA. Virology 133:258–270

Almendral JM, Almazan F, Blasco R, Viñuela E (1990) Multigene families in African swine fever virus: family 110. J Virol 64:2064–2072

Alonso C, Miskin J, Hernaez B, Fernandez-Zapatero P, Soto L, Canto C, Rodriguez-Crespo I, Dixon L, Escribano JM (2001) African swine fever virus protein p54 interacts with the microtubular motor complex through direct binding to light-chain dynein. J Virol 75:9819–9827

Anderson EC, Williams SM, Fischer-Hoch SF, Wilkinson PJ (1987) Arachidonic acid metabolites in the pathophysiology of thrombocytopenia and haemorrhage in acute African swine fever. Res Vet Sci 42:387–394

Anderson EC, Hutchings GH, Mukarati N, Wilkinson PJ (1998) African swine fever virus infection of the bushpig (Potamochoerus porcus) and its significance in the epidemiology of the disease. Vet Microbiol 62:1–15

Andres G, Simon-Mateo C, Viñuela E (1997) Assembly of African swine fever virus: role of polyprotein pp220. J Virol 71:2331–2341

Andres G, Garcia-Escudero R, Simon-Mateo C, Viñuela E (1998) African swine fever virus is enveloped by a two-membraned collapsed cisterna derived from the endoplasmic reticulum. J Virol 72:8988–9001

Andres G, Alejo A, Simon-Mateo C, Salas ML (2001a) African swine fever virus protease, a new viral member of the SUMO-1-specific protease family. J Biol Chem 276:780–787

Andres G, Garcia-Escudero R, Viñuela E, Salas ML, Rodriguez JM (2001b) African swine fever virus structural protein pE120R is essential for virus transport from assembly sites to plasma membrane but not for infectivity. J Virol 75:6758–6768

Andres G, Alejo A, Salas J, Salas ML (2002) African swine fever virus polyproteins pp220 and pp62 assemble into the core shell. J Virol 76:12473–12482

Angulo A, Viñuela E, Alcami A (1993) Inhibition of African swine fever virus binding and infectivity by purified recombinant virus attachment protein p12. J Virol 67:5463–5471

Arzuza O, Urzainqui A, Diaz-Ruiz JR, Tabares E (1992) Morphogenesis of African swine fever virus in monkey kidney cells after reversible inhibition of replication by cycloheximide. Arch Virol 124:343–354

Bakhtina M, Roettger MP, Kumar S, Tsai MD (2007) A unified kinetic mechanism applicable to multiple DNA polymerases. Biochemistry 46:5463–5472

Barbosa E, Moss B (1978) mRNA(nucleoside-2'-)-methyltransferase from vaccinia virus. Characteristics and substrate specificity. J Biol Chem 253:7698–7702

Baroudy BM, Venkatesan S, Moss B (1982) Incompletely base-paired flip-flop terminal loops link the two DNA strands of the vaccinia virus genome into one uninterrupted polynucleotide chain. Cell 28:315–324

Basto AP, Nix RJ, Boinas F, Mendes S, Silva MJ, Cartaxeiro C, Portugal RS, Leitao A, Dixon LK, Martins C (2006) Kinetics of African swine fever virus infection in *Ornithodoros erraticus* ticks. J Gen Virol 87:1863–1871

Bastos AD, Penrith ML, Cruciere C, Edrich JL, Hutchings G, Roger F, Couacy-Hymann ER, Thomson G (2003) Genotyping field strains of African swine fever virus by partial p72 gene characterisation. Arch Virol 148:693–706

Bastos AD, Penrith ML, Macome F, Pinto F, Thomson GR (2004) Co-circulation of two genetically distinct viruses in an outbreak of African swine fever in Mozambique: no evidence for individual co-infection. Vet Microbiol 103:169–182

Baxter T, Wilkinson MC, Turner PC, Dixon LK (1996) Characterisation of transcription factors in African swine fever virus. Biochem Soc Trans 24:415S

Baylis SA, Dixon LK, Vydelingum S, Smith GL (1992) African swine fever virus encodes a gene with extensive homology to type II DNA topoisomerases. J Mol Biol 228:1003–1010

Baylis S, Banham AAH, Vydelingum S, Dixon LK, Smith GL (1993a) African swine fever virus encodes a serine protein kinase which is packaged into virions. J Virol 67:4549–4556

Baylis SA, Twigg SR, Vydelingum S, Dixon LK, Smith GL (1993b) Three African swine fever virus genes encoding proteins with homology to putative helicases of vaccinia virus. J Gen Virol 74:1969–1974

Beard WA, Wilson SH (2001) DNA polymerases lose their grip. Nat Struct Biol 8:915–917

Blasco R, Aguero M, Almendral JM, Viñuela E (1989a) Variable and constant regions in African swine fever virus DNA. Virology 168:330–338

Blasco R, de la Vega I, Almazan F, Aguero M, Viñuela E (1989b) Genetic variation of African swine fever virus: variable regions near the ends of the viral DNA. Virology 173:251–257

Blasco R, Lopez-Otin C, Munoz M, Bockamp EO, Simon-Mateo C, Vinuela E (1990) Sequence and evolutionary relationships of African swine fever virus thymidine kinase. Virology 178:301–304

Borca MV, Irusta P, Carrillo C, Afonso CL, Burrage TG, Rock DL (1994a) African swine fever virus structural protein p72 contains a conformational neutralizing epitope. Virology 201:413–418

Borca MV, Kutish GK, Afonso CL, Irusta P, Carrillo C, Brun A, Sussman M, Rock DL (1994b) An African swine fever virus gene with similarity to the T-lymphocyte surface antigen CD2 mediates hemadsorption. Virology 199:463–468

Borca MV, Irusta PM, Kutish GF, Carillo C, Afonso CL, Burrage AT, Neilan JG, Rock DL (1996) A structural DNA binding protein of African swine fever virus with similarity to bacterial histone-like proteins. Arch Virol 141:301–313

Borca MV, Carrillo C, Zsak L, Laegreid WW, Kutish GF, Neilan JG, Burrage TG, Rock DL (1998) Deletion of a CD2-like gene, 8-DR, from African swine fever virus affects viral infection in domestic swine. J Virol 72:2881–2889

Boshoff CI, Bastos AD, Gerber LJ, Vosloo W (2007) Genetic characterisation of African swine fever viruses from outbreaks in southern Africa (1973–1999). Vet Microbiol 121:45–55

Boursnell M, Shaw K, Yanez RJ, Viñuela E, Dixon L (1991) The sequences of the ribonucleotide reductase genes from African swine fever virus show considerable homology with those of the orthopoxvirus, vaccinia virus. Virology 184:411–416

Breese SSJ, DeBoer JC (1966) Electron microscope observations of African swine fever virus in tissue culture cells. Virology 28:420–428

Brookes SM, Dixon LK, Parkhouse RM (1996) Assembly of African swine fever virus: quantitative ultrastructural analysis in vitro and in vivo. Virology 224:84–92

Brookes SM, Hyatt AD, Wise T, Parkhouse RM (1998a) Intracellular virus DNA distribution and the acquisition of the nucleoprotein core during African swine fever virus particle assembly: ultrastructural in situ hybridisation and DNase-gold labelling. Virology 249:175–188

Brookes SM, Sun H, Dixon LK, Parkhouse RM (1998b) Characterization of African swine fever virion proteins j5R and j13L: immuno-localization in virus particles and assembly sites. J Gen Virol 79:1179–1188

Bulimo WD, Miskin JE, Dixon LK (2000) An ARID family protein binds to the African swine fever virus encoded ubiquitin conjugating enzyme, UBCv1. FEBS Lett 471:17–22

Burrage TG, Lu Z, Neilan JG, Rock DL, Zsak L (2004) African swine fever virus multigene family 360 genes affect virus replication and generalization of infection in *Ornithodoros porcinus* ticks. J Virol 78:2445–2453

Caeiro F, Meireles M, Ribeiro G, Costa JV (1990) In vitro DNA replication by cytoplasmic extracts from cells infected with African swine fever virus. Virology 179:87–94

Camacho A, Viñuela E (1991) Protein p22 of African swine fever virus: an early structural protein that is incorporated into the membrane of infected cells. Virology 181:251–257

Carpintero R, Alonso C, Pineiro M, Iturralde M, Andres M, Le Potier MF, Madec F, Alava MA, Pineiro A, Lampreave F (2007) Pig major acute-phase protein and apolipoprotein A-I responses correlate with the clinical course of experimentally induced African swine fever and Aujeszky's disease. Vet Res 38:741–753

Carrasco L, de Lara FC, Martin de las Mulas J, Gomez-Villamandos JC, Perez J, Wilkinson PJ, Sierra MA (1996) Apoptosis in lymph nodes in acute African swine fever. J Comp Pathol 115:415–428

Carrascosa AL, Sastre I, Viñuela E (1991) African swine fever virus attachment protein. J Virol 65:2283–2289

Carrascosa AL, Saastre I, Gonzalez P, Viñuela E (1993) Localization of the African swine fever virus attachment protein P12 in the virus particle by immunoelectron microscopy. Virology 193:460–465

Carrascosa AL, Sastre I, Viñuela E (1995) Production and purification of recombinant African swine fever virus attachment protein p12. J Biotechnol 40:73–86

Carrascosa AL, Bustos MJ, Galindo I, Viñuela E (1999) Virus-specific cell receptors are necessary, but not sufficient, to confer cell susceptibility to African swine fever virus. Arch Virol 144:1309–1321

Carrascosa AL, Bustos MJ, Nogal ML, González de Buitrago G, Revilla Y (2002) Apoptosis induced in an early step of African swine fever virus entry into Vero cells does not require virus replication. Virology 294:372–382

Carrascosa JL, Carazo JM, Carrascosa AL, García N, Santisteban A, Viñuela E (1984) General morphology and capsid fine structure of African swine fever virus particles. Virology 132:160–172

Carrascosa JL, Del Val M, Santaren JF, Viñuela E (1985) Purification and properties of African swine fever virus. J Virol 54:337–344

Carrascosa JL, Gonzalez P, Carrascosa AL, Garcia-Barreno B, Enjuanes L, Viñuela E (1986) Localization of structural proteins in African swine fever virus particles by immunoelectron microscopy. J Virol 58:377–384

Carrillo C, Borca MV, Afonso CL, Onisk DV, Rock DL (1994) Long-term persistent infection of swine monocytes/macrophages with African swine fever virus. J Virol 68:580–583

Cartwright JL, Safrany ST, Dixon LK, Darzynkiewicz E, Stepinski J, Burke R, McLennan AG (2002) The g5R (D250) gene of African swine fever virus encodes a Nudix hydrolase that preferentially degrades diphosphoinositol polyphosphates. J Virol 76:1415–1421

Carvalho ZG, Rodrigues-Pousada C (1986) African swine fever virus gene expression in infected Vero cells. J Gen Virol 67:1343–1350

Carvalho ZG, De Matos AP, Rodrigues-Pousada C (1988) Association of African swine fever virus with the cytoskeleton. Virus Res 11:175–192

Chang AC, Zsak L, Feng Y, Mosseri R, Lu Q, Kowalski P, Zsak A, Burrage TG, Neilan JG, Lu Z, Laegreid W, Rock DL, Cohen SN (2006) Phenotype-based identification of host genes required for replication of African swine fever virus. J Virol 80:8705–8717

Childerstone A, Takamatsu H, Yang H, Denyer M, Parkhouse RM (1998) Modulation of T cell and monocyte function in the spleen following infection of pigs with African swine fever virus. Vet Immunol Immunopathol 62:281–296

Cistue C, Tabares E (1992) Expression in vivo and in vitro of the major structural protein (VP73) of African swine fever virus. Arch Virol 123:111–124

Cobbold C, Whittle JT, Wileman T (1996) Involvement of the endoplasmic reticulum in the assembly and envelopment of African swine fever virus. J Virol 70:8382–8390

Cobbold C, Wileman T (1998) The major structural protein of African swine fever virus, p73, is packaged into large structures, indicative of viral capsid or matrix precursors, on the endoplasmic reticulum. J Virol 72:5215–5223

Cobbold C, Brookes SM, Wileman T (2000) Biochemical requirements of virus wrapping by the endoplasmic reticulum: involvement of ATP and endoplasmic reticulum calcium store during envelopment of African swine fever virus. J Virol 74:2151–2160

Cobbold C, Windsor M, Wileman T (2001) A virally encoded chaperone specialized for folding of the major capsid protein of African swine fever virus. J Virol 75:7221–7229

Cobbold C, Windsor M, Parsley J, Baldwin B, Wileman T (2007) Reduced redox potential of the cytosol is important for African swine fever virus capsid assembly and maturation. J Gen Virol 88:77–85

Coggins L (1974) African swine fever virus. Pathogenesis. Prog Med Virol 18:48–63

Colgrove GS, Haelterman EO, Coggins L (1969) Pathogenesis of African swine fever in young pigs. Am J Vet Res 30:1343–1359

Conceicao JM (1949) Estudo das zoonoses porcinas de ngola; primeiro relatorio. A zoonose porcina africana de virus filtravel. Pecuaria 1:217–245

Creig A, Plowright W (1970) The excretion of two virulent strains of African swine fever virus by domestic pigs. J Hyg, Cambridge 68:673–682

Cunha CV, Costa JV (1992) Induction of ribonucleotide reductase activity in cells infected with African swine fever virus. Virology 187:73–83

de la Vega I, Viñuela E, Blasco R (1990) Genetic variation and multigene families in African swine fever virus. Virology 179:234–246

de la Vega I, Gonzalez A, Blasco R, Calvo V, Viñuela E (1994) Nucleotide sequence and variability of the inverted terminal repetitions of African swine fever virus DNA. Virology 201:152–156

de Matos AP, Carvalho ZG (1993) African swine fever virus interaction with microtubules. Biol Cell 78:229–234

DeKock G, Robinson EM, Keppel JJG (1994) Swine fever in South Africa. Onderstepoort J Vet Sci Animal Industry 14:31–93

del Val M, Viñuela E (1987) Glycosylated components induced in African swine fever (ASF) virus-infected Vero cells. Virus Res 7:297–308

del Val M, Carrascosa JL, Viñuela E (1986) Glycosylated components of African swine fever virus particles. Virology 152:39–49

Detray DE (1957) Persistence of viremia and immunity in African swine fever. Am J Vet Res 18:811–816

Detray DE (1963) African swine fever. Adv Vet Sci Comp Med 8:299–333

Detray DE, Scott GR (1957) Blood changes in swine with African swine fever. Am J Vet Res 18:484–490

Dixon LK, Wilkinson PJ (1988) Genetic diversity of African swine fever virus isolates from soft ticks (Ornithodoros moubata) inhabiting warthog burrows in Zambia. J Gen Virol 69:2981–2993

Dixon LK, Bristow C, Wilkinson PJ, Sumption KJ (1990) Identification of a variable region of the African swine fever virus genome that has undergone separate DNA rearrangements leading to expansion of minisatellite-like sequences J Mol Biol 216:677–688

Dixon LK, Twigg SR, Baylis SA, Vydelingum S, Bristow C, Hammond JM,Smith GL (1994)
    Nucleotide sequence of a 55 kbp region from the right end of the genome of a pathogenic
    African swine fever virus isolate (Malawi LIL20/1). J Gen Virol 75:1655–1684
Dixon LK, Costa JV, Escribano JM, Rock DL, Viñuela E, Wilkinson PJ (2000) Family
    Asfarviridae. Virus taxonomy: seventh report of the International Committee on Taxonomy of
    Viruses. In: van Regenmortel MHV, Fanquet CM, Bishop DHL, Carstens EB, Estes MK,
    Lemon SM, Maniloff J, Mayo MA, McGeohh DL, Pringle CR, Wickner RB (eds), Academic,
    San Diego, pp 159–165
Dixon LK, Abrams CC, Bowick G, Goatley LC, Kay-Jackson PC, Chapman D, Liverani E, Nix
    R, Silk R, Zhang F (2004) African swine fever virus proteins involved in evading host defence
    systems. Vet Immunol Immunopathol 100:117–134
Edwards JF (1983) The pathogenesis of thrombocytopenia and haemorrhage in African swine
    fever. PhD Thesis, Cornell University. Ithaca, New York
Edwards JF, Dodds WJ, Slauson DO (1985) Mechanism of thrombocytopenia in African swine
    fever. Am J Vet Res 46:2058–2063
Endris RG, Hess WR (1994) Attempted transovarial and venereal transmission of African swine
    fever virus by the Iberian soft tick Ornithodoros (Pavlovskyella) marocanus (Acari: Ixodoidea:
    Argasidae). J Med Entomol 31:373–381
Enjuanes L, Carrascosa AL, Viñuela E (1976) Isolation and properties of the DNA of African
    swine fever (ASF) virus. J Gen Virol 32:479–492
Epifano C, Krijnse-Locker J, Salas ML, Rodriguez JM, Salas J (2006a) The African swine fever
    virus nonstructural protein pB602L is required for formation of the icosahedral capsid of the
    virus particle. J Virol 80:12260–12270
Epifano C, Krijnse-Locker J, Salas ML, Salas J, Rodriguez JM (2006b) Generation of filamentous
    instead of icosahedral particles by repression of African swine fever virus structural protein
    pB438L. J Virol 80:11456–11466
Escribano JM, Tabares E (1987) Proteins specified by African swine fever virus. V. Identification
    of immediate-early, early and late proteins. Arch Virol 92:221–238
Estevez A, Marquez MI, Costa JV (1986) Two-dimensional analysis of African swine fever virus
    proteins and proteins induced in infected cells. Virology 152:192–206
Esteves A, Ribeiro G, Costa JV (1987) DNA-binding proteins specified by African swine fever
    virus. Virology 161:403–409
Eulalio A, Nunes-Correia I, Carvalho AL, Faro C, Citovsky V, Simoes S, Pedroso de Lima MC
    (2004) Two African swine fever virus proteins derived from a common precursor exhibit dif-
    ferent nucleocytoplasmic transport activities. J Virol 78:9731–9739
Eulalio A, Nunes-Correia I, Carvalho AL, Faro C, Citovsky V, Salas J, Salas ML, Simoes S, de
    Lima MC (2006) Nuclear export of African swine fever virus p37 protein occurs through two
    distinct pathways and is mediated by three independent signals. J Virol 80:1393–1404
Eulalio A, Nunes-Correia I, Salas J, Salas ML, Simoes S, Pedroso de Lima MC (2007) African
    swine fever virus p37 structural protein is localized in nuclear foci containing the viral DNA
    at early post-infection times. Virus Res 130:18–27
Fernández de Marco M, Salguero FJ, Bautista MJ, Núñez A, Sánchez-Cordón PJ, Gómez-
    Villamandos JC (2007 ) An immunohistochemical study of the tonsils in pigs with acute
    African swine fever virus infection. Res Vet Sci 83:198–203
Forman AJ, Wardley RC, Wilkinson PJ (1982) The immunological response of pigs and guinea
    pigs to antigens of African swine fever virus. Arch Virol 74:91–100
Freije JM, Lain S, Viñuela E, Lopez-Otin C (1993) Nucleotide sequence of a nucleoside triphos-
    phate phosphohydrolase gene from African swine fever virus. Virus Res 30:63–72
Galindo I, Almazan F, Bustos MJ, Viñuela E, Carrascosa AL (2000a) African swine fever virus
    EP153R open reading frame encodes a glycoprotein involved in the hemadsorption of infected
    cells. Virology 266:340–351
Galindo I, Viñuela E, Carrascosa AL (2000b) Characterization of the African swine fever virus
    protein p49: a new late structural polypeptide. J Gen Virol 81:59–65

Garcia-Beato R, Salas ML, Viñuela E, Salas J (1992) Role of the host cell nucleus in the replication of African swine fever virus DNA. Virology 188:637–649

Garcia-Escudero R, Viñuela E (2000) Structure of African swine fever virus late promoters: requirement of a TATA sequence at the initiation region. J Virol 74:8176–8182

Garcia-Escudero R, Andres G, Almazan F, Vinuela E (1998) Inducible gene expression from African swine fever virus recombinants: analysis of the major capsid protein. J Virol 72:3185–3195

Garcia-Escudero R, Garcia-Diaz M, Salas ML, Blanco L, Salas J (2003) DNA polymerase X of African swine fever virus: insertion fidelity on gapped DNA substrates and AP lyase activity support a role in base excision repair of viral DNA. J Mol Biol 326:1403–1412

Geraldes A, Valdeira ML (1985) Effect of chloroquine on African swine fever virus infection. J Gen Virol 66:1145–1148

Gil S, Spagnuolo-Weaver M, Canals A, Sepúlveda N, Oliveira J, Aleixo A, Allan G, Leitão A, Martins CL (2003) Expression at mRNA level of cytokines and A238L gene in porcine blood-derived macrophages infected in vitro with African swine fever virus (ASFV) isolates of different virulence. Arch Virol 148:2077–2097

Goatley LC, Marron MB, Jacobs SC, Hammond JM, Miskin JE, Abrams CC, Smith GL, Dixon LK (1999) Nuclear and nucleolar localization of an African swine fever virus protein, I14L, that is similar to the herpes simplex virus-encoded virulence factor ICP34.5. J Gen Virol 80:525–535

Goatley LC, Twigg SR, Miskin JE, Monaghan P, St-Arnaud R, Smith GL, Dixon LK (2002) The African swine fever virus protein j4R binds to the alpha chain of nascent polypeptide-associated complex. J Virol 76:9991–9999

Gomez-Puertas P, Rodriguez F, Oviedo JM, Ramiro-Ibañez F, Ruiz-Gonzalvo F, Alonso C, Escribano JM (1996) Neutralizing antibodies to different proteins of African swine fever virus inhibit both virus attachment and internalization. J Virol 70:5689–5694

Gomez-Puertas P, Rodriguez F, Oviedo JM, Brun A, Alonso C, Escribano JM (1998) The African swine fever virus proteins p54 and p30 are involved in two distinct steps of virus attachment and both contribute to the antibody-mediated protective immune response. Virology 243:461–471

Gomez-Villamandos JC, Hervas J, Mendez A, Carrasco L, de las Mulas JM, Villeda CJ, Wilkinson PJ, Sierra MA (1995) Experimental African swine fever: apoptosis of lymphocytes and virus replication in other cells J Gen Virol 76:2399–2405

Gomez del Moral M, Ortuno E, Fernandez-Zapatero P, Alonso F, Alonso C, Ezquerra A, Dominguez J (1999) African swine fever virus infection induces tumor necrosis factor alpha production: implications in pathogenesis. J Virol 73:2173–2180

Gonzalez A, Almendral JM, Viñuela E (1986) Hairpin loop structure of African swine fever virus DNA. Nucleic Acids Res 14:6835–6844

Gonzalez A, Calvo V, Almazan F, Almendral JM, Ramirez JC, de la Vega I, Blasco R, Viñuela E (1990) Multigene families in African swine fever virus: family 360. J Virol 64:2073–2081

Goorha R (1982) Frog virus 3 DNA replication occurs in two stages. J Virol 43:519–528

Granja AG, Nogal ML, Hurtado C, Salas J, Salas ML, Carrascosa AL, Revilla Y (2004a) Modulation of p53 cellular function and cell death by African swine fever virus. J Virol 78:7165–7174

Granja AG, Nogal ML, Hurtado C, Vila V, Carrascosa AL, Salas ML, Fresno M, Revilla Y (2004b) The viral protein A238L inhibits cyclooxygenase-2 expression through a nuclear factor of activated T cell-dependent transactivation pathway. J Biol Chem 279:53736–53746

Granja AG, Nogal ML, Hurtado C, Del Aguila C, Carrascosa AL, Salas ML, Fresno M, Revilla Y (2006a) The viral protein A238L inhibits TNF-alpha expression through a CBP/p300 transcriptional coactivators pathway. J Immunol 176:451–462

Granja AG, Sabina P, Salas ML, Fresno M, Revilla Y (2006b) Regulation of inducible nitric oxide synthase expression by viral A238L-mediated inhibition of p65/RelA acetylation and p300 transactivation. J Virol 80:10487–10496

Greig A (1972) The localization of African swine fever virus in the tick Ornithodoros moubata porcinus. Arch Gesamte Virusforsch 39:240–247

Hamdy FM, Dardiri AH (1984) Clinic and immunologic responses of pigs to African swine virus isolated from the Western hemisphere. Am J Vet Res 45:711–714

Hammond JM, Dixon LK (1991) Vaccinia virus-mediated expression of African swine fever virus genes. Virology 181:778–782

Hammond JM, Kerr SM, Smith GL, Dixon LK (1992) An African swine fever virus gene with homology to DNA ligases. Nucleic Acids Res 20:2667–2671

Haresnape JM, Wilkinson PJ, Mellor PS (1988) Isolation of African swine fever virus from ticks of the Ornithodoros moubata complex (Ixodoidea: Argasidae) collected within the African swine fever enzootic area of Malawi. Epidemiol Inf 101:173–185

He B, Gross M, Roizman B (1998) The gamma134.5 protein of herpes simplex virus 1 has the structural and functional attributes of a protein phosphatase 1 regulatory subunit and is present in a high molecular weight complex with the enzyme in infected cells. J Biol Chem 273:20737–20743

Heath CM, Windsor M, Wileman T (2001) Aggresomes resemble sites specialized for virus assembly. J Cell Biol 153:449–455

Heath CM, Windsor M, Wileman T (2003) Membrane association facilitates the correct processing of pp220 during production of the major matrix proteins of African swine fever virus. J Virol 77:1682–1690

Hernaez B, Diaz-Gil G, Garcia-Gallo M, Ignacio-Quetglas J, Rodriguez-Crespo I, Dixon LK, Escribano JM, Alonso C (2004) The African swine fever virus dynein-binding protein p54 induces infected cell apoptosis. FEBS Lett 569:224–228

Hernaez B, Escribano JM, Alonso C (2006) Visualization of the African swine fever virus infection in living cells by incorporation into the virus particle of green fluorescent protein-p54 membrane protein chimera. Virology 350:1–14

Hess WR, Endris RG, Lousa A, Caiado JM (1989) Clearance of African swine fever virus from infected tick (Acari) colonies. J Med Entomol 26:314–317

Heuschele WP, Coggins L (1969) Epizootiology of African swine fever in warthogs. Bull Epizoot Dis Afr 17:179–183

Hingamp PM, Arnold JE, Mayer RJ, Dixon LK (1992) A ubiquitin conjugating enzyme encoded by African swine fever virus EMBO J 11:361–366

Hingamp PM, Leyland ML, Webb J, Twigger S, Mayer RJ, Dixon LK (1995) Characterization of a ubiquitinated protein which is externally located in African swine fever virions. J Virol 69:1785–1793

Hurtado C, Granja AG, Bustos MJ, Nogal ML, Gonzalez de Buitrago G, de Yebenes VG, Salas ML, Revilla Y, Carrascosa AL (2004) The C-type lectin homologue gene (EP153R) of African swine fever virus inhibits apoptosis both in virus infection and in heterologous expression. Virology 326:160–170

Irusta PM, Borca MV, Kutish GF, Lu Z, Caler E, Carrillo C, Rock DL (1996) Amino acid tandem repeats within a late viral gene define the central variable region of African swine fever virus. Virology 220:20–27

Iyer LM, Aravind L, Koonin EV (2001) Common origin of four diverse families of large eukaryotic DNA viruses. J Virol 75:11720–11734

Iyer LM, Balaji S, Koonin EV, Aravind L (2006) Evolutionary genomics of nucleo-cytoplasmic large DNA viruses. Virus Res 117:156–184

Jezewska MJ, Marcinowicz A, Lucius AL, Bujalowski W (2006) DNA polymerase X from African swine fever virus: quantitative analysis of the enzyme-ssDNA interactions and the functional structure of the complex. J Mol Biol 356:121–141

Jezewska MJ, Bujalowski PJ, Bujalowski W (2007) Interactions of the DNA polymerase X from African swine fever virus with gapped DNA substrates. Quantitative analysis of functional structures of the formed complexes. Biochemistry 46:12909–12924

Jouvenet N, Wileman T (2005) African swine fever virus infection disrupts centrosome assembly and function. J Gen Virol 86:589–594

Jouvenet N, Monaghan P, Way M, Wileman T (2004) Transport of African swine fever virus from assembly sites to the plasma membrane is dependent on microtubules and conventional kinesin. J Virol 78:7990–8001

Jouvenet N, Windsor M, Rietdorf J, Hawes P, Monaghan P, Way M, Wileman T (2006) African swine fever virus induces filopodia-like projections at the plasma membrane. Cell Microbiol 8:1803–1811

Kay-Jackson PC, Goatley LC, Cox L, Miskin JE, Parkhouse RM, Wienands J, Dixon LK (2004) The CD2v protein of African swine fever virus interacts with the actin-binding adaptor protein SH3P7. J Gen Virol 85:119–130

Keck JG, Baldick CJJ, Moss B (1990) Role of DNA replication in vaccinia virus gene expression: a naked template is required for transcription of three late trans-activator genes. Cell 61:801–809

Kihm U, Ackerman M, Mueller H, Pool R (1987) Approaches to vaccination. In: African swine fever. Becker Y (ed) Martinus Nijhoff, Boston, pp 127–144

Kleiboeker SB, Burrage TG, Scoles GA, Fish D, Rock DL (1998) African swine fever virus infection in the argasid host, Ornithodoros porcinus porcinus. J Virol 72:1711–1724

Kleiboeker SB, Scoles GA, Burrage TG, Sur J (1999) African swine fever virus replication in the midgut epithelium is required for infection of Ornithodoros ticks. J Virol 73:8587–8598

Konno S, Taylor WD, Dardiri AH (1971a) Acute African swine fever. Proliferative phase in lymphoreticular tissue and the reticuloendothelial system. Cornell Vet 61:71–84

Konno S, Taylor WD, Hess WR, Heuschele WP (1971b) Liver pathology in African swine fever. Cornell Vet 61:125–150

Konno S, Taylor WD, Hess WR, Heuschele WP (1972 ) Spleen pathology in African swine fever. Cornell Vet 62:486–506

Kumar S, Lamarche BJ, Tsai MD (2007) Use of damaged DNA and dNTP substrates by the error-prone DNA polymerase X from African swine fever virus. Biochemistry 46:3814–3825

Kuznar J, Salas ML, Viñuela E (1980) DNA-dependent RNA polymerase in African swine fever virus. Virology 101:169–175

Kuznar J, Salas ML, Viñuela E (1981) Nucleoside triphosphate phosphohydrolase activities in African swine fever virus. Arch Virol 69:307–310

Lamarche BJ, Tsai MD (2006) Contributions of an endonuclease IV homologue to DNA repair in the African swine fever virus. Biochemistry 45:2790–2803

Lamarche BJ, Showalter AK, Tsai MD (2005) An error-prone viral DNA ligase. Biochemistry 44:8408–8417

Lamarche BJ, Kumar S, Tsai MD (2006) ASFV DNA polymerase X is extremely error-prone under diverse assay conditions and within multiple DNA sequence contexts. Biochemistry 45:14826–14833

Lewis T, Zsak L, Burrage TG, Lu Z, Kutish GF, Neilan JG, Rock DL (2000) An African swine fever virus ERV1-ALR homologue, 9GL, affects virion maturation and viral growth in macrophages and viral virulence in swine. J Virol 74:1275–1285

Ley V, Almendral JM, Carbonero P, Beloso A, Viñuela E, Talavera A (1984) Molecular cloning of African swine fever virus DNA. Virology 133:249–257

Li SJ, Hochstrasser M (1999) A new protease required for cell-cycle progression in yeast. Nature 398:246–251

Lopez-Otin C, Freije JM, Parra F, Mendez E, Viñuela E (1990) Mapping and sequence of the gene coding for protein p72, the major capsid protein of African swine fever virus. Virology 175:477–484

Lopez-Otin C, Simon C, Mendez E, Viñuela E (1988) Mapping and sequence of the gene encoding protein p37, a major structural protein of African swine fever virus. Virus Genes 1:291–303

Lopez-Otin C, Simon-Mateo C, Martinez L, Viñuela E (1989) Gly-Gly-X, a novel consensus sequence for the proteolytic processing of viral and cellular proteins. J Biol Chem 264:9107–9110

Lu Z, Kutish GF, Sussman MD, Rock DL (1993) An African swine fever virus gene with a similarity to eukaryotic RNA polymerase subunit 6. Nucleic Acids Res 21:2940

Lubisi BA, Bastos ADS, Dwarka RM, Vosloo W (2003) Genotyping African swine fever virus strains from East Africa. Proceedings of the Conference of Southern African Society of Veterinary Epidemiology Preventive Medicine. Roodevallei, pp 10–14

Lubisi BA, Bastos AD, Dwarka RM, Vosloo W (2007) Intra-genotypic resolution of African swine fever viruses from an East African domestic pig cycle: a combined p72-CVR approach. Virus Genes 35:729–735

Maciejewski MW, Shin R, Pan B, Marintchev A, Denninger A, Mullen MA, Chen K, Gryk MR, Mullen GP (2001) Solution structure of a viral DNA repair polymerase. Nat Struct Biol 8:936–941

Manso Ribeiro J, Rosa Azevedo F (1961) Réapparition de la peste porcine africaine (PPA) au Portugal. Bull Off Int Epizoot 55:88–106

Martin Hernandez AM, Tabares E (1991) Expression and characterization of the thymidine kinase gene of African swine fever virus. J Virol 65:1046–1052

Martinez-Pomares L, Simon-Mateo C, Lopez-Otin C, Viñuela E (1997) Characterization of the African swine fever virus structural protein p14.5: a DNA binding protein. Virology 229:201–211

Martins A, Ribeiro G, Marques MI, Costa JV (1994) Genetic identification and nucleotide sequence of the DNA polymerase gene of African swine fever virus. Nucleic Acids Res 22:208–213

Maurer FD, Griesemer RA, Jones FC (1958) The pathology of African swine fever – a comparison with hog cholera. Am J Vet Res 19:517–539

McCrossan M, Windsor M, Ponnambalam S, Armstrong J, Wileman T (2001) The trans Golgi network is lost from cells infected with African swine fever virus. J Virol 75:11755–11765

Mebus CA (1988) African swine fever. Adv Virus Res 35:251–269

Meireles M, Costa JV (1994) Nucleotide sequence of the telomeric region of the African swine fever virus genome. Virology 203:193–196

Mendes AM (1961) Considérations sur le diagnostic et la prophylaxie de la peste porcine africaine. Bull Off Int Epizoot 57:591–600

Miskin JE, Abrams CC, Goatley LC, Dixon LK (1998) A viral mechanism for inhibition of the cellular phosphatase calcineurin. Science 281:562–565

Miskin JE, Abrams CC, Dixon LK (2000) African swine fever virus protein A238L interacts with the cellular phosphatase calcineurin via a binding domain similar to that of NFAT. J Virol 74:9412–9420

Montgomery RE (1921) On a form of swine fever occurring in British East Africa (Kenya Colony). J Comp Pathol 34:159–191, 243–262

Moore DM, Zsak L, Neilan JG, Lu Z, Rock DL (1998) The African swine fever virus thymidine kinase gene is required for efficient replication in swine macrophages and for virulence in swine. J Virol 72:10310–10315

Moreno MA, Carrascosa AL, Ortin J, Viñuela E (1978) Inhibition of African swine fever (ASF) virus replication by phosphonoacetic acid. J Gen Virol 39:253–258

Moss B, Ahn BY, Amegadzie B, Gershon PD, Keck JG (1991) Cytoplasmic transcription system encoded by vaccinia virus. J Biol Chem 266:1355–1358

Moulton J, Coggins L (1968a) Comparison of lesions in acute and chronic African swine fever. Cornell Vet 58:364–388

Moulton J, Coggins L (1968b) Synthesis and cytopathogenesis of African swine fever virus in porcine cell cultures. Am J Vet Res 29:219–232

Moura Nunes JF, Vigario JD, Terrinha AM (1975) Ultrastructural study of African swine fever virus replication in cultures of swine bone marrow cells. Arch Virol 49:59–66

Munoz M, Freije JM, Salas ML, Viñuela E, Lopez-Otin C (1993) Structure and expression in E. coli of the gene coding for protein p10 of African swine fever virus. Arch Virol 130:93–107

Neilan JG, Lu Z, Afonso CL, Kutish GF, Sussman MD, Rock DL (1993a) An African swine fever virus gene with similarity to the proto-oncogene bcl-2 and the Epstein-Barr virus gene BHRF1. J Virol 67:4391–4394

Neilan JG, Lu Z, Kutish GF, Sussman MD, Roberts PC, Yozawa T, Rock DL (1993b) An African swine fever virus gene with similarity to bacterial DNA binding proteins, bacterial integration host factors, and the Bacillus phage SPO1 transcription factor, TF1. Nucleic Acids Res 21:1496

Neilan JG, Lu Z, Kutish GF, Zsak L, Burrage TG, Borca MV, Carrillo C, Rock DL (1997a) A BIR motif containing gene of African swine fever virus, 4CL, is nonessential for growth in vitro and viral virulence. Virology 230:252–264

Neilan JG, Lu Z, Kutish GF, Zsak L, Lewis TL, Rock DL (1997b) A conserved African swine fever virus IKB homolog, 5EL, is nonessential for growth in vitro and virulence in domestic pigs. Virology 235:377–385

Neilan JG, Borca MV, Lu Z, Kutish GF, Kleiboeker SB, Carrillo C, Zsak L, Rock DL (1999) An African swine fever virus ORF with similarity to C-type lectins is non-essential for growth in swine macrophages in vitro and for virus virulence in domestic swine. J Gen Virol 80:2693–2697

Neilan JG, Zsak L, Lu Z, Kutish GF, Afonso CL, Rock DL (2002) Novel swine virulence determinant in the left variable region of the African swine fever virus genome. J Virol 76:3095–3104

Neilan JG, Zsak L, Lu Z, Burrage TG, Kutish GF, Rock DL (2004) Neutralizing antibodies to African swine fever virus proteins p30, p54, and p72 are not sufficient for antibody-mediated protection. Virology 319:337–342

Netherton CL, Parsley JC, Wileman T (2004a) African swine fever virus inhibits induction of the stress-induced proapoptotic transcription factor CHOP/GADD153. J Virol 78:10825–10828

Netherton C, Rouiller I, Wileman T (2004b) The subcellular distribution of multigene family 110 proteins of African swine fever virus is determined by differences in C-terminal KDEL endoplasmic reticulum retention motifs. J Virol 78:3710–3721

Netherton CL, McCrossan MC, Denyer M, Ponnambalam S, Armstrong J, Takamatsu HH, Wileman TE (2006) African swine fever virus causes microtubule-dependent dispersal of the trans-Golgi network and slows delivery of membrane protein to the plasma membrane. J Virol 80:11385–11392

Nix RJ, Gallardo C, Hutchings G, Blanco E, Dixon LK (2006) Molecular epidemiology of African swine fever virus studied by analysis of four variable genome regions. Arch Virol 151:2475–2494

Nogal ML, Gonzalez de Buitrago G, Rodriguez C, Cubelos B, Carrascosa AL, Salas ML, Revilla Y (2001) African swine fever virus IAP homologue inhibits caspase activation and promotes cell survival in mammalian cells. J Virol 75:2535–2543

Nunes Petisca JL (1965) Etudes anatomo-pathologiques et histopathologiques sur la peste porcine africaine (Virose L) au Portugal. Bull Off Int Epizoot 63:103–142

Oliveros M, Yanez RJ, Salas ML, Salas J, Viñuela E, Blanco L (1997) Characterization of an African swine fever virus 20-kDa DNA polymerase involved in DNA repair. J Biol Chem 272:30899–30910

Oliveros M, Garcia-Escudero R, Alejo A, Viñuela E, Salas ML, Salas J (1999) African swine fever virus dUTPase is a highly specific enzyme required for efficient replication in swine macrophages. J Virol 73:8934–8943

Onisk DV, Borca MV, Kutish GF, Kramer E, Irusta P, Rock DL (1994) Passively transferred African swine fever virus antibodies protect swine against lethal infection. Virology 198:350–354

Ortin J, Viñuela E (1977) Requirement of cell nucleus for African swine fever virus replication in Vero cells. J Virol 21:902–905

Ortin J, Enjuanes L, Viñuela E (1979) Cross-links in African swine fever virus DNA. J Virol 31:579–583

Oura CA, Powell PP, Anderson E, Parkhouse RMJ (1998a) The pathogenesis of African swine fever in the resistant bushpig. Gen Virol 79:1439–1443

Oura CA, Powell PP, Parkhouse RME (1998b) Detection of African swine fever virus in infected pig tissues by immunocytochemistry and in situ hybridisation. J Virol Methods 72:205–217

Oura CAL, Powell PP, Parkhouse RME (1998c) African swine fever: a disease characterized by apoptosis. J Gen Virol 79:1427–1438

Oura CAL, Denyer MS, Takamatsu H, Parkhouse RME (2005) In vivo depletion of CD8+ T lymphocytes abrogates protective immunity to African swine fever virus. J Gen Virol 86:2445–2450

Pan IC, Shimizu M, Hess WR (1980) Replication of African swine fever virus in cell cultures. Am J Vet Res 41:1357–1367

Parker J, Plowright W, Pierce MA (1969) The epizootiology of African swine fever in Africa. Vet Rec 85:668–674

Parrish S, Resch W, Moss B (2007) Vaccinia virus D10 protein has mRNA decapping activity, providing a mechanism for control of host and viral gene expression. Proc Natl Acad Sci U S A 104:2139–2144

Pena L, Yanez RJ, Revilla Y, Viñuela E, Salas ML (1993) African swine fever virus guanylyltransferase. Virology 193:319–328

Phologane SB, Bastos AD, Penrith ML (2005) Intra- and inter-genotypic size variation in the central variable region of the 9RL open reading frame of diverse African swine fever viruses. Virus Genes 31:357–360

Pini A (1977) Strains of African swine fever virus isolated from domestic pigs and from the tick Ornithodoros moubata in South Africa. DVSc Thesis, University of Pretoria

Plowright W (1977) Vector transmission of African swine fever. In: Liess B (ed) Seminar on hog cholera/classical swine fever and African swine fever. EUR5904 EN. Commission of the European Communities

Plowright W (1981) African swine fever. In: Infectious Diseases of wild animals, 2nd edn. Davis JW, Karstad LH, Trainer DO (eds) Iowa University Press, Ames, IA

Plowright W, Parker J, Peirce MA (1969a) African swine fever virus in ticks (Ornithodoros moubata, Murray) collected from animal burrows in Tanzania. Nature 221:1071–1073

Plowright W, Parker J, Pierce MA (1969b) The epizootiology of African swine fever in Africa. Vet Rec 85:668–674

Plowright W, Perry CT, Peirce MA (1970a) Transovarian infection with African swine fever virus in the tick, Ornithodoros moubata porcinus, Walton. Res Vet Sci 11:582–584

Plowright W, Perry CT, Peirce MA, Parker J (1970b) Experimental infection of the argasid tick, Ornithodoros moubata porcinus, with African swine fever virus. Arch Gesamte Virusforsch 31:33–50

Plowright W, Perry CT, Greig A (1974) Sexual transmission of African swine fever virus in the tick, Ornithodoros moubata porcinus, Walton. Res Vet Sci 17:106–113

Plowright W, Thomson GR, Neser JA (1994) African swine fever. Infectious diseases of livestock. In: Coetzer JAW, Thomson GR, Tustin RC (eds) Oxford University Press, Capetown pp. 568–599

Polatnick J, Hess W (1970) Altered thymidine kinase activity in culture cells inoculated with African swine fever virus. Am J Vet Res 31:1609–1613

Polatnick J, Hess WR (1972) Increased deoxyribonucleic acid polymerase activity in African swine fever virus-infected culture cells. Brief report. Arch Gesamte Virusforsch 38:383–385

Polatnick J, Pan IC, Gravell M (1974) Protein kinase activity in African swine fever virus. Arch Gesamte Virusforsch 44:156–159

Powell PP, Dixon LK, Parkhouse RME (1996) An IKB homolog encoded by African swine fever virus provides a novel mechanism for downregulation of proinflammatory cytokine responses in host macrophages. J Virol 70:8527–8533

Ramiro-Ibañez F, Ortega A, Brun A, Escribano JM, Alonso C (1996) Apoptosis: a mechanism of cell killing and lymphoid organ impairment during acute African swine fever virus infection. J Gen Virol 77:2209–2219

Ravanello MP, Hruby DE (1994) Conditional lethal expression of the vaccinia virus L1R myristoylated protein reveals a role in virion assembly. J Virol 68:6401–6410

Redrejo-Rodriguez M, Garcia-Escudero R, Yanez-Muñoz RJ, Salas ML, Salas J (2006) African swine fever virus protein pE296R is a DNA repair apurinic/apyrimidinic endonuclease required for virus growth in swine macrophages. J Virol 80:4847–4857

Rennie L, Wilkinson PJ, Mellor PS (2000) Effects of infection of the tick *Ornithodoros moubata* with African swine fever virus. Med Vet Entomol 14:355–360

Rennie L, Wilkinson PJ, Mellor PS (2001) Transovarial transmission of African swine fever virus in the argasid tick *Ornithodoros moubata*. Med Vet Entomol 15:140–146

Revilla Y, Cebrian A, Baixeras E, Martinez C, Viñuela E, Salas ML (1997) Inhibition of apoptosis by the African swine fever virus Bcl-2 homologue: role of the BH1 domain. Virology 228:400–404

Revilla Y, Callejo M, Rodriguez JM, Culebras E, Nogal ML, Salas ML, Viñuela E, Fresno M (1998) Inhibition of nuclear factor kappaB activation by a virus-encoded IkappaB-like protein. J Biol Chem 273:5405–5411

Rivera J, Abrams C, Hernáez B, Alcázar A, Escribano JM, Dixon L, Alonso C (2007) The MyD116 African swine fever virus homologue interacts with the catalytic subunit of protein phosphatase 1 and activates its phosphatase activity. J Virol 81:2923–2929

Roberts PC, Lu Z, Kutish GF, Rock DL (1993) Three adjacent genes of African swine fever virus with similarity to essential poxvirus genes. Arch Virol 132:331–342

Rodriguez CI, Nogal ML, Carrascosa AL, Salas ML, Fresno M, Revilla Y (2002) African swine fever virus IAP-like protein induces the activation of nuclear factor kappa B. J Virol 76:3936–3942

Rodriguez F, Alcaraz C, Eiras A, Yanez RJ, Rodriguez JM, Alonso C, Rodriguez JF, Escribano JM (1994) Characterization and molecular basis of heterogeneity of the African swine fever virus envelope protein p54. J Virol 68:7244–7252

Rodriguez F, Ley V, Gomez-Puertas P, Garcia R, Rodriguez JF, Escribano JM (1996) The structural protein p54 is essential for African swine fever virus viability. Virus Res 40:161–167

Rodriguez I, Redrejo-Rodriguez M, Rodriguez JM, Alejo A, Salas J, Salas ML (2006) African swine fever virus pB119L protein is a flavin adenine dinucleotide-linked sulfhydryl oxidase. J Virol 80:3157–3166

Rodriguez JM, Salas ML, Viñuela E (1992) Genes homologous to ubiquitin-conjugating proteins and eukaryotic transcription factor SII in African swine fever virus. Virology 186:40–52

Rodriguez JM, Yanez RJ, Almazan F, Viñuela E, Rodriguez JF (1993a) African swine fever virus encodes a CD2 homolog responsible for the adhesion of erythrocytes to infected cells. J Virol 67:5312–5320

Rodriguez JM, Yanez RJ, Rodriguez JF, Viñuela E, Salas ML (1993b) The DNA polymerase-encoding gene of African swine fever virus: sequence and transcriptional analysis. Gene 136:103–110

Rodriguez JM, Yanez RJ, Pan R, Rodriguez JF, Salas ML, Viñuela E (1994) Multigene families in African swine fever virus: family 505. J Virol 68:2746–2751

Rodriguez JM, Salas ML, Viñuela E (1996) Intermediate class of mRNAs in African swine fever virus. J Virol 70:8584–8589

Rodriguez JM, Salas ML, Santaren JF (2001) African swine fever virus-induced polypeptides in porcine alveolar macrophages and in Vero cells: two-dimensional gel analysis. Proteomics 1:1447–1456

Rodriguez JM, Garcia-Escudero R, Salas ML, Andres G (2004) African swine fever virus structural protein p54 is essential for the recruitment of envelope precursors to assembly sites. J Virol 78:4299–4313

Rojo G, Chamorro M, Salas ML, Viñuela E, Cuezva JM, Salas J (1998) Migration of mitochondria to viral assembly sites in African swine fever virus-infected cells. J Virol 72:7583–7588

Rojo G, Garcia-Beato R, Viñuela E, Salas ML, Salas J (1999) Replication of African swine fever virus DNA in infected cells. Virology 257:524–536

Rosales R, Harris N, Ahn BY, Moss B (1994) Purification and identification of a vaccinia virus-encoded intermediate stage promoter-specific transcription factor that has homology to eukaryotic transcription factor SII (TFIIS) and an additional role as a viral RNA polymerase subunit. J Biol Chem 269:14260–14267

Rouiller I, Brookes SM, Hyatt AD, Windsor M, Wileman T (1998) African swine fever virus is wrapped by the endoplasmic reticulum. J Virol 72:2373–2387

Rubio D, Alejo A, Rodriguez I, Salas ML (2003) Polyprotein processing protease of African swine fever virus: purification and biochemical characterization. J Virol 77:4444–4448

Ruiz-Gonzalvo F, Carnero ME, Bruyel V (1981) Immunological responses of pigs to partially attenuated ASF and their resistance to virulent homologous and heterologous viruses. In: Wilkinson PJ (ed) FAO/CEC Expert Consultation in ASF Research, Rome, pp 206–216

Salas ML, Kuznar J, Viñuela E (1981) Polyadenylation, methylation, and capping of the RNA synthesized in vitro by African swine fever virus. Virology 113:484–491

Salas ML, Kuznar J, Viñuela E (1983) Effect of rifamycin derivatives and coumermycin A1 on in vitro RNA synthesis by African swine fever virus. Brief report. Arch Virol 77:77–80

Salas ML, Rey-Campos J, Almendral JM, Talavera A, Viñuela E (1986) Transcription and translation maps of African swine fever virus. Virology 152:228–240

Salas J, Salas ML, Viñuela E (1988a) Effect of inhibitors of the host cell RNA polymerase II on African swine fever virus multiplication. Virology 164:280–283

Salas ML, Salas J, Viñuela E (1988b) Phosphorylation of African swine fever virus proteins in vitro and in vivo. Biochimie 70:627–635

Salguero FJ, Ruiz-Villamor E, Bautista MJ, Sanchez-Cordon PJ, Carrasco L, Gomez-Villamandos JC (2002) Changes in macrophages in spleen and lymph nodes during acute African swine fever: expression of cytokines. Vet Immunol Immunopathol 90:11–22

Salguero FJ, Sánchez-Cordón PJ, Sierra MA, Jover A, Núñez A, Gómez-Villamandos JC (2004) Apoptosis of thymocytes in experimental African swine fever virus infection. Histol Histopathol 19:77–84

Salguero FJ, Sánchez-Cordón PJ, Núñez A, Fernández de Marco M, Gómez-Villamandos JC (2005) Proinflammatory cytokines induce lymphocyte apoptosis in acute African swine fever infection. J Comp Pathol 132:289–302

Sampoli Benitez BA, Arora K, Schlick T (2006) In silico studies of the African swine fever virus DNA polymerase X support an induced-fit mechanism. Biophys J 90:42–56

Sanchez-Cordon PJ, Ceron JJ, Nunez A, Martinez-Subiela S, Pedrera M, Romero-Trevejo JL, Garrido MR, Gomez-Villamandos JC (2007) Serum concentrations of C-reactive protein, serum amyloid A, and haptoglobin in pigs inoculated with African swine fever or classical swine fever viruses. Am J Vet Res 68:772–777

Sánchez-Torres C, Gomez-Puertas P, Gomez-del-Moral M, Alonso F, Escribano JM, Ezquerra A, Dominguez J (2003) Expression of porcine CD163 on monocytes/macrophages correlates with permissiveness to African swine fever infection. Arch Virol 148:2307–2323

Sánchez Vizcaino JM, Slauson DO, Ruiz Gonzalvo F, Valero MS (1981) Lymphocyte function and cell-mediated immunity in pigs with experimentally induced African swine fever. Am J Vet Res 42:1335–1341

Santaren JF, Viñuela E (1986) African swine fever virus-induced polypeptides in Vero cells. Virus Res 5:391–405

Santurde G, Ruiz Gonzalvo F, Carnero ME, Tabares E (1988) Genetic stability of African swine fever virus grown in monkey kidney cells. Brief report. Arch Virol 98:117–122

Sanz A, Garcia-Barreno B, Nogal ML, Viñuela E, Enjuanes L (1985) Monoclonal antibodies specific for African swine fever virus proteins. J Virol 54:199–206

Schloer GM (1985) Polypeptides and structure of African swine fever virus. Virus Res 3:295–310

Senkevich TG, White CL, Koonin EV, Moss B (2002) Complete pathway for protein disulfide bond formation encoded by poxviruses. Proc Natl Acad Sci U S A 99:6667–6672

Showalter AK, Byeon IJ, Su MI, Tsai MD (2001) Solution structure of a viral DNA polymerase X and evidence for a mutagenic function. Nat Struct Biol 8:942–946

Showalter AK, Tsai MD (2001) A DNA polymerase with specificity for five base pairs. J Am Chem Soc 123:1776–1777

Shuman S (1989) Functional domains of vaccinia virus mRNA capping enzyme. Analysis by limited tryptic digestion. J Biol Chem 264:9690–9695

Silk RN, Bowick GC, Abrams CC, Dixon LK (2007) African swine fever virus A238L inhibitor of NF-kappaB and of calcineurin phosphatase is imported actively into the nucleus and exported by a CRM1-mediated pathway. J Gen Virol 88:411–419

Simon-Mateo C, Andres G, Almazan F, Viñuela E (1997) Proteolytic processing in African swine fever virus: evidence for a new structural polyprotein, pp62. J Virol 71:5799–5804

Simon-Mateo C, Andres G, Viñuela E (1993) Polyprotein processing in African swine fever virus: a novel gene expression strategy for a DNA virus. EMBO J 12:2977–2987

Simon-Mateo C, Freije JM, Andres G, Lopez-Otin C, Viñuela E (1995) Mapping and sequence of the gene encoding protein p17, a major African swine fever virus structural protein. Virology 206:1140–1144

Sogo JM, Almendral JM, Talavera A, Viñuela E (1984) Terminal and internal inverted repetitions in African swine fever virus DNA. Virology 133:271–275

Stefanovic S, Windsor M, Nagata K, Inagaki M, Wileman T (2005) Vimentin rearrangement during African swine fever virus infection involves retrograde transport along microtubules and phosphorylation of vimentin by calcium calmodulin kinase II. J Virol 79:11766–11775

Steyn DG (1928) Preliminary report on a South African virus disease amongst pigs. 13th and 14th Reports of the Director of Veterinary Education and Research, Union of South Africa, pp 415–428

Steyn DG (1932) East African virus disease in pigs. 18th Report of the Director of Veterinary Services and Animal Industry, Union of South Africa 1:99–109

Sumption KJ, Hutchings GH, Wilkinson PJ, Dixon LK (1990) Variable regions on the genome of Malawi isolates of African swine fever virus. J Gen Virol 71:2331–2340

Sun H, Jacobs SC, Smith GL, Dixon LK, Parkhouse RM (1995) African swine fever virus gene j13L encodes a 25–27 kDa virion protein with variable numbers of amino acid repeats J Gen Virol 76:1117–1127

Sun H, Jenson J, Dixon LK, Parkhouse ME (1996) Characterization of the African swine fever virion protein j18L. J Gen Virol 77:941–946

Sussman MD, Lu Z, Kutish GF, Afonso CL, Roberts P, Rock DL (1992) Identification of an African swine fever virus gene with similarity to a myeloid differentiation primary response gene and a neurovirulence-associated gene of herpes simplex virus. J Virol 66:5586–5589

Tabares E, Sanchez Botija C (1979) Synthesis of DNA in cells infected with African swine fever virus. Arch Virol 61:49–59

Tabares E, Marcotegui MA, Fernandez M, Sanchez-Botija C (1980a) Proteins specified by African swine fever virus. I. Analysis of viral structural proteins and antigenic properties. Arch Virol 66:107–117

Tabares E, Martinez J, Ruiz Gonzalvo F, Sanchez-Botija C (1980b) Proteins specified by African swine fever virus. II. Analysis of proteins in infected cells and antigenic properties. Arch Virol 66:119–132

Tabares E, Martinez J, Martin E, Escribano JM (1983) Proteins specified by African swine fever virus. IV. Glycoproteins and phosphoproteins. Arch Virol 77:167–180

Tabares E, Olivares I, Santurde G, Garcia MJ, Martin E, Carnero ME (1987) African swine fever virus DNA: deletions and additions during adaptation to growth in monkey kidney cells. Arch Virol 97:333–346

Tait SW, Reid EB, Greaves DR, Wileman TE, Powell PP (2000) Mechanism of inactivation of NF-kappa B by a viral homologue of I kappa b alpha. Signal-induced release of I kappa b alpha results in binding of the viral homologue to NF-kappa B. J Biol Chem 275:34656–34664

Takamatsu H, Denyer MS, Oura C, Childerstone A, Andersen JK, Pullen L, Parkhouse RM (1999) African swine fever virus: a B cell-mitogenic virus in vivo and in vitro. J Gen Virol 80:1453–1461

Thomson GR (1985) The epidemiology of African swine fever: the role of free-living hosts in Africa. Onderstepoort J Vet Res 52:201–209

Thomson GR, Gainaru MD, Van Dellen AF (1979) African swine fever: Pathogenicity and immunogenicity of two non-haemadsorbing viruses. Onderstepoort J Vet Res 46:149–154

Thomson GR, Gainaru M, Lewis A, Biggs H, Nevill E, van Der Pypekamp M, Gerbes L, Esterhuysen J, Bengis R, Bezuidenhout D, Condy J (1983) The relationship between ASFV, the warthog and *Ornithodoros* species in southern Africa. In: Wilkinson PJ (ed) African swine fever. ASF, EUR 8466 EN, Procceedings of CEC/FAO Research Seminar, Sardinia, Italy, September 1981, Commission of the European Communities, Rome, pp 85–100

Urzainqui A, Tabares E, Carrasco L (1987) Proteins synthesized in African swine fever virus-infected cells analyzed by two-dimensional gel electrophoresis. Virology 160:286–291

Valdeira ML, Geraldes A (1985) Morphological study on the entry of African swine fever virus into cells. Biol Cell 55:35–40

Valdeira ML, Bernardes C, Cruz B, Geraldes A (1998) Entry of African swine fever virus into Vero cells and uncoating. Vet Microbiol 60:131–140

Vigario JD, Relvas ME, Ferraz FP, Ribeiro JM, Pereira CG (1967) Identification and localization of genetic material of African swine fever virus by autoradiography. Virology 33:173–175

Vilanova M, Ferreira P, Ribeiro A, Arala-Chaves M (1999) The biological effects induced in mice by p36, a proteinaceous factor of virulence produced by African swine fever virus, are mediated by interleukin-4 and also to a lesser extent by interleukin-10. Immunology 96:389–395

Viñuela E (1985) African swine fever virus. Curr Top Microbiol Immunol 116:151–170

Vydelingum S, Baylis SA, Bristow C, Smith GL, Dixon LK (1993) Duplicated genes within the variable right end of the genome of a pathogenic isolate of African swine fever virus. J Gen Virol 74:2125–2130

Wambura PN, Masambu J, Msami H (2006) Molecular diagnosis and epidemiology of African swine fever outbreaks in Tanzania. Vet Res Commun 30:667–672

Wang S, Sakai H, Wiedmann M (1995) NAC covers ribosome-associated nascent chains thereby forming a protective environment for regions of nascent chains just emerging from the peptidyl transferase center. J Cell Biol 130:519–528

Wardley RC, Wilkinson PJ (1977) The association of African swine fever virus with blood components of infected pigs. Arch Virol 55:327–334

Wardley RC, Wilkinson PJ (1980) Lymphocyte responses to African swine fever virus infection. Res Vet Sci 28:185–189

Webb JH, Mayer RJ, Dixon LK (1999) A lipid modified ubiquitin is packaged into particles of several enveloped viruses. FEBS Lett 444:136–139

Wesley RD, Pan IC (1982) African swine fever virus DNA: restriction endonuclease cleavage patterns of wild-type, Vero cell-adapted and plaque-purified virus. J Gen Virol 63:383–391

Wesley RD, Tuthill AE (1984) Genome relatedness among African swine fever virus field isolates by restriction endonuclease analysis. Prev Vet Med 2:53–62

Wilkinson PJ (1989) African swine fever virus. In: Pensaert MB (ed) Virus infections of porcines. Elsevier, Amsterdam, pp 17–35

Wilkinson PJ, Pegram RG, Berry BD, Lemche J, Schels HF (1988) The distribution of African swine fever virus isolated from *Ornithodoros moubata* in Zambia. Epidemiol Inf 101:547–564

Yanez RJ, Viñuela E (1993) African swine fever virus encodes a DNA ligase. Virology 193:531–536

Yanez RJ, Boursnell M, Nogal ML, Yuste L, Viñuela E (1993a) African swine fever virus encodes two genes which share significant homology with the two largest subunits of DNA-dependent RNA polymerases. Nucleic Acids Res 21:2423–2427

Yanez RJ, Rodriguez JM, Boursnell M, Rodriguez JF, Viñuela E (1993b) Two putative African swine fever virus helicases similar to yeast 'DEAH' pre-mRNA processing proteins and vaccinia virus ATPases D11L and D6R. Gene 134:161–174

Yanez RJ, Rodriguez JM, Rodriguez JF, Salas ML, Viñuela E (1993c) African swine fever virus thymidylate kinase gene: sequence and transcriptional mapping. J Gen Virol 74:1633–1638

Yanez RJ, Rodriguez JM, Nogal ML, Yuste L, Enriquez C, Rodriguez JF, Viñuela E (1995) Analysis of the complete nucleotide sequence of African swine fever virus. Virology 208:249–278

Yates PR, Dixon LK, Turner PC (1995) Promoter analysis of an African swine fever virus gene encoding a putative elongation factor. Biochem Soc Trans 23:139S

Yozawa T, Kutish GF, Afonso CL, Lu Z, Rock DL (1994) Two novel multigene families, 530 and 300, in the terminal variable regions of African swine fever virus genome. Virology 202:997–1002

Zhang F, Hopwood P, Abrams CC, Downing A, Murray F, Talbot R, Archibald A, Lowden S, Dixon LK (2006) Macrophage transcriptional responses following in vitro infection with a highly virulent African swine fever virus isolate. J Virol 80:10514–10521

Zsak L, Onisk DV, Afonso CL, Rock DL (1993) Virulent African swine fever virus isolates are neutralized by swine immune serum and by monoclonal antibodies recognizing a 72-Kda viral protein. Virology 196:596–602

Zsak L, Lu Z, Kutish GF, Neilan JG, Rock DL (1996) An African swine fever virus virulence-associated gene NL-S with similarity to the herpes simplex virus ICP34.5 gene. J Virol 70:8865–8871

Zsak L, Caler E, Lu Z, Kutish GF, Neilan JG, Rock DL (1998) A nonessential African swine fever virus gene UK is a significant virulence determinant in domestic swine. J Virol 72:1028–1035

Zsak L, Lu Z, Burrage TG, Neilan JG, Kutish GF, Moore DM, Rock DL (2001) African swine fever virus multigene family 360 and 530 genes are novel macrophage host range determinants. J Virol 75:3066–3076

# Mimivirus

J.-M. Claverie (✉), C. Abergel, H. Ogata

## Contents

**Abstract** *Acanthamoeba polyphaga* Mimivirus, the first representative and proto-type member of the *Mimiviridae*, is the latest addition to the menagerie of lesser-known big DNA viruses. Due to the size of its particle—a fiber-covered icosahedral protein capsid with a diameter of 0.7 μm—Mimivirus was initially mistaken for an intracellular parasitic bacteria. Its 1.2-Mb genome sequence was then found to encode more than 900 proteins, many of them associated with functions never before encountered in a virus, such as four aminoacyl-tRNA synthetases. The finding of Mimivirus-encoded central components of the protein translation apparatus thought to be the signature of cellular organisms revived the debate about the origin

J.-M. Claverie
Structural and Genomic Information Laboratory, Parc Scientifique de Luminy,
Case 934 13288 Marseille cedex 09, France
Jean-Michel.Claverie@univmed.fr

James L. Van Etten (ed.) *Lesser Known Large dsDNA Viruses.*
Current Topics in Microbiology and Immunology 328.
© Springer-Verlag Berlin Heidelberg 2009

of DNA viruses and their possible role in the emergence of the eukaryotic cell. Despite the many features making it unique in the viral world, Mimivirus is nevertheless phylogenetically close to other large DNA viruses, such as phycodnaviruses and iridoviruses, and most likely share a common ancestry with all nucleocytoplasmic large DNA viruses. Postgenomic studies have now started in various laboratories, slowly shedding some light on the physiology of the largest and most complex virus isolated to date. This chapter summarizes our present knowledge on Mimivirus.

## Introduction

The discovery of Mimivirus (for mimicking microbe virus) (La Scola et al. 2003), a double-stranded DNA virus infecting the common ameba *Acanthamoeba polyphaga*, followed by the sequencing and analysis of its genome (Raoult et al. 2004) sent a shock wave through the community of virologists and evolutionists. By its record particle size (750 nm in diameter) and genome length (1.2 million bp), the complexity of its gene repertoire (911 protein coding genes) as well as its particle composition (it contains products of more than 130 virus genes), Mimivirus blurred the established boundaries between viruses and parasitic cellular organisms. Beyond these quantitative aspects, Mimivirus has many types of genes never before encountered in a virus, most noticeably genes encoding central components of the protein translation machinery, previously thought to be the signature of cellular organisms. These exceptional genes include those encoding four aminoacyl-tRNA synthetases. The revolutionary finding of a partial protein translation apparatus in a virus, together with the presence of components of other pathways previously unique to cellular organisms, came at the right time to lend support to several bold theories linking ancestral viruses to the emergence of the eukaryotic domain (Claverie 2006). As more researchers are becoming involved in the study of Mimivirus, experimental information is now slowly accumulating, although very little is known about its physiology. This article reviews some of the recent progress, mostly including individual protein characterization, electron microscopy, and proteomics.

## The Serendipitous Discovery of Mimivirus

In 1992, a pneumonia outbreak in the West Yorkshire mill town of Bradford (England), triggered an investigation for *Legionella* (a pneumonia causing intracellular parasitic bacterium) in the water of a nearby cooling tower. This investigation was conducted by Timothy Rowbotham, the officer in charge of Britain's Public Health Laboratory Service. Instead of the expected Gram-negative bacillus-like *Legionella*, he discovered a microorganism resembling a small Gram-positive coccus (initially called Bradfordcoccus) (Fig. 1).

After unsuccessful cultivation attempts and the failure of molecular identification using universal 16S rDNA bacterial primers, the mysterious sample was stored

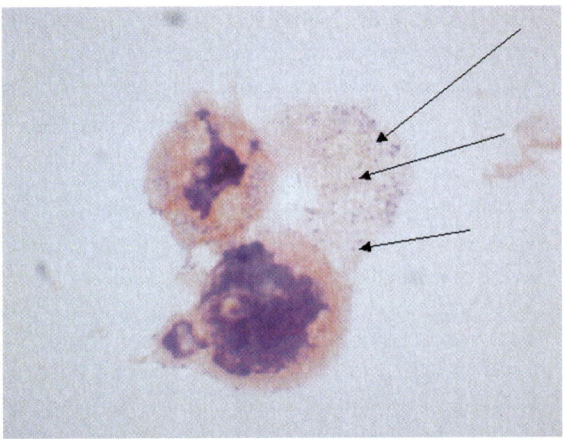

**Fig. 1** Light microscopy appearance of Mimivirus particles in infected amoeba, following Gram coloration. The three *arrows* point to individual particles (*purple-blue*) within the ameba cytoplasm (*pink*). From La Scola et al. (2003)

in a freezer for approximately 10 years. It was then brought to the Rickettsia Unit at the school of Medicine in Marseille, France by Dr. Richard Birtle. There, following additional characterization attempts, electron microscopy of infected *Acanthamoeba polyphaga* cells provided the first hint that Bradfordcoccus was in fact a giant virus, with mature icosahedral particles approximately 0.7 µm in diameter, a size comparable to that of mycoplasma cells (La Scola et al. 2003). The viral nature of the agent was further established by the demonstration of an eclipse phase during its replication, and the analysis of several gene sequences exhibiting a clear phylogenetic affinity with nucleocytoplasmic large DNA viruses (NCLDV), a group of viruses including the *Poxviridae*, the *Iridoviridae*, the *Phycodnaviridae*, and the *Asfarviridae*. This new virus was named *A. polyphaga* Mimivirus and is now classified by ICTV as the first and prototype member species of the *Mimiviridae*, a new family within the NCLDV. The size of its particles makes Mimivirus the largest virus ever described. Mimivirus does not pass through a 0.3-µm pore filter, a usual experimental procedure to separate bacterial cells from viruses. Following such filtering steps, prevalent in environmental microbiology studies (for instance, metagenomics), Mimivirus is retained in the pool of prokaryotic cellular organisms (Ghedin and Claverie 2005).

## Potential Source of Other Mimiviridae

Phylogenetic analyses of the most conserved genes common to all NCLDVs (the so-called NCLDV core genes, Iyer et al. 2001) consistently places Mimivirus in an independent lineage, between the *Phycodnaviridae* (algal viruses) and *Iridoviridae* (predominantly fish viruses). This resemblance of Mimivirus with viruses found in aqueous environments prompted Ghedin and Claverie (2005) to search for evidence of other *Mimiviridae* in the environmental microbial DNA sequences gathered in the Sargasso Sea (Venter et al. 2004). This in silico search was successful since 15% of the Mimivirus 911-predicted protein sequences had their closest homologs in this metagenomic data set rather than in viral sequences of known origin. Furthermore, 43% of Mimivirus core genes had their closest homologs in the Sargasso Sea data set. It is thus very likely that other species of *Mimividae* remain to be isolated from the marine environment, probably infecting microalgae (Monier et al. 2008b) or heterotrophic protozoans. Interestingly, the Sargasso Sea data set where Mimivirus sequence homologs are detected correspond to bacteria-sized organisms that passed through 3-μm pore-sized filters and were retained by 0.2-μm pore-sized filters. Mimivirus-like particles (0.75 μm in diameter) are in this range. Our preliminary analysis of the latest oceanic metagenomic sequence data brought back by the *Sorcerer II* Global Ocean Sampling Expedition (Rusch et al. 2007) confirmed the presence of Mimivirus relatives at various oceanic locations around the globe (Monier et al. 2008a). These results thus predict that the marine environment contains Mimivirus relatives that are abundant enough to be randomly sampled from sea water. Thus it is only a matter of time before new *Mimiviridae* members are found in aqueous environments.

## Host Range and Pathogenicity

Only cells from species of the *Acanthamoeba* genus have been productively infected by a cell-free viral suspension, among a large number of primary or established cell lines from vertebrates or invertebrates that were tested for their ability to support Mimivirus infection and replication (Suzan-Monti et al. 2006).

Upon infection of *A. polyphaga* cells, Mimivirus has a typical viral replication cycle with an eclipse phase until 5 h postinfection (p.i.), followed by the steady appearance of newly synthesized virions in the cytoplasm, leading to the clustered accumulation of viral particles filling up most of the intracellular space (Fig. 2A), until infected amebae start to lyse after 14 h p.i. The burst size is larger than 300 particles per cell. In a recent study, Suzan-Monti et al. (2007) described the assembly of the virion within and around very large cytoplasmic virus factories. Given that Mimivirus particles contain a rather complete transcription system (see Sect. 4.2) as do poxviruses, the entire replication cycle might occur in the cytoplasm. The presence of a highly conserved promoter motif in 50% of the Mimivirus genes (see Sect. 3.4) suggests that there are two categories of genes, some with promoters recognized by the viral RNA polymerase and some lacking the conserved promoter element that may be transcribed by the host RNA polymerase recruited from the ameba nucleus.

The mechanism of delivery of the particle content into the ameba cytoplasm also remains to be clarified, but seems to require the partial digestion of the sturdy fibril layer surrounding the mature particles (Figs. 2B and 3).

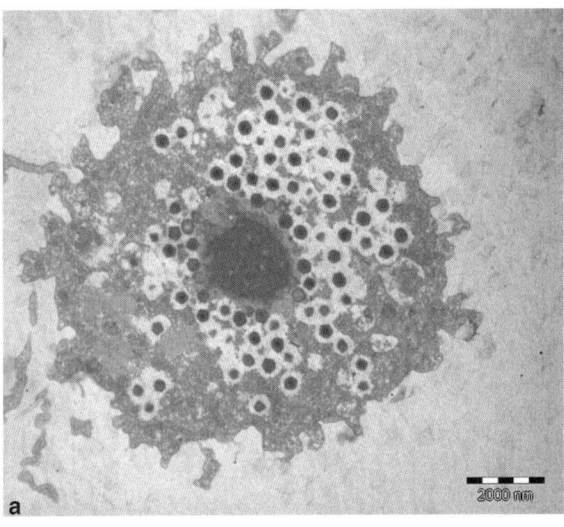

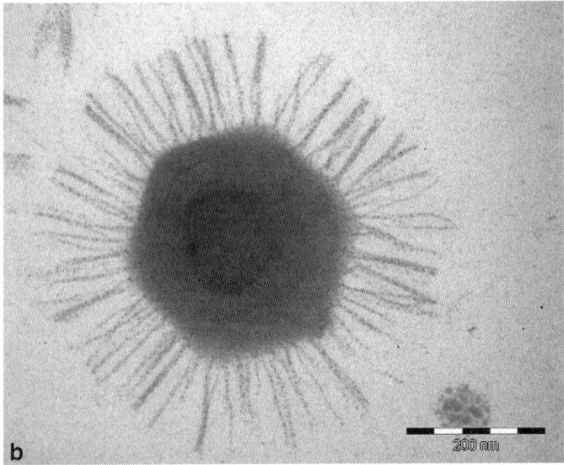

**Fig. 2** Transmission electron microscopy of Mimivirus particles. **a** Mimivirus-infected *A. poly-phaga* at 8 h p.i. shows intracytoplasmic accumulation of virus particles (Bar = 2 μ m). The *central dark nucleus-like region* is a cytoplasmic virus factory. **b** Mimivirus particles, purified from the supernatant of infected cells, appear as nonenveloped icosahedral virions surrounded by fibrils (Bar = 100 nm) (From Raoult et al. 2007)

The combination of genomic, proteomic, and ultrastructural analyses suggests the following infection scenario:

1. Free virus particles mimicking bacteria (by their size and perhaps a lipopolysac-charide [LPS]-like layer surrounding the capsid) are taken up as food by the ameba.

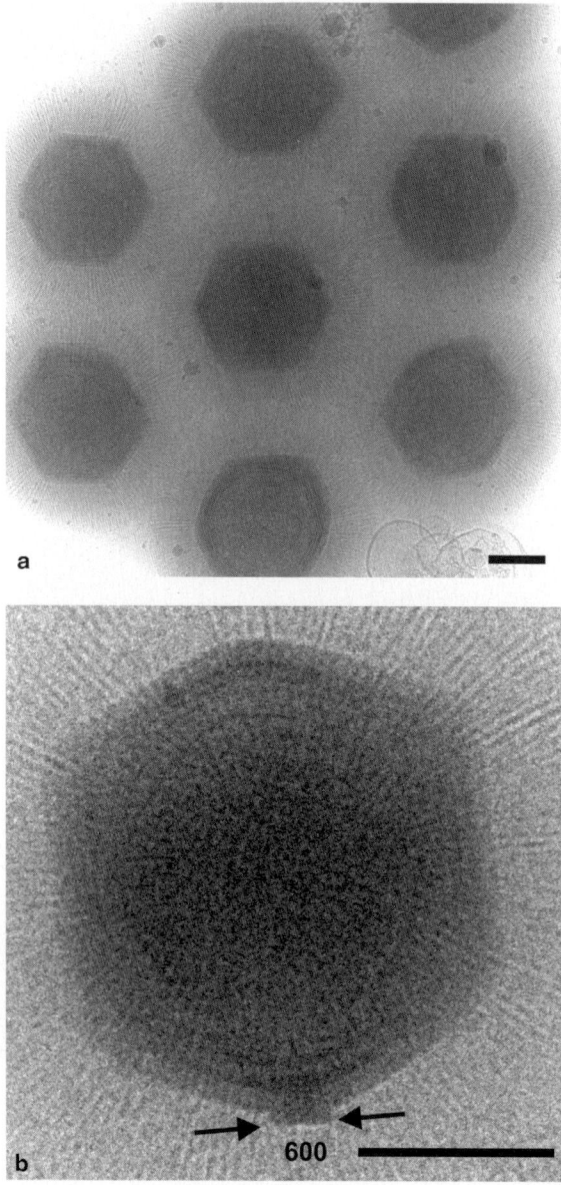

**Fig. 3** Cryo-EM highquality images of Mimivirus particles. A Cluster of mature particles, exhibiting a solid and compact fiber layers. **b** Close-up of one particle (0.75 μm across) exhibiting a densely packed layer of crosslinked fibers and a single vertex. (From Xiao et al. 2005). Note the difference in fibril density with Fig. 2B

2. The LPS-like fibril layer is partially digested within the ameba endocytic vacuole, making the surface of the capsid accessible for interaction with the vacuole membrane.
3. The content of the capsid is then discharged into the ameba cytoplasm, probably through a fusion between the virus internal lipid membrane and the phagosome membrane, leaving the empty particle in the endocytic vacuole.
4. Transcription of early and late-early genes then occurs in the cytoplasm, most likely under the control of the Mimivirus highly conserved promoter using the virus-encoded and virion-packaged transcription machinery.

The experimentally determined narrow range of Mimivirus host cells, restricted to protozoans belonging to the *Acanthamoeba* genus, conflicts with reports suggesting that Mimivirus might be a human pathogen. La Scola et al. (2005) reported the presence of Mimivirus-specific antibodies in the sera of patients with community- or hospital-acquired pneumonia (see also Berger et al. 2006). In contrast, no evidence of Mimivirus infection was found in hospitalized children in Austria (Larcher et al. 2006), nor in a large CDC-led analysis of respiratory specimens from 496 pneumonia cases (Dare et al. 2008). An isolated case of laboratory infection of a technician by Mimivirus has been reported (Raoult et al. 2006). The patient's serum reacted strongly with several Mimivirus proteins. However, isolation of Mimivirus from the infected patient did not formally link the virus with the disease. Finally, mice experimentally inoculated (via intracardiac route) with Mimivirus developed histopathological features of pneumonia (Khan et al. 2006), but again, no virus was recovered from the lung tissues. In summary, it is not clear whether Mimivirus should be considered a potential pneumonia agent or is simply highly immunogenic (perhaps due to the unique LPS-like layer surrounding its protein capsid) or cross-reacts with a common bacterial species. As a precautionary measure, it is probably best to treat Mimivirus as a biosafety class 2 pathogen. In this context, it is worth remembering that Mimivirus particles remain infectious for at least 1 year when stored at 4–32°C in a neutral buffer.

# Genomics of Mimivirus

## *Overall Genome Structure*

Mimivirus genome sequencing revealed a single linear dsDNA molecule of 1,181,404 bp. A combination of bioinformatic methods led to the initial prediction that the virus had 911 protein-encoding genes and six tRNA genes. The exact number of protein-encoding genes may change slightly in the future due to the difficulty of identifying introns in some of the genes that lack relatives in the databases. Proteomic and resequencing data has already led to the correction of some annotations (see Sect. 4.2). With an overall coding percentage of 90.5% and an average intergenic distance of 157 nt, Mimivirus exhibits the genome compaction observed in other DNA viruses. The large size of the Mimivirus genome is therefore not due to the accumulation of noncoding junk DNA. The overall nucleotide composition is 72% (A+T), leading both to strong positive bias in the usage of A+T-rich synonymous codons and to an increased abundance of amino-acid residues with A+T-rich codons (Raoult et al. 2004). For instance, isoleucine (9.9%), aspargine (8.9%) and tyrosine (5.4%) are twice as frequent in Mimivirus proteins as in ameba or human proteins. On the other hand, alanine (encoded by GCN codons) is a rare amino acid (3.1%). Such surprising flexibility, despite the constraints imposed by the necessity to maintain protein 3D structure, solubility and function, was previously noted for other viruses with even higher A+T (82.2%) contents such as *Amsacta moorei* entomopoxvirus (Bawden et al. 2000). The two strands of the Mimivirus DNA molecule encode roughly the same number of genes (450 R genes vs. 465 L genes). However, both the gene excess and the A+C excess profiles exhibit a clear slope reversal (around nucleotide position 400,000) as found in bacterial genomes; this reversal is usually associated with the origin (or terminus) of replication. Mimivirus genes are preferentially transcribed away from this location (578 leading strand ORFs vs. 333 lagging strand ORFs).

Mimivirus genome termini do not have the large terminal inverted repeats (up to 2 kb) found in Phycodnaviruses (Chlorovirus [Yamada et al. 2006] or *Ectocarpus siliculosus* virus [Delaroque et al. 2003]), its closest NCLDV relatives, but also a conserved feature in Poxviruses and Asfarviruses. The putative circular DNA molecules generated by pairing these repeats might be important during DNA replication. Interestingly, the closest phylogenetic relative of Mimivirus, *Emiliania huxleyi* virus (EhV) 86 appears to replicate as a circular molecule (Wilson et al. 2005). In place of inverted terminal repeats, the Mimivirus genome has a quasi-perfect (616/617) inverted repetition of a 617-bp sequence, beginning at nucleotide position 22,515 and its unique complementary counterpart near the end of the chromosome beginning at nucleotide position 1,180,529. As these regions are intergenic and are not flanked by paralogous genes, their extreme conservation suggests a strong functional constraint related to their perfect base-pairing (Fig. 4). Pairing these inverted repeats leads to a putative Q-like form for the Mimivirus genome, with a long (22,514-bp) and a short (259-bp) tail (Fig. 4). The short tail does not overlap with any ORFs. The long tail has a lower coding density than the rest of the genome (75% vs 90.5%), with larger intergenic distances (435 nt vs 157 nt in average).

This long tail region encodes 12 proteins as follows:

- R1 (795 aa) corresponds to a predicted replication origin binding protein (OBP), homologous to the herpesvirus core gene UL9. Its N-terminal DEAD-like helicase domain is 48% identical to the one found in the R8 (1052 aa). The R1 protein shares its C-terminal domain with the products of nearby genes R8 (38% identity), R9 (49% identity), and R10 (26% identity). This domain of unknown function is also found in other predicted viral OBPs (from Herpesvirus and Asfarvirus). These four proteins might be involved in the DNA replication priming process.

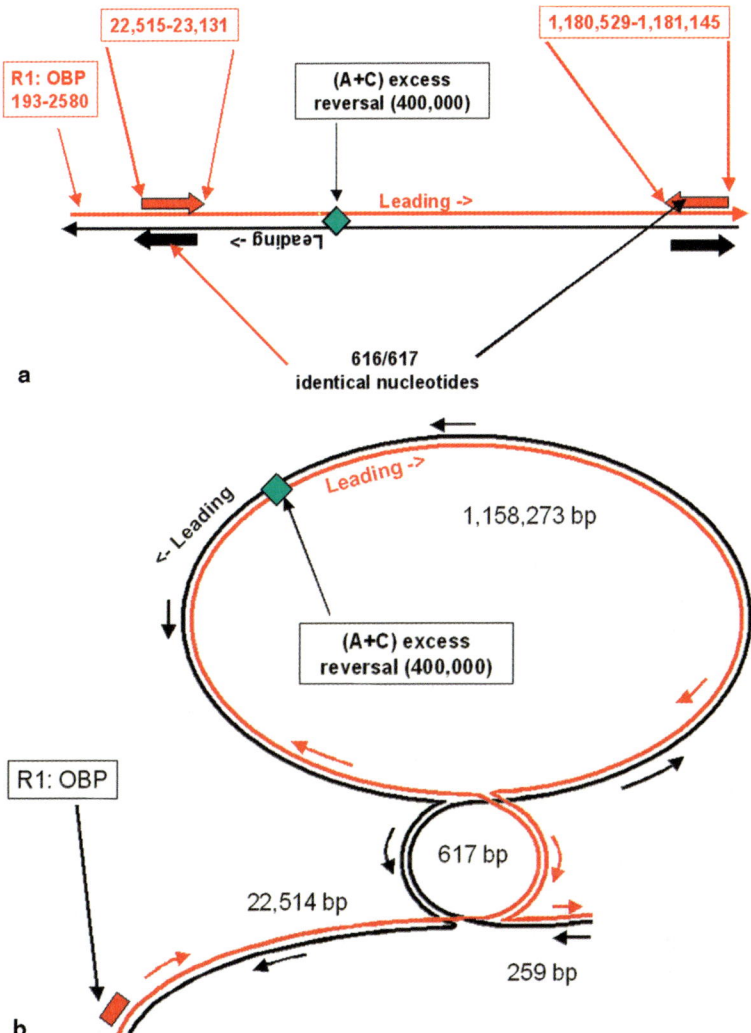

**Fig. 4** Schematic structure of Mimivirus chromosome. **a** Remarkable features along the linear chromosome. **b** Putative circularized Q-like form obtained by pairing the largest perfect intergenic inverted repeat in the genome. The first ORF (R1) has a clear similarity to an origin of replication binding protein (OBP)

- L2 (246 aa) has a BRO family, N-terminal domain, which is associated with DNA binding.
- L3 (666 aa) encodes a homolog of chromosome segregation ATPases (COG1196).
- L4 (454 aa) encodes a predicted DNA binding protein of the N1R/P28 type.
- L5 (461 aa) encodes a protein with no functional attribute.
- L6 (218 aa) and L7 (155 aa) are 50% identical, but have no functional attribute or homolog in the database except for the Mimivirus L57 gene product.

- R8 (1052 aa), R9 (376 aa) and R10 (376 aa) are related to R1 and each other as described above.
- R11 (267 aa) encodes a protein of unknown function.
- L12 (487 aa) encodes a protein with no known function, but is 45% identical to L5.

Among the proteins encoded by these 12 genes, a putative function can be attributed to seven of them, all of which are related to DNA replication or binding. These statistics suggest that this clustering of genes encoding DNA replication components at one extremity of the viral chromosome may have functional significance. This makes the proteins of unknown function—L5, L6, L7, R11 and L12—all the more interesting because they may be involved in unknown DNA replication events.

| Region | | Identity | Overlapping ORFs |
|---|---|---|---|
| 98,340–99,316 | 1,114,002–1,113,026 | 951 /977bp | L79–R854 (transposase) |
| 1,007,267–1,008,591 | 1,112,548–1,113,872 | 1320/1325 bp | L770–R854 (transposase) |

Other remarkable regions of the genomes include two inverted repeats:
These regions may be the result of recent transposase-mediated duplications and thus may be devoid of topological significance.

## Mimivirus as a Bona Fide NCLDV

Iyer and collaborators (Iyer et al. 2001) performed a detailed comparative analysis of the protein sequences encoded in the genomes of four families of large DNA viruses (collectively abbreviated as NCLDV) that replicate, completely or partly, in the cytoplasm of eukaryotic cells (poxviruses, asfarviruses, iridoviruses, and phycodnaviruses). They identified nine genes (class 1 core genes) that are shared by all these viruses and 22 more genes that are found in at least three of these four viral families (Class 2 and 3 core genes). Our analysis of the Mimivirus genome unambiguously identified homologs of the nine Class 1 core genes, and 17 of the 22 other core genes. The phylogenetic analysis of a concatenation of the protein sequences encoded by the Class 1 core genes robustly places Mimivirus in an independent lineage (the *Mimiviridae*) among the NCLDVs, in between the *Iridoviridae* and *Phycodnaviridae* (Fig. 5). Together with its morphological (icosahedral capsid) and ultrastructure (nonenveloped particle with an internal lipid membrane; see Sect. 4) characteristics, Mimivirus is clearly a bona fide member of the NCLDVs, despite being much larger and three times genetically more complex than the phycodnavirus EhV86, its closest relative and the NCLDV with the second largest genome.

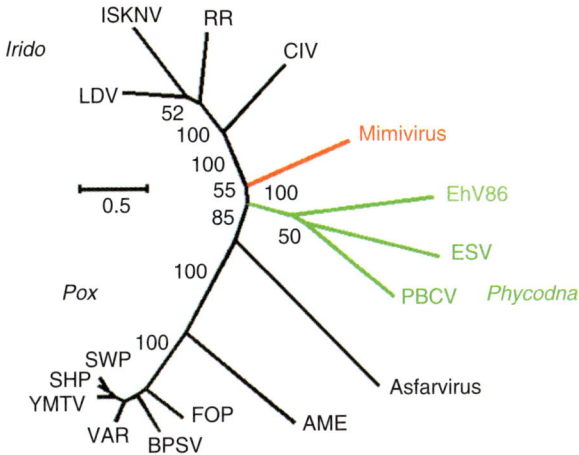

**Fig. 5** Phylogenetic position of Mimivirus among established NCLDV families. Viral species are as follows: Iridoviridae (*CIV* chilo iridescent virus, *RR Regina ranavirus*, *LDV*, lymphocystis disease virus type 1, *ISKNV* Infectious spleen and kidney necrosis virus), *Poxviridae* (*SWP* swinepox virus, *SHP* sheeppox virus, *YMTV* Yaba monkey tumor virus, *VAR* variola virus, *BSPV* bovine popular stomatitis virus, *FOP* fowl pox virus, *AME Amsacta mooreii* entomopoxvirus), Asfarvirus: African swine fever virus, *Phycodnaviridae* (*PBCV Paramecium bursaria* chlorella virus 1, *ESV Ectocarpus siliculosus* virus, *EhV86 Emiliania huxleyi* virus 86). This tree was built using maximum likelihood and based on the concatenated sequenced of the proteins sequences encoded by the NCLDV Class 1 core genes. Bootstrap percentages are shown along the branches (except for the Pox families where they all are close to 100)

Interestingly, prior to the sequencing of EhV86, phycodnaviruses (algal virus, such as virus-infecting chlorella) were the closest relatives to Mimivirus. However, despite their large genomes (>300 kb) they lack a virus-encoded RNA polymerase, which made unique among the NCLDVs (Van Etten 2003) and at odds with the plethoric gene content of Mimivirus. This apparent paradox is now alleviated with the finding of a complete transcription apparatus encoded in EhV86, the largest phycodnavirus genome sequenced to date. As it is unlikely that the many different genes required to constitute a functional transcription apparatus were independently acquired by EhV86, their presence strongly suggests that all extant NCLDV families (*Poxviridae*, *Asfarviridae*, *Iridoviridae*, *Phycodnaviridae*, and *Mimiviridae*) share a common ancestor (a mostly cytoplasmic DNA virus) that might have been even more complex than today's Mimivirus. The smaller phycodnavirus (Chloroviruses and Phaeoviruses) genomes probably underwent lineage-specific gene losses leading to the (still surprising) disappearance of their host-independent transcription apparatus. Similarly, Iridoviruses, Asfarviruses, and Poxviruses are all missing a few of the NCLDV core genes (Raoult et al. 2004), which may correspond to lineage-specific losses. Despite its huge genome, Mimivirus is not immune to this phenomenon, for instance with the puzzling absence of a dUTPase homolog, a universal enzyme that is required to avoid the incorporation of deoxyuridine into its thymidine-rich DNA. An equivalent activity might be performed by the protein product of ORF L479 that has a MazG-type nucleotide pyrophosphohydrolase domain, probably derived from a bacterial source (Iyer et al. 2006). Similarly, the ATP-dependent DNA ligase present in all other NCLDVs is replaced by an NAD-dependent version of the enzyme in Mimivirus (R303). A succession of gene losses, some of them compensated by nonorthologous gene replacements via horizontal transfers, might explain the small number of recognizable orthologous genes shared by today's NCLDVs.


## *Mimivirus as a Unique Giant Virus: Virally Encoded Translation Components*

If, on one hand, Mimivirus exhibits many of the features characteristic of previously described NCLDVs, on the other hand its 1.2-Mb genome encodes many unique genes not previously found in a virus (Raoult et al. 2004). For instance, Mimivirus possesses a complete set of DNA repair enzymes capable of correcting nucleotide mismatches as well as errors induced by oxidation, UV irradiation and alkylating agents. Mimivirus is also the only virus to encode the three major types of topoisomerases (the usual type IIA, the poxvirus-like type Ib, and the first viral type Ia). In addition, Mimivirus uniquely possesses a number of polysaccharide, amino-acid and lipid manipulating enzymes. Such metabolic capabilities, although covering a broader biochemical spectrum in Mimivirus, also exist in other NCLDVs, specially the phycodnaviruses (Van Etten 2003), where they often differ from one species to the next, suggesting their involvement in specific virus–host relationships. EhV86, for instance, encodes many components of the ceramide biosynthesis pathway (Wilson et al. 2005) that are thought to interfere with its host apoptosis-like cell death pathway.

Probably the most spectacular discovery in the Mimivirus genome was finding ten homologs of proteins with functions central to protein translation: four aminoacyl-tRNA synthetases, a mRNA cap-binding protein (eukaryotic initiation factor eIF4E, ORF L496), translation initiation factor eEF-1 (GTP-binding translocation factor, ORF R624), translation initiation factor SUI1/eIF1 (ORF R464), translation initiation factor eIF4A (an ATP-dependent RNA helicase, ORF R458), and peptide chain release factor eRF1 (ORF R726). In addition, Mimivirus encodes a homolog (ORF R405) of the tRNA (uracil-5-)-methyltransferase, the tRNA-modifying enzyme whose *Escherichia coli* counterpart catalyzes the methylation of the invariant tRNA uracil at position 54, thus defining the T-loop (TΨC arm) in all tRNAs. This region of the tRNA serves as a recognition site for the ribosome.

The four aminoacyl-tRNA synthetases (aaRS), all from Class I aaRSs, are specific for tyrosine (TyrRS), arginine (ArgRS), cysteine (CysRS) and methionine (MetRS). Finding these components of the translation apparatus in Mimivirus clearly violated the dogma that viruses rely entirely on the host translation machinery for protein synthesis.

Genes encoding tRNAs were previously described in a few viruses, including bacteriophage T4, herpes virus 4, and the chlorella viruses (Van Etten 2003). Similarly, Mimivirus encodes six tRNA-like genes, albeit mostly unrelated to the above aaRSs: three $tRNA_{leu}$ (2 TTA, 1 TTG), one $tRNA_{trp}$ (TGG), one $tRNA_{his}$ (CAC), and one $tRNA_{cys}$ (TGG). Although Mimivirus exhibits a codon usage that is fairly distinct from its ameba host, the above tRNAs or aaRSs are not related to the most conspicuous differences. One exception to this statement is TyrRS, which may help incorporate tyrosine into Mimivirus proteins where its frequency (5.4%) is twice that observed in ameba proteins.

The presence of many translation machinery components encoded in the genome of Mimivirus can be explained by two opposing hypotheses. On one hand, the traditional view that viruses capture genes from their environment and their host predicts that these translation components were acquired from cellular organisms (Moreira and Lopez-Garcia 2005). However, this hypothesis suffers from a lack of phylogenetic evidence (Ogata et al. 2005a). Also, the independent random acquisition (and retention) of so many translation-related genes is unlikely, given their lack of usefulness as individual components of an incomplete system. On the other hand, one may interpret the translation components found in Mimivirus as the remains of an even more complex ancestral genome that encoded a complete and functional translation apparatus, as occurs in cellular organisms. Such a genome reduction scenario is consistent with the hypothesis that NCLDVs originated from the primitive nucleus of ancestral eukaryotes (Claverie 2006). A genome reduction process, akin to the one observed in intracellular parasitic bacteria (Blanc et al. 2007) may have led to the present day Mimivirus. However, this evolutionary process appears to have stopped as Mimivirus shows no signs of ongoing genome degradation such as pseudogenes, repeat accumulation, or reduced coding density (Claverie et al. 2006). Presumably, there is a strong selective advantage for Mimivirus to retain its incomplete translation system, given the high evolutionary rates usually associated with viruses. In this context, assessing the biochemical and cellular function of Mimivirus-encoded translation components during the infection process becomes important (see Sect. 5).

## *Other Remarkable Features of the Mimivirus Genome*

### Intein and Introns

Inteins are protein-splicing domains encoded by mobile intervening sequences. They catalyze their own excision from the host protein. Although found in all domains of life (Eukarya, Archaea and Eubacteria) their distribution is sporadic. Mimivirus is one of the few dsDNA viruses containing an intein, inserted in its DNA polymerase B protein (Ogata et al. 2005b). The Mimivirus intein is closely related to one found in the DNA polymerase of *Heterosigma akashiwo* virus (HaV) a phycodnavirus that infects the single-cell bloom-forming raphidophyte (golden brown alga) *H. akashiwo* (Nagasaki et al. 2005). Both inteins appear monophyletic to archaeal inteins. Two additional inteins have recently been reported in chlorovirus proteins (Fitzgerald et al. 2007). Type I introns are self-splicing intervening sequences that are excised at the mRNA level. One type IB intron has been identified in several chlorella viruses, but they are rare in viruses infecting eukaryotes. Mimivirus has six self-excising introns: one in the largest RNA polymerase subunit gene and the other three in the second-largest RNA polymerase subunit gene. Two introns were recently discovered in the gene encoding the major capsid protein (L425, now corrected in the UniProt Q5UQL7 entry). Given that introns are mostly detected when they interrupt the coding sequence of know proteins, additional introns located within anonymous Mimivirus ORFs might exist.

## A Uniquely Conserved Promoter Signal

An exhaustive search for overrepresented "words" in the Mimivirus genome led to the discovery of an "AAAATTGA" octamer within the 150-nt upstream region of 403 of the 911 (45%) predicted protein-coding genes. A search for more sophisticated signals (Bailey and Gribskov 1998) led to a very similar result with 446 genes (49%) showing a conserved upstream motif. The location of this motif at positions ranging from −80 to −50 before the initiator codon is consistent with the short average size (157 ±113 nt) of the intergenic region in Mimivirus, and the compact promoter/5′ UTR structure (as well as 3′UTR) known for some ameba protists (Vanacova et al. 2003). Suhre et al. (2005) proposed that the AAAATTGA octamer might correspond to a TATA box-like core promoter element. The finding of such a strongly conserved sequence motif in front of nearly half of the Mimivirus genes is one more unique feature of this virus because eukaryotic (as well as viral) promoters usually lack clear consensus sequences.

There is a significant correlation between the upstream AAAATTGA motif and genes transcribed from the predicted leading strand (54% vs 40%). Finally, Suhre et al. (2005) noted that this motif was not common in the available ameba genome sequences. Applying this same analysis to the genomes of other large DNA viruses, confirmed that the homogeneity of this promoter sequence is unique to Mimivirus.

Based on the predicted function of the proteins encoded by the genes possessing the AAAATTGA motif in their upstream region, this putative promoter element appears to correlate with functions required for the early (or late-early) phase of viral infection. Suhre et al. (2005) also proposed that this Mimivirus TATA box-like signal might have co-evolved with the virus-encoded transcription preinitiation complex consisting of two RNA polymerase II subunits and a TFIID initiation factor homolog. This late-early promoter may thus be recognized by the Mimivirus encoded-transcription machinery, while the genes lacking this signal could be transcribed by the host RNA polymerase. This hypothesis received additional support from the proteomic analysis of Mimivirus particles showing that: (i) the virus-encoded transcription machinery is associated with the particle (and thus accessible immediately after infection), and (ii) only a small fraction (approximately 10%) of the late genes encoding proteins associated with the virion have the AAAATTGA promoter element. Finally, it is worth noting that the exact same sequence was shown to function as a promoter for the *Chilo* iridescent virus DNA polymerase gene (Nalçacioglu et al. 2007).

## Gene and Genome Duplication in Mimivirus

The Mimivirus genome is roughly 2.4 times larger than the second largest virus (Phage G) and ten times larger than the average DNA viruses (www.giantvirus. org). This observation raises the question of the mechanisms by which such a viral genome might have occurred, and of the evolutionary forces allowing such an anomalous genome to be maintained. DNA viruses vary widely in DNA content as well as in their genetic complexity. Larger genomes may result primarily from the accumulation of noncoding DNA. An extreme example is the *Cotesia congregata* Bracovirus, a *Polydnaviridae* with a 568-kb genome, but encoding a mere 156 proteins, for an overall 27% coding density (Espagne et al. 2004). Closer to Mimivirus, some sequenced members of the *Iridoviridae* (Zhang et al. 2004) or the *Baculoviridae* (Cheng et al. 2002) have coding densities below 69%.

With a coding density above 90%, the size of Mimivirus genome cannot be explained by a propensity to accumulate junk DNA. After a thorough analysis of the gene content, Suhre (2005) identified two main mechanisms contributing to the Mimivirus genome size. First, a segmental duplication of about 200 kb is at the origin of the telomeric regions of the Mimivirus linear genome. Second, many tandem gene duplications exist in various positions in the genome, sometimes generating large paralogous families of up to 66 members (these are ankyrin-domain containing ORFs). A perfect tandem expansion of 12 paralogous ORFs occurs from L174 to L185. Depending on the similarity threshold used, approximately 35% of the Mimivirus genes have at least one homolog in the virus's genome (E value $<10^{-5}$). This fraction lies well within the range of values encountered throughout the three domains of life: for example 17% for *Haemophilus influenzae*, 44% for *Mycoplasma pneumoniae*, 30% for *Saccharomyces cerevisiae*, and 65% for *Arabidopsis thaliana* (Suhre 2005 and references therein). From a different perspective, Ogata et al. (2005a) showed that horizontal gene transfer from its host, or other exogenous sources does not account for much of the Mimivirus genome. Despite their crude methodology, overestimating horizontal gene transfers, Filee et al. (2007) identified less than 10% of Mimivirus genes as originating from bacteria. Overall, these studies do not indicate that Mimivirus is quantitatively different from cellular organisms or other large DNA viruses (Monier et al. 2007) with respect to the various mechanisms leading to genome expansion. If a fraction of the Mimivirus genome can be attributed to segmental or tandem duplications, as well as to exogenous sources, this leaves roughly 400 unique genes that might be part of Mimivirus lineage, dating back to its origin.

# The Mimivirus Particle

## *Morphology and Ultrastructure*

Despite its unprecedented size, the icosahedral symmetry of the Mimivirus particle was good enough to allow Xiao et al. (2005) to make a computer-generated
3D reconstruction at a resolution of approximately 75 Å from series of cryo-
electron microscopy (cryoEM) images. According to the reconstruction, the
Mimivirus capsid has a diameter of approximately 0.5 μm, and is covered by
0.125 μm long, closely packed fibers (Fig. 3). The total diameter of a free particle
is thus roughly 0.75 μm, consistent with its visibility in the light microscope
(Fig. 1). Mimivirus has a pseudo-triangulation number of approximately 1180,
predicting that the capsid contains approximately 70,000 individual molecules of
the L425 ORF-encoded major capsid protein. Inside the 70-Å-thick protein shell,
the cryoEM images reveal two 40-Å-thick lipid membranes, a structure also
found in some other NCLDVs such as African swine fever virus (Asfarvirus); the
phycodnaviruses and iridoviruses have a single membrane inside their capsids
(Xiao et al. 2005).

The chemical nature of the fibers projecting from the outer layer of the particle
is unknown. Mimivirus encodes eight large proteins with triple-helix forming collagen repeats: L71 (945 aa, seven repeats), R196 (1595 aa, nine repeats), R238
(441 aa, one repeat), R239 (939 aa, eight repeats), R240 (817 aa, six repeats), R241
(812 aa, three repeats), L668 (1387 aa, six repeats), L669 (1937 aa, ten repeats).
Furthermore, Mimivirus possesses a homolog of the procollagen-lysine, 2-oxoglutarate-
5-dioxygenase that catalyzes the posttranslational formation of hydroxylysine in
X-Lys-Gly sequences, and one putative prolyl-4-hydroxylase (L593) that forms 4-
hydroxyproline in -X-Pro-Gly sequences. Hydroxylysines are involved in the inter-
molecular crosslinking of collagen molecules, and hydroxyproline plays a central
role in collagen folding and stability. In addition, a fraction of collagen hydroxylysine residues are the target of O-glycosylation. Consequently, it is tempting to
speculate that the fiber layer surrounding the Mimivirus particle is made of a dense
mesh of crosslinked and glycosylated collagen-like gene products.

Paradoxically, none of the proteins predicted to constitute the fiber layer (the
above-mentioned ORF products with collagen repeats) were detected in the particle
proteome (Table 1; Renesto et al. 2006). This result might be the consequence of
their heavy crosslinking, making them irreversibly insoluble and excluding them
from gel electrophoresis and subsequent mass-spectrometry analysis. The same
explanation may apply to the three paralogs of the major capsid protein L425
(R439, R440, R441) that are not detected in the proteomic analysis. Another unique
feature of Mimivirus particles is the presence of a pentagonal star-shaped structure
centered at a single vertex of the icosahedral capsid (Fig. 6). This feature, nick-
named stargate, was proposed to play a central role in initiating the viral–phagosome
membrane fusion by Dr. Nathan Zauberman (see www.weizmann.ac.il/ Organic_
Chemistry/minsky/nathan/mimivirus.shtml).

**Table 1** Mimivirus particle proteins

| ORF | Protein annotation | Identification by Renesto et al. (2006) | Identification by this study (E-value) |
|---|---|---|---|
| R1 | Replication origin binding protein | – | <0.1 |
| L3 | Unknown | – | <0.1 |
| R10 | Unknown | – | <0.1 |
| L12 | Unknown | – | <0.1 |
| L18 | Unknown | – | <0.1 |
| L48 | Unknown | 2D | – |
| L56 | Ankyrin-containing protein | – | <0.01 |
| L65 | Virion-associated membrane protein | 1D | – |
| L66 | Ankyrin-containing protein | – | <0.01 |
| R69 | Unknown | – | <0.01 |
| L86 | Ankyrin-containing protein | – | <0.1 |
| L90 | Unknown | – | <0.1 |
| L98 | Unknown | – | <0.1 |
| L98b | New short ORF Unknown [37 aa; complement (124358..124471)] overlapping L98 | – | <0.1 |
| L116 | Unknown | – | <0.1 |
| L122 | Ankyrin-containing protein | – | <0.1 |
| R135 | Choline dehydrogenase or related protein | 1D–2D | <0.01 |
| L137 | Glycosyl-transferase domain | – | <0.1 |
| L145 | Unknown | – | <0.1 |
| R160 | Unknown | 1D | – |
| R161 | Unknown | 1D | – |
| L164 | Cysteinyl-tRNA synthetase | – | <0.1 |
| L172 | Unknown (L cluster) | – | <0.1 |
| L173 | Unknown (L cluster) | – | <0.1 |
| L177 | Unknown (L cluster) | – | <0.1 |
| L180 | Unknown (L cluster) | – | <0.1 |
| L183 | Unknown (L cluster) | – | <0.1 |
| R186 | Putative transposase | – | <0.1 |
| R188 | Unknown, similar to AAQ60770 from *Chromobacterium violaceum* | 1D–2D | <0.01 |
| R194 | Topoisomerase I (pox-like) | 1D | – |
| R195 | Glutaredoxin (ESV128 type) | 1D | – |
| L208 | Unknown | 1D | <0.1 |
| L221 | Topoisomerase I (bacterial type) | 1D | <0.01 |
| R225 | Unknown | – | <0.1 |
| L228 | Unknown | – | <0.01 |
| L230 | Procollagen-lysine,2-oxoglutarate 5-dioxygenase | – | <0.01 |
| L232 | Protein kinase domain | – | <0.1 |
| L235 | RNA polymerase subunit 5 | 1D–2D | <0.01 |
| L244 | RNA polymerase II second Largest submit (Rpb2) | ID – | – |
| R253 | Unknown | 1D–2D | <0.01 |
| L264 | Unknown | 1D | – |
| L269 | Unknown | – | <0.1 |
| L271 | Ankyrin-containing protein | – | <0.1 |

(continued)

**Table 1** (continued)

| ORF | Protein annotation | Identification by Renesto et al. (2006) | Identification by this study (E-value) |
|-----|---------|----------|----------|
| L274 | Unknown | 1D | – |
| L279 | Unknown | – | <0.1 |
| L293 | Unknown | 1D | – |
| L294 | Unknown | 1D–2D | <0.01 |
| URF130 | Unknown (149 aa; 382733..383182) | – | <0.1 |
| R301 | Uncharacterized protein (Chilo iridescent virus 380R) | 1D | – |
| R307 | Protein phosphatase 2C domain | 1D | – |
| L309 | Unknown | 1D | – |
| R311 | BIR domain (Chilo iridescent virus 193R) | – | <0.1 |
| L318 | DNA polymerase family X | 1D | – |
| R322 | DNA polymerase (B family) | – | <0.01 |
| R326 | Unknown | 1D | – |
| R327 | Unknown | 1D | – |
| L330 | Unknown | 1D–2D | <0.01 |
| L334 | Unknown | – | <0.1 |
| R341 | Putative polyadenylate polymerase | 1D–2D | <0.01 |
| R345 | Unknown | 1D–2D | <0.01 |
| R347 | Unknown | 1D | – |
| R349 | Unknown | – | <0.1 |
| R350 | Putative transcription termination factor, VV D6R helicase | 1D | <0.01 |
| L352 | Unknown | 1D | – |
| R355 | Unknown | 1D | – |
| L357 | Unknown | – | <0.1 |
| R362 | Thioredoxin domain | 1D–2D | <0.1 |
| R366 | Helicase domain | – | <0.01 |
| L376 | Unknown | 1D–2D | <0.01 |
| L377 | Putative NTPase I | 1D | – |
| R382 | mRNA capping enzyme | 1D | – |
| R383 | Unknown | 1D | – |
| R387 | Unknown | 1D–2D | <0.01 |
| L389 | Unknown | 1D | <0.1 |
| L394 | Unknown | – | <0.1 |
| R395 | Similar to EsV-1–87 (*Ectocarpus siliculosus* virus) | – | <0.1 |
| L396 | VV A18 helicase | – | <0.1 |
| R398 | Calcineurin-like phosphoesterase domain | 1D | – |
| L399 | Unknown | 1D | – |
| R400 | S/T protein kinase, similar to PBCV–1 A617R | 1D | – |
| R402 | Unknown | 1D–2D | <0.01 |
| R403 | Unknown | 1D | – |
| R406 | Alkylated DNA repair | – | <0.1 |
| R407 | tRNA (uracil–5-)-methyltransferase | 2D | <0.1 |
| L410 | Similar to poxvirus P4B major core protein | 1D–2D | <0.01 |
| L417 | Unknown | 1D | – |
| R423 | Unknown | – | <0.1 |
| L425 | Capsid protein 1, [SWISS-PROT: Q5UQL7], complement(join | 1D–2D | <0.01 |

(continued)

**Table 1** (continued)

| ORF | Protein annotation | Identification by Renesto et al. (2006) | Identification by this study (E-value) |
|---|---|---|---|
| | (557530..559233,559658.. 559681,560873..560926)) | | |
| R429 | PBCV1-A494R-like | – | <0.1 |
| L437 | VV A32 virion packaging ATPase | – | <0.1 |
| L442 | Unknown | 1D–2D | <0.01 |
| R443 | Thioredoxin domain | 1D–2D | <0.01 |
| L446 | Patatin-like phospholipase (463L) | – | <0.1 |
| R449b | New short ORF Unknown (68 aa; 592148..592354) (overlapping R449) | – | <0.1 |
| L452 | Unknown | 1D | – |
| L454 | Unknown | 1D | – |
| R457 | Unknown | 1D | – |
| R459 | Unknown | 1D–2D | <0.01 |
| R463 | Unknown | 1D | – |
| R470 | DNA-directed RNA polymerase subunit L | 1D–2D | <0.01 |
| R472 | Unknown | 1D | <0.01 |
| R476 | ATPase domain | – | <0.1 |
| R480 | Topoisomerase II | – | <0.1 |
| L484 | Ankyrin-containing protein | 1D | – |
| L485 | Unknown | 2D | <0.01 |
| R486 | Two PAN domains | 1D | – |
| L488 | Unknown | 1D | <0.1 |
| R489 | Unknown | 1D–2D | <0.01 |
| L492 | Unknown | 1D | <0.1 |
| L498 | Zn-dependent alcohol dehydrogenase | 1D | – |
| R501 | RNA polymerase II largest subunit (Rpb1) | 1D | – |
| R510 | Putative replication factor C subunit | – | <0.1 |
| L515 | Unknown | 1D | <0.1 |
| L516 | Unknown | 1D | – |
| R526 | Putative triacylglycerol lipase | 1D–2D | <0.01 |
| R528 | Unknown | 1D | – |
| L532 | Cytochrome p450 domain | 2D | <0.1 |
| L533 | Unknown | 1D | – |
| L538 | Helicase conserved C-terminal domain (PFAM) | 1D | – |
| L540 | VVI8 helicase | 1D | <0.1 |
| L544 | Transcription initiation factor TFIIB | 1D | – |
| L550 | Unknown | 1D–2D | <0.01 |
| R553 | Unknown | 1D–2D | <0.01 |
| R557 | Unknown | 1D | – |
| R559b | New short ORF Unknown (90 aa; 750051..750323) overlapping R559 | – | <0.01 |
| R563 | Helicase conserved C-terminal domain | 1D | – |
| R566 | Unknown | – | <0.01 |
| L567 | Unknown | 1D–2D | <0.01 |
| R571 | Patatin-like phospholipase (similar to Chilo iridescent virus 463L) | – | <0.1 |

(continued)

**Table 1** (continued)

| ORF | Protein annotation | Identification by Renesto et al. (2006) | Identification by this study (E-value) |
|---|---|---|---|
| L581 | Unknown | – | <0.01 |
| R584 | Unknown | 1D–2D | <0.01 |
| L585 | Unknown | 1D | – |
| L591 | Unknown | 1D–2D | – |
| R592 | Helicase conserved C-terminal domain | – | <0.1 |
| L593 | Prolyl 4-hydroxylase | 1D | – |
| R596 | Thiol oxidoreductase E10R | 1D–2D | <0.01 |
| R607 | Unknown | – | <0.01 |
| R607b | New short ORF Unknown (31 aa; 801314..801409) overlapping R607 | – | <0.1 |
| R610 | Proline-rich protein | 1D–2D | <0.1 |
| L611 | Unknown | – | <0.01 |
| L612 | Mannose–6P isomerase | 1D–2D | <0.01 |
| R639 | Methionyl-tRNA synthetase | – | <0.1 |
| R641 | Unknown | – | <0.01 |
| R642 | Unknown | – | <0.1 |
| R644 | Putative phosphatidylethanolamine-binding protein | 1D | – |
| R646 | Unknown | 1D | – |
| L647 | Unknown | 1D–2D | – |
| R648 | Unknown | 1D–2D | <0.01 |
| R653 | Unknown | 1D–2D | <0.01 |
| R658 | Unknown | 1D | – |
| R661 | Unknown | – | <0.1 |
| R663 | Arginyl-tRNA synthetase | – | <0.1 |
| L670 | Protein kinase domain and Cyclin N-terminal domain | – | <0.1 |
| R679 | Unknown | 1D | – |
| URF277 | Unknown (439 aa; complement [904734..906053]) | – | <0.01 |
| L687 | Endonuclease for the repair of UV-irradiated DNA | 1D | – |
| L688 | Unknown | 1D–2D | – |
| L690 | Unknown | 1D–2D | <0.01 |
| R691 | Unknown | 1D | – |
| L692b | New short ORF Unknown (78 aa; complement [911840..912076]) overlapping R692 | – | <0.1 |
| R692 | Unknown | 1D–2D | <0.01 |
| R695 | Unknown | 1D | – |
| L701 | Unknown | 1D | – |
| R705 | Unknown | 1D | – |
| R706 | Unknown | 1D | – |
| R710 | Unknown | 1D | <0.1 |
| L720 | Hydrolysis of DNA containing ring-opened N7 methylguanine | – | <0.1 |
| R721 | Similar to CheD, chemotaxis protein | 1D–2D | <0.1 |
| R722 | Unknown | 1D | – |
| L724 | Unknown | 1D–2D | <0.01 |
| L725 | Unknown | 1D–2D | <0.01 |
| R727 | Unknown | 1D | – |
| R741 | Unknown | – | <0.01 |

(continued)

**Table 1** (continued)

| ORF | Protein annotation | Identification by Renesto et al. (2006) | Identification by this study (E-value) |
|---|---|---|---|
| R745 | Unknown | – | <0.1 |
| R753 | Unknown | – | <0.01 |
| L754 | Unknown | – | <0.01 |
| R756 | Similar to predicted Fe-S-cluster redox enzyme | – | <0.01 |
| L763 | Unknown | – | <0.1 |
| L766 | Unknown | – | <0.1 |
| L767 | Unknown | – | <0.01 |
| L774 | Unknown | – | <0.1 |
| R776 | Unknown | – | <0.1 |
| L778 | Unknown | 1D | – |
| R787 | Ankyrin-containing protein | – | <0.1 |
| L794b | New short ORF Unknown (66 aa; complement [1033747..1033947]) overlapping L694 | – | <0.1 |
| R811 | Unknown | – | <0.01 |
| R826 | Two protein kinase domains | – | <0.01 |
| L829 | Unknown | 1D–2D | <0.01 |
| L834 | Unknown | – | <0.1 |
| R841 | Ankyrin-containing protein | – | <0.01 |
| R842 | Unknown | – | <0.1 |
| L851 | Unknown | 1D | – |
| L872 | Unknown | 1D–2D | <0.01 |
| R877 | Putative outer membrane lipoprotein | 1D | – |
| R878b | New short ORF Unknown (72 aa; 1142194..1142412) overlapping R878 | --- | <0.1 |
| L893 | Putative oxidoreductase (C-term) | 1D | – |
| L894 | Putative oxidoreductase (N-term) | 1D | – |
| L899 | Unknown | 1D | – |
| R903 | Unknown | – | <0.1 |
| L909 | Unknown | – | <0.01 |

Twenty-three new proteins identified at the high confidence level (E-value<0.01) are highlighted in grey.

Mimivirus encodes several enzymes usually involved in the synthesis of complex reticulated polysaccharides such as perosamine, found in the O-antigen moiety of the lipopolysaccharide (LPS) of various bacteria. The outer layer of the Mimivirus particle may resemble a bacterial cell wall, explaining its retention of the Gram stain. The presence of this polysaccharide layer could also make it palatable for its ameba host, the phagocytosis of which is both triggered by bacterial-sized particles (>0.6 μm) and enhanced by the recognition of surface sugar moieties (reviewed in Claverie et al. 2006). It is likely that Mimivirus is packaged in this spore-like structure, and that the digestion of the fiber outer layer by the ameba endocytic vacuole is a prerequisite to a productive infection. The virus-host specificity (*Acanthamoeba*) might be in part dictated by the presence of the necessary enzymes in the phagosomes of various ameba species (Weekers et al. 1995). Electron micrographs of Mimivirus in the phagocytic vacuole of its host suggest that a significant disruption of the particle's outer layer occurs at this time of the infection (Fig. 7).

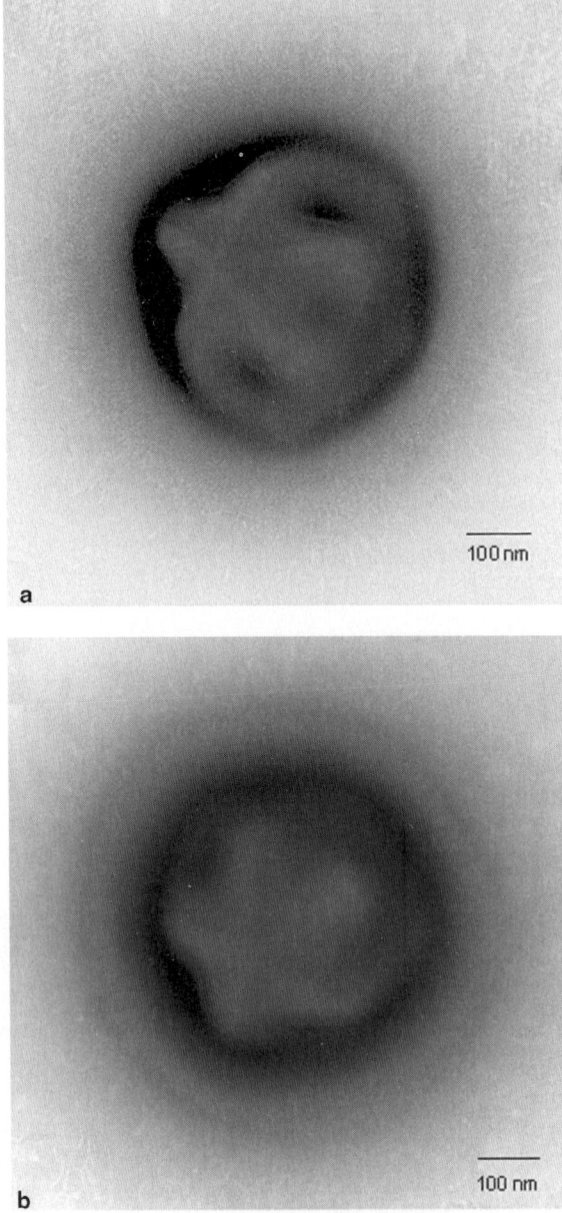

**Fig. 6a,b** TEM pictures (200 kV with FEI CM 200, ×50,000 ) of negatively stained (2% uranyl acetate) Mimivirus particles exhibiting a single pentagonal star-shaped structure. Also note the dense layer of fibers covering the protein capsid. Courtesy of Dr. Wai Li Ling and Dr. Jorge Navaza, Institut de Biologie Structurale, Grenoble

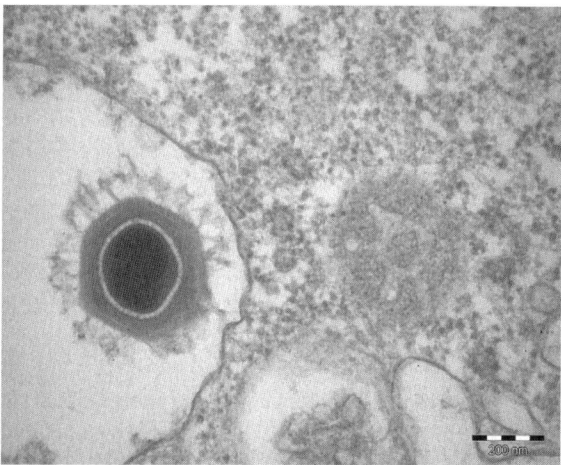

**Fig. 7** Mimivirus particle within an *Acanthamoeba polyphaga* phagocytic vacuole. Note the disorganization of the fiber outer layer compared to its appearance around free particles (Fig. 2B). From Raoult et al. (2007)

Like bacteria, the LPS-like antigens at the surface of the virus particles might constitute the dominant Mimivirus epitope. Gold-labeled antibody molecules against Mimivirus bind to the outer layer of the virus (La Scola et al. 2005). In this context, the frequent seroconversion of pneumonia patients attributed to Mimivirus infection, although no Mimiviruses were isolated from these patients (Berger et al. 2006), could be the result of crossreactivity with a cell-wall antigen from a common bacterial human pathogen.

## Proteomics of the Virion

The composition of purified virions was analyzed by using total extracts, and 1D and 2D gel electrophoresis followed by MALDI-TOF mass-spectrometry analysis of the in-gel trypsin digested bands. Proteins encoded by 114 Mimivirus genes were identified (Renesto et al. 2006). The function of over half of these virion-associated proteins are unknown. The 2D gels revealed numerous isoforms, probably due to posttranslational modifications such as glycosylation, acetylation and phosphorylation.

In addition to the expected major structural components (e.g., the major capsid protein L425 and core L410 protein), transcription enzymes and factors (12 gene products) constitute the largest functional category associated with the viral particles. This set includes all five predicted DNA-directed RNA polymerase subunits, two helicases (R350, L540), the mRNA capping enzyme, and four transcription factors (L377, L538, L544, R563), including a TATA box-like binding protein. The completeness of the transcription machinery components in the Mimivirus particles resembles the poxviruses (Zachertowska et al. 2005; Yoder et al. 2006; Resch et al. 2007) and is expected for a DNA virus that replicates predominantly or exclusively in the cytoplasm.

The next largest functional group contains nine gene products associated with oxidative pathways. These enzymes might help the virus cope with the oxidative stress generated by the host defense. The protein/lipid modification functional category is also well represented, including a phosphoesterase and a lipase, which are eventually used for digesting the cell (vacuole) membrane, two protein kinases, and a protein phosphatase. Finally, five proteins associated with DNA topology and damage repair are in the virion, including topoisomerases IA and IB and a DNA UV damage repair endonuclease (Renesto et al. 2006).

We used this review article to reanalyze the previously generated tryptic peptide mass lists from 487 spots in the 2D gels (Renesto et al 2006), using less stringent, albeit statistically sound criteria. Our main purpose is to eventually identify small exons (or ORFs) that might have been overlooked in the initial analysis of the Mimivirus genome. Each of the mass data was searched against a hybrid database containing all the previously annotated sequences (911 standard ORFs and 347 downgraded URFs) complemented with all other small ORFs (30–99 codons; 6,393 ORFs in total) delineated in the Mimivirus genome. The statistical significance of peptide identifications was assessed by randomizing every sequence in the hybrid database 100 times and repeating the same searches against this randomized database. We obtained an E-value (expected number of protein hits) for a given number of identified peptides in a protein of a given size range. Table 1 shows all 200 ORFs that were reported by Renesto et al. (2006) plus those identified here with an E-value less than 0.1; 23 new proteins identified at the high confidence level (E-value <0.01) are highlighted in grey. They include the B-type DNA polymerase (R322), several ankyrin repeat-containing proteins (L56, L66, R841), the procollagen-lysine hydroxylase (L230), a helicase (R366), a kinase (R826), 13 proteins of unknown functions, URF277, and a short ORF not previously annotated (overlapping with R559). Less confident identification includes the replication origin binding protein (R1), topoisomerase II (R480), putative replication factor C (R510), and three of the four tRNA synthetases (Met-, Arg-, CysRS). In addition, we putatively identified two URFs (URF130, URF277) and seven small ORFs (31–90 aa) not annotated in the original work. This analysis also allowed us to correct the sequence of the major capsid protein (L425, Uniprot Q5UQL7), two exons of which were previously overlooked (see Sect. 3.3.1).

## Experimentally Validated Mimivirus Genes

Most of our knowledge on Mimivirus is derived from bioinformatic analyses of its genome sequence, proteomics, or electron microscopy studies, all of which are subject to overinterpretation. Following the initial excitement of the discovery of this exceptional virus, a few groups initiated postgenomic studies on the biochemistry and physiology of Mimivirus. The various genes for which molecular studies are in progress are summarized in this section.

## Mimivirus-Encoded Components of the Translation Machinery

Genes encoding aminoacyl-tRNA synthetases were probably the most unexpected finding in Mimivirus. Therefore, they became the immediate focus of functional and structural studies in our laboratory. The status of these studies follows:

- The cysteinyl-tRNA synthetase (L164) was expressed in *E. coli* and the protein obtained in a soluble form.
- The methionyl-tRNA synthetase (R639) was expressed in *E. coli* and purified. Its enzymatic function has been characterized and it is specific for both Met and the eukaryotic tRNA$_{Met}$ (Abergel et al. 2007).
- Studies on the tyrosyl-tRNA synthetase (L124) are the most advanced. Following its production in *E. coli*, its enzymatic function was characterized and its specificity for the Tyr amino-acid and the eukaryotic tRNA$_{Tyr}$ validated using a panel of mutant tRNA$_{Tyr}$s. In addition, the crystal structure of the enzyme was determined at 2.2-Å resolution (Abergel et al. 2005, 2007). Mimivirus tyrosyl-tRNA synthetase has unique characteristics in its anti-codon recognizing regions and homodimer organization.

## Mimivirus-Encoded Nucleotide Metabolism and DNA Replication Enzymes

Our laboratory also initiated a systematic characterization of the nucleotide metabolism enzymes. Enzymes being studied include:

- The deoxynucleotide monophosphate kinase (DNK, R512) has been expressed and purified. It behaves as a dimer. It is active with the two substrates dCMP and dGMP. Bacteriophage T4 DNK is the only member of this family of enzymes that recognizes three structurally dissimilar nucleotides: dGMP, dTMP and 5-hydroxymethyl-dCMP, while excluding dCMP and dAMP. The mimivirus homolog has 29% amino acid identity over a region of 120 residues with the T4 enzyme.
- The 584-aa R341 gene product, initially annotated as unknown, contains a polyadenylate polymerase domain at its N-terminus. The protein is associated with Mimivirus particles (Table 1) and African swine fever virus has a homolog (C475L). The polyadenylate polymerase is responsible for adding a poly A tail to the 3Î end of mRNA in eukaryotes. Its identification in Mimivirus adds to a number of virally encoded proteins involved in transcription. The R341 gene product has been expressed, purified and crystals diffracting at 4-Å resolution have been obtained.
- Studies on the nucleoside diphosphate kinase (NDK, R418), the first virus-encoded protein of its kind, are the most advanced. This enzyme usually catalyzes the synthesis of nucleoside triphosphates (NTPs) other than ATP. A detailed characterization of its enzymatic activity showed that the Mimivirus enzyme has a strong preference for deoxypyrimidine nucleotides (Jeudy et al. 2006). This

property might represent an adaptation to the production of the limiting TTP deoxynucleotide required for the replication of the large A+T-rich (72%) Mimivirus genome. The viral NDK might also assume a role in dUTP detoxification to compensate for the surprising absence of a Mimivirus dUTPase (deoxyuridine triphosphate pyrophosphatase), an important enzyme conserved in most viruses. The crystal structure of the enzyme in complex with various ligands has been obtained at 2.2-Å resolution (PDB: 2B8P, 2B8Q). Additional structural studies are in progress on the NDK to investigate the role of its shorter Kpn-loop and other specific Mimivirus features in the active site on its substrate specificity (Jeudy et al. 2005).

Two additional Mimivirus enzymes associated with DNA replication have been characterized outside our laboratory:

- The topoisomerase IB (R194) was characterized by Benarroch et al. (2006). The mimivirus enzyme was functionally more similar to the poxvirus enzyme than to its bacterial homolog, despite its greater sequence similarity to the latter.
- Instead of the ATP-dependent DNA ligase that is present in most NCLDVs, Mimivirus has a NAD⁺-dependent DNA ligase (R303), which is found in bacteria and entomopoxvirus. Benarroch and Shuman (2006) validated the predicted function of the gene product, but found significant differences in its enzymatic behavior compared with both the bacterial and the entomopoxvirus enzymes. They proposed that the Mimivirus enzyme is an intermediate evolutionary stage between the bacterial and entomopoxvirus form of the NAD⁺-dependent DNA ligase, suggesting a horizontal transfer in an ancestral ameba host.

## Characterization of Other Mimivirus Gene Products

ORF L276 is predicted to encode the first viral mitochondrial substrate carrier. In eukaryotic organisms, these proteins are located in the inner mitochondrial membrane or are integral to the membrane of other eukaryotic organelles. Monné et al. (2007) produced milligram quantities of the L276 gene product in *Lactococcus lactis* that were used for a detailed functional characterization. The protein transports dATP and dTTP, suggesting that Mimivirus might target its host mitochondria for obtaining the necessary deoxynucleotide required for replication of its A+T-rich genome.

We successfully produced the R355 gene product, predicted to encode a polyprotein protease potentially involved in the regulation of sumoylation. Crystals (diffracting up to 1.5-Å resolution) of the recombinant protein have been obtained.

The L678 gene product (homologous with a histone methyltransferase) has also been produced, but not in a soluble form. Finally, the product of the L222 gene, a member of a family of 12 close Mimivirus paralogs, has been produced in a soluble form. This protein, of unknown function, is predicted to interact with RNA.

Finally, the predicted DNA glycosylase activities (Endonuclease VIII) of ORF L315 and ORF L720 gene products were experimentally characterized (Bandaru et al. 2007).

# Mimivirus in the Tree of Life

## *Phylogenetic Analysis Using Components of the Translational Apparatus*

Building the Tree of Life, in other words, reconstructing the phylogenetic relationships among all living organisms, is one of the fundamental challenges in biology. Numerous attempts to derive such a tree have been published (see Delsuc et al. 2005). They all involve a comparison of universal genes present in all organisms from the three domains Eubacteria, Eukarya and Archaea. Besides the two major subunits of the DNA-directed RNA polymerase, these universal genes all belong to the protein translation apparatus including ribosomal proteins (~20), a handful of transcription factors, and the aminoacyl-tRNA synthetases (~20). Given the absence of virus-encoded translational machinery, as well as the nonuniversality of virus-encoded RNA polymerases, DNA viruses were always excluded from Tree of Life constructions. However, the numerous components of the translational machinery coded by Mimivirus allow, for the first time, a DNA virus to be included in a Tree of Life analysis. Including the same set of genes used for cellular organisms a phylogenetic analysis indicates that Mimivirus branches near the origin of the Eukarya domain (Fig. 8).

Considering that the central position that Mimivirus occupies among the NCLDVs (Fig. 5), it is tempting to speculate that the ancestor of these large DNA

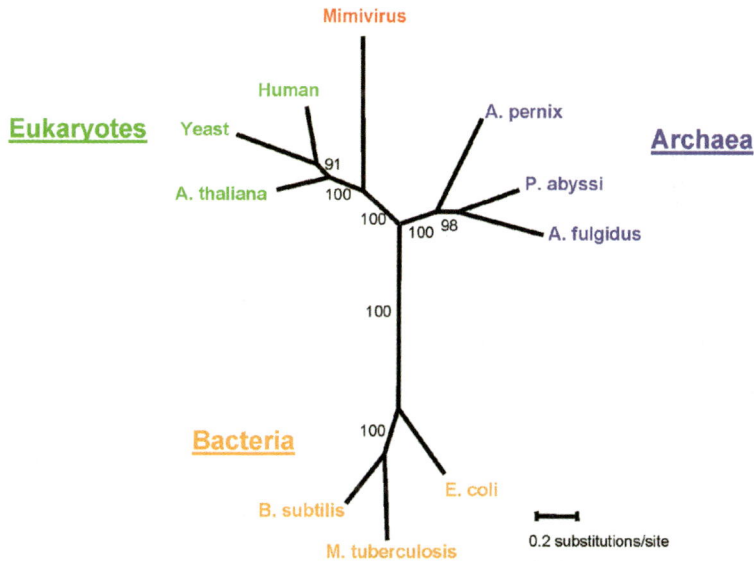

**Fig. 8** Mimivirus in the Tree of Life. This tree was built using the concatenated sequences of seven universally conserved protein sequences (ArgRS, MetRS, TyrRS, the two RNA polymerase subunits, PCNA and a 5'-3' exonuclease. From Raoult et al. (2004)

viruses infecting eukaryotes could predate the radiation of the various eukaryotic kingdoms, or even the emergence of eukaryotes. Such a proposition, highly controversial at the time it was first proposed (Raoult et al. 2004), has since gained some acceptance since it fits nicely with previously published hypotheses linking primitive viruses to the origin of the eukaryotic nucleus (Villarreal and DeFilippis 2000; Takemura 2001; Bell 2001). An alternative hypothesis taking into account the reversibility and flexibility of the gene flow between ancestral viruses and the genome of primitive eukaryotes was more recently proposed (Claverie 2006). According to this scenario, ancestral viruses and pro-nucleus might have exchanged their roles iteratively during the predarwinian era, accounting for the diversity of extant DNA virus families and their partial monophyletic character.

## *Phylogenetic Analysis Using Clamp Loaders*

Despite its high bootstrap values, a common criticism of the phylogenetic tree shown in Fig. 8 is the long branch connecting Mimivirus to the tree trunk, indicating that the sequences used to build the tree are quite divergent. We identified the clamp loader (replication factor C subunits) proteins as an alternative set of sequences exhibiting minimal divergence across the three domains of life, and present in a few DNA viruses, including Mimivirus. Clamp loaders use ATP hydrolysis to load the ring-shaped sliding-clamp made of PCNA subunits around the DNA molecule at the time of replication, promoting processivity. Clamp loader homologs remain remarkably similar in sequence (>25% identity over more than 250 residues) across the three domains of life. Mimivirus encodes its own PCNA molecule, and is again unique among viruses in possessing four clamp loader small subunits (R395, L499, R510, L478) and one large subunit (R411), as found in cellular eukaryotes. In contrast, the archaeal functional homolog (from which the eukaryote clamp loader is thought to have evolved) is usually composed of one small subunit and one large subunit (with the exception of *Methanosarcina acetivorans* that has two similar small subunits and one large subunit) (Chen et al. 2005). Eubacterial clamp loaders are made of two different small subunits and one large subunit (Majka and Burgers 2004). Robust phylogenetic trees encompassing the three domains of life can be made from the multiple alignment of one Mimivirus clamp loader paralogs with its most similar homologs (reciprocal best match) in cellular organisms, as shown in Fig. 9.

In this reconstruction using the small subunit R395 clamp loader protein sequence, Mimivirus is positioned near its fellow NCLDV *Ectocarpus siliculosus* virus (EsV), both of them at the very root of the branch leading to all eukaryotes. Note that the Mimivirus and EsV sequences are positioned in between the Eubacteria and Prokarya domains. Similarly, the phage SPM2 sequence is positioned in between the Archea and Eubacteria domains. Taken at face value, these positions definitely suggest that ancestors of these DNA viruses were present at the time the three major forms of cellular organisms were individualized and are consistent with the hypothesis that they provided the DNA biochemistry and the necessary replication machinery to emerging cellular microorganisms (Forterre 2006; Claverie 2006).

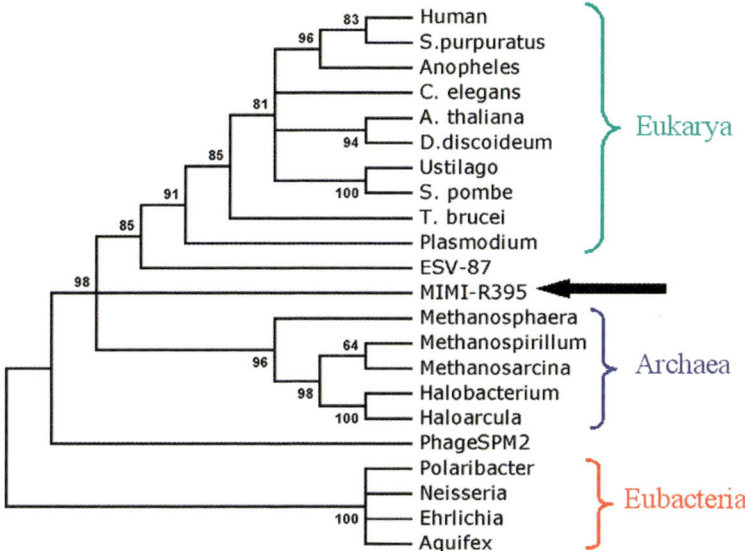

**Fig. 9** Mimivirus in the Tree of Life. The tree was built using an alignment (145 positions retained) of the sequence of the Mimivirus clamp loader protein (R395) with its best reciprocal homologs in the indicated species. The server at www.phylogeny.fr was used with defaults parameters (rooting at midpoint)

## Evidence of Lateral Gene Transfer Between Mimivirus and Corals

As noted above, a large fraction of the genes now constituting the Mimivirus genome have an ancestry predating the emergence of eukaryotes. However, it is also clear that several genes found their way into the Mimivirus genome through horizontal transfers, probably facilitated by the concentration of bacteria or other viruses within the ameba host during its normal feeding process. Together with the metagenomic analyses, identifying the putative origin of the laterally transferred genes might provide hints on ecological niches to look for additional Mimiviridae members. The MutS (mismatch repair) gene provides an intriguing example.

Mimivirus is the sole virus known to have a MutS family protein gene (L359). MutS proteins function in DNA mismatch repair and recombination. These enzymes are ubiquitous in bacteria and eukaryotes and are also found in several Archaea (e.g., *Halobacterium* spp. and *Methanosarcina mazei*). Eukaryotes have at least six major paralogous groups of MutS proteins (MSH1 to MSH6/7), as well as an additional isolated paralogous group encoded in the gorgonian coral mitochondrial DNA (mtMSH). Gorgonians are the sole known eukaryotes exhibiting mitochondrial DNA-encoded mtMutS family proteins (Pont-Kingdon et al. 1998). Interestingly, the Mimivirus MutS sequence is clearly related to the gorgonian mtMSH (Fig. 10). Furthermore, *Sulfurimonas denitrificans* (formerly *Thiomicrospira denitrificans*),

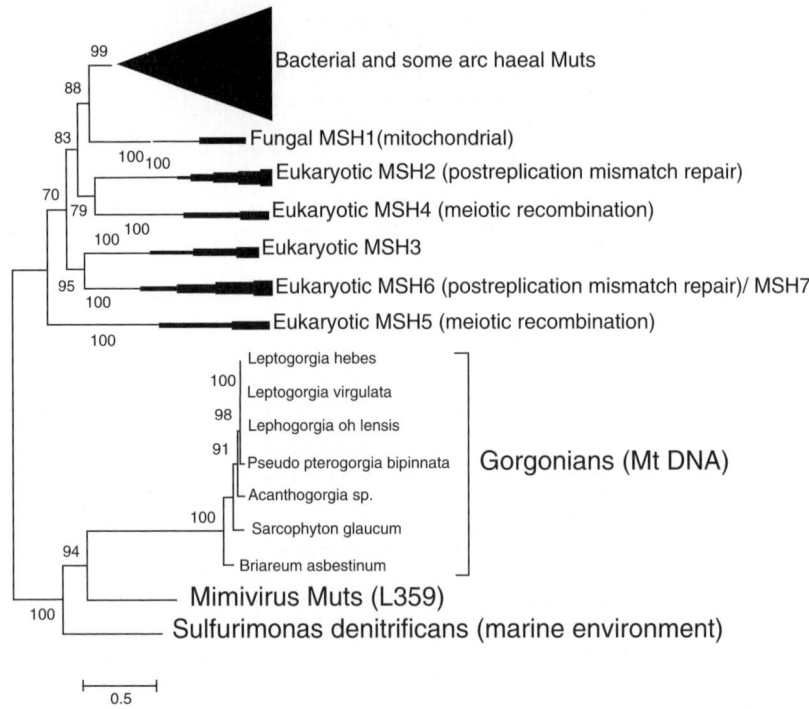

**Fig. 10** Phylogenetic tree of the MutS family proteins. The tree was built using the PhyML server (www.phylogeny.fr) with the option of T-coffee for the alignment program. Bootstrap percentages above 70% are indicated

the marine environmental epsilonproteobacterium, has a MutS homolog related to the Mimivirus MutS and the mtMSH. These results strongly suggest that horizontal transfers of the MutS protein genes occurred (eventually within an ameba host) between an ancestor of Mimivirus, corals and environmental epsilonproteobacteria. Together with the identification of Mimivirus relatives in the Sargasso Sea data set, this last result is one more incentive to look for new *Mimiviridae* species in association with marine protists.

**Acknowledgements** We thank Jim Van Etten for his correction of the original version of this chapter, as well as Dr. Wai Li Ling and Dr. Jorge Navaza Jorge for allowing us to present micrographs of the stargate structure (Fig. 6) prior to publication.

# References

Abergel C, Chenivesse S, Byrne D, Suhre K, Arondel V, Claverie JM (2005) Mimivirus TyrRS: preliminary structural and functional characterization of the first amino-acyl tRNA synthetase found in a virus. Acta Crystallograph Sect F Struct Biol Cryst Commun. 61:212–215

Abergel C, Rudinger-Thirion J, Giege R, Claverie JM (2007) Virus-encoded aminoacyl-tRNA synthetases: structural and functional characterization of Mimivirus TyrRS and MetRS. J Virol81:12406–12417

Bailey TL, Gribskov M (1998) Combining evidence using p-values: application to sequence homology searches. Bioinformatics 14:48–54

Bandaru V, Zhao X, Newton MR, Burrows CJ, Wallace SS (2007) Human endonuclease VIII-like (NEIL) proteins in the giant DNA Mimivirus. DNA Repair 6:1629–1641

Bawden AL, Glassberg KJ, Diggans J, Shaw R, Farmerie W, Moyer RW (2000) Complete genomic sequence of the *Amsacta moorei* entomopoxvirus: analysis and comparison with other poxviruses. Virology 274:120–139

Bell PJ (2001) Viral eukaryogenesis: was the ancestor of the nucleus a complex DNA virus? J Mol Evol 53:251–256

Benarroch D, Shuman S (2006) Characterization of mimivirus NAD+-dependent DNA ligase. Virology 353:133–143

Benarroch D, Claverie JM, Raoult D, Shuman S (2006) Characterization of mimivirus DNA topoisomerase Ib suggests horizontal gene transfer between eukaryal viruses and bacteria. J Virol 80: 314–321

Berger P, Papazian L, Drancourt M, La Scola B, Auffray JP, Raoult D (2006) Ameba-associated microorganisms and diagnosis of nosocomial pneumonia. Emerg Infect Dis 12:248–255

Blanc G, Ogata H, Robert C, Audic S, Suhre K, Vestris G, Claverie JM, Raoult D (2007) Reductive genome evolution from the mother of rickettsia. PLoS Genet 3:e14

Chen YH, Kocherginskaya SA, Lin Y, Sriratana B, Lagunas AM, Robbins JB, Mackie RI, Cann IK (2005) Biochemical and mutational analyses of a unique clamp loader complex in the archaeon *Methanosarcina acetivorans*. J Biol Chem 280:41852–1863

Cheng CH, Liu SM, Chow TY, Hsiao YY, Wang DP, Huang JJ, Chen HH (2002) Analysis of the complete genome sequence of the Hz-1 virus suggests that it is related to members of the Baculoviridae. J Virol 76:9024–9034

Claverie JM (2006) Viruses take center stage in cellular evolution. Genome Biol 7:110

Claverie JM, Ogata H, Audic S, Abergel C, Suhre K, Fournier PE (2006) Mimivirus and the emerging concept of "giant" virus. Virus Res 117:133–144

Dare RK, Chittaganpitch M, Erdman DD (2008) Screening pneumonia patients for mimivirus. Emerg Infect Dis 14:465–467

Delaroque N, Boland W, Muller DG, Knippers R (2003) Comparisons of two large phaeoviral genomes and evolutionary implications. J Mol Evol 57:613–622

Delsuc F, Brinkmann H, Philippe H (2005) Phylogenomics and the reconstruction of the tree of life. Nat Rev Genet 6:361–375

Espagne E, Dupuy C, Huguet E, Cattolico L, Provost B, Martins N, Poirie M, Periquet G, Drezen JM (2004) Genome sequence of a polydnavirus: insights into symbiotic viral evolution. Science 306:286–289

Filee J, Siguier P, Chandler M (2007) I am what I eat and I eat what I am: acquisition of bacterial genes by giant viruses. Trends Genet 23:10–15

Fitzgerald LA, Graves MV, Li X, Feldblyum T, Nierman WC, Van Etten JL (2007) Sequence and annotation of the 369-kb NY-2A and the 345-kb AR158 viruses that infect Chlorella NC64A. Virology 358:472–484

Forterre P (2006) Three RNA cells for ribosomal lineages and three DNA viruses to replicate their genomes: a hypothesis for the origin of cellular domain. Proc Natl Acad Sci U S A 103:3669–3674

Ghedin E, Claverie JM (2005) Mimivirus relatives in the Sargasso sea. Virol J 2:62

Iyer LM, Aravind L, Koonin EV (2001) Common origin of four diverse families of large eukaryotic DNA viruses. J Virol 75:11720–11734

Iyer LM, Balaji S, Koonin EV, Aravind L (2006) Evolutionary genomics of nucleo-cytoplasmic large DNA viruses. Virus Res 117:156–184

Jeudy S, Coutard B, Lebrun R, Abergel C (2005) *Acanthamoeba polyphaga* mimivirus NDK: preliminary crystallographic analysis of the first viral nucleoside diphosphate kinase. Acta Crystallograph Sect F Struct Biol Cryst Commun. 61:569–572

Jeudy S, Claverie JM, Abergel C (2006) The nucleoside diphosphate kinase from mimivirus: a peculiar affinity for deoxypyrimidine nucleotides. J Bioenerg Biomembr 38:247–254

Khan M, La Scola B, Lepidi H, Raoult D (2006) Pneumonia in mice inoculated experimentally with *Acanthamoeba polyphaga mimivirus*. Microb Pathog 42:56–61

La Scola B, Audic S, Robert C, Jungang L, de Lamballerie X, Drancourt M, Birtles R, Claverie JM, Raoult D (2003) A giant virus in amoebae. Science 299:2033

La Scola B, Marrie TJ, Auffray JP, Raoult D (2005) *Mimivirus* in pneumonia patients. Emerg Inf Dis 11:449–452

Larcher C, Jeller V, Fischer H, Huemer HP (2006) Prevalence of respiratory viruses, including newly identified viruses, in hospitalised children in Austria. Eur J Clin Microbio Infect Dis 25:681–886

Majka J, Burgers PM (2004) The PCNA-RFC families of DNA clamps and clamp loaders. Prog Nucleic Acid Res Mol Biol 78:227–260

Monier A, Claverie JM, Ogata H (2007) Horizontal gene transfer and nucleotide compositional anomaly in large DNA viruses. BMC Genomics 8:456

Monier A, Claverie JM, Ogata H (2008a) Taxonomic distribution of large DNA viruses in the sea. Genome Biol 9:R106

Monier A, Larsen JB, Sandaa RA, Bratbak G, Claverie JM, Ogata H (2008b) Marine mimivirus relatives are probably large algal viruses. Virol J 5:12

Monne M, Robinson AJ, Boes C, Harbour ME, Fearnley IM, Kunji ER (2007) The mimivirus genome encodes a mitochondrial carrier that transports dATP and dTTP. J Virol 81:3181–3186

Moreira D, Lopez-Garcia P (2005) Comment on "The 1.2-megabase genome sequence of Mimivirus". Science 308:1114

Nagasaki K, Shirai Y, Tomaru Y, Nishida K, Pietrokovski S (2005) Algal viruses with distinct intraspecies host specificities include identical intein elements. Appl Environ Microbiol 71:3599–3607

Nalçacioglu R, Ince IA, Vlak JM, Demirbag Z, van Oers MM (2007) The Chilo iridescent virus DNA polymerase promoter contains an essential AAAAT motif. J Gen Virol 88:2488–2494

Ogata H, Abergel C, Raoult D, Claverie J-M (2005a) Response to Comment on "The 1.2-megabase genome sequence of Mimivirus". Science 308:1114

Ogata H, Raoult D, Claverie JM (2005b) A new example of viral intein in Mimivirus. Virol J 2:8

Pont-Kingdon G, Okada NA, Macfarlane JL, Beagley CT, Watkins-Sims CD, Cavalier-Smith T, Clark-Walker GD, Wolstenholme DR (1998) Mitochondrial DNA of the coral *Sarcophyton glaucum* contains a gene for a homologue of bacterial MutS: a possible case of gene transfer from the nucleus to the mitochondrion. J Mol Evol 46:419–431

Raoult D, Audic S, Robert C, Abergel C, Renesto P, Ogata H, La Scola B, Suzan M, Claverie JM (2004) The 1.2-megabase genome sequence of Mimivirus. Science 306:1344–1350

Raoult D, Renesto P, Brouqui P (2006) Laboratory infection of a technician by mimivirus. Ann Intern Med 144:702–703

Raoult D, La Scola B, Birtles R (2007) The discovery and characterization of Mimivirus, the largest known virus and putative pneumonia agent. Clin Infect Dis 45:95–102

Resch W, Hixson KK, Moore RJ, Lipton MS, Moss B (2007) Protein composition of the vaccinia virus mature virion. Virology 358:233–247

Renesto P, Abergel C, Decloquement P, Moinier D, Azza S, Ogata H, Fourquet P, Gorvel JP, Claverie JM (2006) Mimivirus giant particles incorporate a large fraction of anonymous and unique gene products. J Virol 80:11678–11685

Rusch DB, Halpern AL, Sutton G, Heidelberg KB, Williamson S et al (2007) The *Sorcerer II* Global Ocean Sampling Expedition: Northwest Atlantic through Eastern Tropical Pacific. PLoS Biol 5:e77

Suhre K (2005) Gene and genome duplication in Acanthamoeba polyphaga Mimivirus. J Virol 79:14095–14101

Suhre K, Audic S, Claverie JM (2005) Mimivirus gene promoters exhibit an unprecedented conservation among all eukaryotes. Proc Natl Acad Sci U S A 102:14689–14693

Suzan-Monti M, La Scola B, Raoult D (2006) Genomic and evolutionary aspects of Mimivirus. Virus Res 117:145–150

Suzan-Monti M, La Scola B, Barrassi L, Espinosa L, Raoult D (2007) Ultrastructural characterization of giant volcano-like virus factory of *A. polyphaga* Mimivirus. PloS ONE 3: e328

Takemura M (2001) Poxviruses and the origin of the eukaryotic nucleus. J Mol Evol 52:419–425

Vanacova S, Liston DR, Tachezy J, Johnson PJ (2003) Molecular biology of the amitochondriate parasites, *Giardia intestinalis*, *Entamoeba histolytica* and *Trichomonas vaginalis*. Int J Parasitol 33:235–255

Van Etten JL (2003) Unusual life style of giant chlorella viruses. Annu Rev Genet 37:153–195

Venter JC. Remington K, Heidelberg JF, Halpern AL, Rusch D, Eisen JA, Wu D, Paulsen I, Nelson KE, Nelson W, Fouts DE, Levy S, Knap AH, Lomas MW, Nealson K, White O, Peterson J, Hoffman J, Parsons R, Baden-Tillson H, Pfannkoch C, Rogers YH, Smith HO (2004) Environmental genome shotgun sequencing of the Sargasso Sea. Science 304: 66–74

Villarreal LP, DeFilippis VR (2000) A hypothesis for DNA viruses as the origin of eukaryotic replication proteins. J Virol 74:7079–7084

Weekers PH, Engelberts AM, Vogels GD (1995) Bacteriolytic activities of the free-living soil amoebae, *Acanthamoeba castellanii*, *Acanthamoeba polyphaga* and *Hartmannella vermiformis*. Antonie Van Leeuwenhoek 68:237–243

Wilson WH, Schroeder DC, Allen MJ, Holden MT, Parkhill J, Barrell BG, Churcher C, Hamlin N, Mungall K, Norbertczak H, Quail MA, Price C, Rabbinowitsch E, Walker D, Craigon M, Roy D, Ghazal P (2005) Complete genome sequence and lytic phase transcription profile of a Coccolithovirus. Science 309:1090–1092

Xiao C, Chipman PR, Battisti AJ, Bowman VD, Renesto P, Raoult D, Rossmann MG (2005) Cryo-electron microscopy of the giant Mimivirus. J Mol Biol 353:493–496

Yamada T, Onimatsu H, Van Etten JL (2006) Chlorella viruses. Adv Virus Res 66:293–336

Yoder JD, Chen TS, Gagnier CR, Vemulapalli S, Maier CS, Hruby DE (2006) Pox proteomics: mass spectrometry analysis and identification of Vaccinia virion proteins. Virol J 3:10

Zachertowska A, Brewer D, Evans DH (2005) Characterization of the major capsid proteins of myxoma virus particles using MALDI-TOF mass spectrometry. J Virol Methods 132:1–12

Zhang QY, Xiao F, Xie J, Li ZQ, Gui JF (2004) Complete genome sequence of lymphocystis disease virus isolated from China. J Virol 78:6982–6994

# Family *Iridoviridae*: Poor Viral Relations No Longer

V.G. Chinchar(✉), A. Hyatt, T. Miyazaki, T. Williams

**Contents**

**Abstract** Members of the family *Iridoviridae* infect a diverse array of invertebrate and cold-blooded vertebrate hosts and are currently viewed as emerging pathogens of fish and amphibians. Iridovirid replication is unique and involves both nuclear and cytoplasmic compartments, a circularly permuted, terminally redundant genome that, in the case of vertebrate iridoviruses, is also highly methylated, and the efficient shutoff of host macromolecular synthesis. Although initially neglected largely due to the perceived lack of health, environmental, and economic concerns, members of the genus *Ranavirus*, and the newly recognized genus *Megalocytivirus*, are rapidly attracting

V.G. Chinchar
Department of Microbiology, University of Mississippi Medical Center, Jackson, MS USA
vchinchar@microbio.umsmed.edu

James L. Van Etten (ed.) *Lesser Known Large dsDNA Viruses.*
Current Topics in Microbiology and Immunology 328.
© Springer-Verlag Berlin Heidelberg 2009

growing interest due to their involvement in amphibian population declines and their
adverse impacts on aquaculture. Herein we describe the molecular and genetic basis
of viral replication, pathogenesis, and immunity, and discuss viral ecology with reference to members from each of the invertebrate and vertebrate genera.

# Introduction

Since their isolation nearly 50 years ago, iridovirids (i.e., members of the family
*Iridoviridae*) have been overshadowed by other DNA viruses of medical or veterinary importance, specifically herpesviruses and poxviruses. Although one family
member, lymphocystis disease virus (LCDV), has been known for over a century
by the wart-like disease it causes in multiple species of salt- and fresh-water fish
(Weissenberg 1965), and study of a second family member, frog virus 3 (FV3), has
elucidated novel events in eukaryotic virus replication, the perceived absence of
commercial, agricultural, medical, or ecological damage resulting from iridovirid
infections limited interest in, and study of, this diverse virus family (Williams et al.
2005). Initially, iridovirids attracted interest because of their unusual biology and
widespread occurrence in amphibians, fish, and insects. However, within the last
20 years the increased recognition of vertebrate iridovirids as important pathogens infecting commercially and ecologically important fish and amphibian species
has attracted the interest of fish pathologists, wildlife biologists, ecologists and others
interested in the impact of infectious disease on ectothermic vertebrates (Hyatt et al.
2000; Chinchar 2002; Williams et al. 2005; Mendelson et al. 2006). For example,
members of the genus *Ranavirus* were identified as the causative agent in approximately half the documented cases of amphibian mortality reported in the United
States between 1996 and 2001 (Green et al. 2002). In addition, viruses in the genus
*Megalocytivirus* have been responsible for numerous outbreaks of severe disease in
fish farming facilities throughout Asia (Nakajima et al. 1998). Given the growing
impact of iridovirus diseases worldwide and the increased use of contemporary
molecular approaches to elucidate the phylogeny and life cycle of iridoviruses, it
appears that this virus family is finally receiving the scientific recognition it
deserves. In this chapter, we provide a summary of iridovirid taxonomy, genetic
organization, and replication strategy, followed by a description of the biology and
ecology of specific genera and viral species. Additional information can be found
in several recent reviews (Williams 1996, 1998; Chinchar 2002; Williams et al.
2005).

# Taxonomy

The family *Iridoviridae* is currently classified into five genera (*Iridovirus,
Chloriridovirus, Ranavirus, Megalocytivirus,* and *Lymphocystivirus*), each consisting of
one or more virus species, tentative species and strains (Table 1) (Chinchar et al. 2005).
In keeping with the recent suggestion of Vetten and Haenni (2006), members of the

**Table 1** Taxonomy of the family *Iridoviridae*

| Genus | Viral species [strains][a] | Tentative species |
|---|---|---|
| *Iridovirus* | *Invertebrate iridescent virus 6* (IIV–6), IIV–1 | Anticarsia gemmatalis iridescent virus (AGIV), IIV–2, –9, –16, –21, –22, –23, –24, 29, –30, –31 |
| *Chloriridovirus* | *Invertebrate iridescent virus* 3 (IIV–3) | |
| *Ranavirus* | *Frog virus 3* (FV3), [tadpole edema virus, TEV; tiger frog virus, TFV] | Singapore grouper iridovirus (SGIV); Grouper iridovirus (GIV) |
| | *Ambystoma tigrinum virus* (ATV), [Regina ranavirus, RRV] | Rana catesbeiana virus-Z (RCV-Z) |
| | *Bohle iridovirus* (BIV) | |
| | *Epizootic haematopoietic necrosis virus* (EHNV) | |
| | *European catfish virus* (ECV), [European sheatfish virus, ESV] | |
| | *Santee-Cooper ranavirus,* [Largemouth bass virus, LMBV; doctor fish virus, DFV; guppy virus 6, GV–6] | |
| *Megalocytivirus* | *Infectious spleen and kidney necrosis virus* (ISKNV) [Red sea bream iridovirus, RSIV; African lampeye iridovirus, ALIV; Orange spotted grouper iridovirus, OSGIV; Rock bream iridovirus, RBIV] | |
| *Lymphocystivirus* | *Lymphocystis disease virus 1* (LCDV–1) | LCDV–2, LCDV-C, LCDV-RF |
| Unclassified | White sturgeon iridovirus (WSIV) | |

[a] Viral species recognized by the ICTV are italicized, whereas strains or isolates are listed within brackets (Chinchar et al. 2005). Common abbreviations are indicated

family *Iridoviridae* will be referred to collectively as iridovirids to distinguish them from members of the genus *Iridovirus*. Morphologically, iridovirids are large, icosahedral viruses (120–200 nm in diameter) that possess an internal lipid membrane located between the viral core and outer capsid. In contrast to other virus families, a viral envelope, present on virions that bud from the plasma membrane, is not required for infectivity, and many virions remain cell-associated and are released as naked particles following cell lysis.

Members of the family possess linear, double-stranded DNA genomes, which vary in size from approximately 140 kbp (genus *Ranavirus*) to over 200 kbp (genus *Iridovirus*). Iridovirid genomes are unique among animal viruses in that they are circularly permuted and terminally redundant (Goorha and Murti 1982; Delius et al. 1984). For example, if the letters of the alphabet represent the viral genome, analysis of linear genomes from individual virus particles would yield sequences such as **ABCDE...UVWXYZABCDE, CDEFG...WXYZABCDEFG, FGHIJ...ZABCEDFGHIJ.** Because the terminal repeat region accounts for 5%–50% of the genome length, the total length of each genome (e.g., 140 kbp for a typical ranavirus) is more than the length of the unique region (e.g., ~105 kbp).

The five iridovirid genera can be partitioned into two groups (that in the future may be classified as subfamilies) based on the hosts they infect and the level of genomic methylation (Chinchar et al. 2005). Members of the genera *Iridovirus* and *Chloriridovirus* infect invertebrates (i.e., insects, crustaceans, etc.) and lack a highly methylated genome. In contrast, members of the *Ranavirus*, *Lymphocystivirus*, and *Megalocytivirus* genera infect cold-blooded vertebrates such as fish, amphibians, and reptiles and possess genomes in which approximately 25% of the cytosine residues are methylated by a virus encoded DNA methyltransferase (Willis and Granoff 1980). However, there is at least one ranavirus, Singapore grouper iridovirus (SGIV), that lacks the DNA methyltransferase gene and cannot methylate its DNA (Song et al. 2004).

The division of the family into genera was initially based on biological properties of the viruses (e.g., host range, GC content of the genome, serology, virion morphology, particle size, histopathology, and clinical signs of disease). GC content varies markedly and ranges from 27%–29% (irido- and lymphocystiviruses) to 48%–55% (chlorirido-, rana- and megalocytiviruses) and does not correspond to either the GC content of the host or the methylation status of the virus. Not unexpectedly, codon usage is influenced by the overall GC content, but the basis for the marked difference in GC content among different viral genera is unknown (Schackelton et al. 2006; Eaton et al. 2007; Tsai et al. 2007). Recent analyses of the amino acid sequences of the major capsid protein (MCP) and other viral proteins confirmed these taxonomic divisions and indicated that species within a genus generally shared high levels of identity/similarity. Typically, members of the same viral genus show more than 70% similarity within the major capsid protein (MCP) at the amino acid level, whereas species from different genera show less than 50% similarity (Do et al. 2005a, 2005b).

Although identification of iridovirid genera has been relatively straightforward, identification of individual viral species has proven to be more difficult because of high levels of sequence identity/similarity within the MCP and other highly conserved proteins among members of the same genus. For example, several ranavirus species show greater than 90% amino acid identity within the highly conserved MCP. Thus, differentiation of viral species is based on multiple criteria including viral protein profiles, DNA restriction fragment length polymorphisms (RFLPs), host species infected, clinical signs (i.e., histopathology and gross pathology), and differences in nucleotide and amino acid sequences (Mao et al. 1997; Chinchar and Mao 2000; Chinchar et al. 2005). Unfortunately, the lack of sequence information from many vertebrate and invertebrate iridovirid isolates continues to hamper taxonomic classification and results in a significant number of tentatively assigned and unclassified isolates. Moreover, because of the aforementioned high level of amino acid conservation within the MCP, differentiation of iridovirid species based on serological differences has not proven useful (Hedrick et al. 1992a; Chinchar et al. 2005). To further confound taxonomic classification, identical, or nearly identical, viruses have been given different names reflecting the host species or the geographic regions from which they were isolated.

As the number of iridovirid isolates increases, a change in the way that new viruses are designated may be in order. In the past, multiple isolates of ostensibly the same viral species were made from different host species and given different names. While some of these may represent novel viral species, others are likely strains or isolates of the same virus. Although the International Committee for the Taxonomy of Viruses (ICTV) does not recognize taxonomic distinctions below the species level, it may be important for working virologists to unambiguously identify clinically or ecologically important isolates in the same way that bacteriologists identify pathogenic strains of *Escherichia coli,* or virologists identify strains of influenza A virus. Based on the latter example, it may be appropriate to utilize a standard nomenclature to designate iridovirid isolates. For example, a novel isolate of *Ambystoma tigrinum virus* would be designated as *Ambystoma tigrinum virus (A. tigrinum*/Fort Collins/1/2005) where, following iden-tification of the viral species, the four identifiers within the parentheses indicate the host species from which the isolate was obtained, the location of the isolate, the isolate number, and the year of isolation. As with influenza virus, once the virus has been clearly identified, it could be referred to by a suitable abbreviation such as ATV/ FC/1/05. While far from perfect, use of this convention may help resolve the confusion resulting when a given virus species displays a range of clinical symptoms, a broad host range, and wide geographical distribution.

# The Viral Genome

## *Genomic Organization*

Because the viral genome is circularly permuted and terminally redundant, no two iridovirid genes consistently occupy the ends of the viral genome. Moreover, gene order within the family is not fixed, and even among members of the same genus, gene order can be quite variable. For example, dot-plot analysis, involving the systematic comparison of nucleotide positions in one genome with nucleotide positions in another genome, revealed that although tiger frog virus (TFV) and ATV displayed considerable co-linearity over large portions of their genomes, two inver-sions were noted, the largest of which involved an approximately 40-kbp fragment (He et al 2002; Jancovich et al. 2003). As expected, more closely related viruses such as FV3 and TFV, which are considered to be strains of the same species, pos-sess nearly co-linear genomes (Tan et al. 2004), whereas more distantly related ranaviruses such as SGIV and FV3, or red sea bream iridovirus (RSIV) and FV3, which are species from different genera, showed, respectively, little or no conserva-tion of gene order (V.G. Chinchar, unpublished observation). These results suggest that the relative positions of genes within iridovirid genomes are not important determinants of gene expression or virion viability. The marked differences in gene order among various viral species may be a consequence of the high rate of recom-bination that has been noted previously (Chinchar and Granoff 1986). Furthermore, these findings support earlier observations that iridovirid genes are not spatially clustered by functional or temporal class (McMillan and Kalmakoff 1994; D'Costa et al. 2004; Tan et al. 2004; Lua et al. 2005).

## Viral Genes: Identity and Function

Although there is still some disagreement on the exact numbers, based on the
genome size, iridovirids contain between roughly 100 (ATV and FV3) and 200
(IIV-6) open reading frames (ORFs) (Tsai et al. 2007; Eaton et al. 2007; see
Table 2). Some reports suggest that more than 400 ORFs are encoded by IIV-6, but
these include ORFs that overlap each other on the same or different DNA strands
and thus likely overestimate the actual number of viral gene products. Consistent
with this lower estimate, D'Costa et al. (2004) detected 137 IIV-6 transcripts (38
immediate-early, 34 delayed-early, and 65 late) by Northern blotting. Among the
viruses sequenced to date, the functions or presumed functions of about 25%–35%
of the putative viral proteins are known or have been inferred by homology to other
viral or cellular proteins. Most of the remaining ORFs match putative proteins
present in one or more iridovirids, suggesting that they play important roles in
iridovirid replication, biogenesis, and survival (Eaton et al. 2007). Viral genes
involved in DNA and RNA synthesis, dTTP synthesis, and the evasion of host
immune responses have been identified. For example, iridovirids encode their own
DNA polymerase, homologs of the two largest subunits of RNA polymerase II, a
RAD2-like repair enzyme, a DNA methyltransferase (vertebrate iridovirids only),
an RNAse III-like protein, helicases, NTPases, and various kinases (reviewed in
detail by Williams et al. 2005). Overall, iridovirids contain a common set of viral
genes that encode viral structural and catalytic proteins that permit replication in a
broad range of cell types (Eaton et al. 2007). However, individual species vary
considerably in their content of non-core genes, likely reflecting differences in their
respective hosts and survival strategies.

**Table 2**  Coding potential of iridovirus genomes

| Genus | Virus | Kbp[a] | No. genes[b] | %GC | GenBank Acc. No. |
|---|---|---|---|---|---|
| *Iridovirus* | IIV–6 | 212,482 | 193 | 29% | AF303741 |
| *Chloriridovirus* | IIV–3 | 191,132 | 135 | 48% | DQ643392 |
| *Lymphocystivirus* | LCDV–1 | 102,653 | 107 | 29% | L63545 |
| | LCDV-C | 186,250 | 160 | 27% | AY380826 |
| *Megalocytivirus* | ISKNV | 111,362 | 121 | 55% | AF371960 |
| | OSGIV | 112,636 | 126 | 54% | Ay894343 |
| | RBIV | 112,080 | 127 | 53% | AY532606 |
| *Ranavirus* | FV3 | 105,903 | 97 | 55% | AY548484 |
| | TFV | 105,057 | 101 | 55% | AF389451 |
| | ATV | 106,332 | 93 | 54% | AY150217 |
| | SGIV | 140,131 | 142 | 49% | AY521625 |
| | GIV | 139,793 | 142 | 49% | AY666015 |

[a] The value shown represents the unique genome size in kilobase pairs minus the length of the terminal repeat

[b] The value shown is an estimate of the total number of nonoverlapping genes encoded by a given virus. It is generally lower than the total number of putative ORFs, which includes putative genes encoded on opposing DNA strands and overlapping genes. The numbers shown are averages based on the estimates of Eaton et al. (2007) and Tsai et al. (2007)

## *Iridovirid Phylogeny*

To date 12 complete genomic sequences, including representatives of each of
the five established genera, as well as partial sequences of numerous other
isolates, have been determined, and phylogenetic trees based on individual viral
protein-coding regions or concatenated sequence sets have been constructed
(Jancovich et al. 2003; Tan et al. 2004; Delhon et al. 2006; Kitamura et al. 2006).
Although the resulting phylogenies generally confirm the current taxonomy
(Fig. 1), representatives of the two genera from invertebrates form what is, essentially, one cluster. Furthermore, while trees similar to those shown in Fig. 1 and
elsewhere (Delhon et al. 2006; Tang et al. 2007) indicate that IIV-6 (genus
*Iridovirus*) and IIV-3 (genus *Chloriridovirus*) are distantly related, several
tentative invertebrate virus species that were presumed to be members of the genus

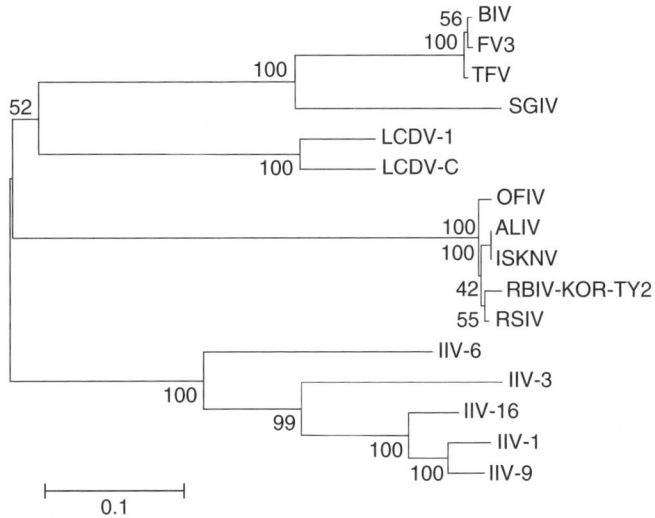

**Fig. 1** Iridovirid phylogeny. The inferred amino acid sequences of the MCP of 16 iridovirids
representing all five currently recognized genera were aligned using the CLUSTAL W program.
Subsequently, a phylogenetic tree was constructed using the Neighbor-Joining algorithm and
Poisson correction within MEGA version 3.1 and validated by 1,000 bootstrap repetitions (Kumar
et al. 2004). Branch lengths are drawn to scale, and a scale bar is shown. The number at each node
indicates bootstrapped percentage values. The sequences used to construct the tree were obtained
from the following viruses: genus *Megalocytivirus—ISKNV infectious skin and kidney necrosis
virus* (AF370008), *ALIV* African lampeye iridovirus (AB109368), *OFIV* olive flounder iridovirus
(AY661546), *RSIV* red sea bream iridovirus (AY310918), *RBIV* rock bream iridovirus (AY533035);
genus *Ranavirus—SGIV* Singapore grouper iridovirus (AF364593), *TFV* tiger frog virus
(AY033630), *BIV Bohle iridovirus* (AY187046), *FV3 Frog virus 3* (U36913); genus
*Lymphocystivirus*: *LCDV-1 Lymphocystis disease virus* (L63545), *LCDV-C* lymphocystis disease
virus—China (AAS47819.1); genus *Iridovirus* —*IIV-6 Invertebrate iridescent virus 6*
(AAK82135.1), *IIV-16* (AF025775), *IIV-1* (M33542), and *IIV-9* (AF025774); genus
*Chloriridovirus—IIV-3* (DQ643392)

*Iridovirus* cluster closer to IIV-3 than to IIV-6. Moreover, the tree shows that iridovirids group into four well-resolved clusters corresponding to the *Ranavirus*, *Megalocytivirus*, *Lymphocystivirus*, and *Iridovirus/Chloriridovirus* genera. However, it is not possible to determine if one genus is descended from another, or if all four genera are derived from a common ancestor. In addition to studies focused solely on iridovirids, several investigators have examined the phylogenetic relationships among ascoviruses, mimiviruses, and four families of nuclear cytoplasmic large DNA-containing viruses (NCLDV), i.e., *Iridoviridae*, *Phycodnaviridae*, *Asfarviridae*, and *Poxviridae*. These analyses suggest that the six families share a common ancestry (Stasiak et al. 2000, 2003; Iyer et al. 2001; Allen et al. 2006). One recent study postulates that iridovirids may be closer to the recently discovered mimiviruses than to other virus families (Allen et al. 2006), but the branching order is far from certain and the authentic phylogeny remains unresolved.

## Replication Cycle

Although iridovirids other than FV3 have been examined by several groups, few of these studies have focused on the mechanisms of replication. Consequently, most of what we know about iridovirid biogenesis is derived from studies of FV3, and the assumption, often unstated, is that other members of the family utilize the same general replication strategy as FV3. While this assumption is likely true among members of the genus *Ranavirus*, experience in other viral families suggests that care must be taken when extending strategies utilized in one genus to members of different genera. Nevertheless, this summary will focus primarily on events in FV3-infected cells, but include findings from other iridovirids where appropriate. The main events in iridovirid replication are shown schematically in Fig. 2.

### Attachment and Uncoating

Little is known about early events in iridovirid-infected cells. Both naked and enveloped FV3 virions are infectious, although the latter have a higher specific infectivity, likely owing to the presence of specific receptor proteins within the viral envelope (Gendrault et al. 1981; Braunwald et al. 1979). The cellular receptor(s) for FV3 is unknown but is thought to be a ubiquitous cellular molecule since the in vitro host range of FV3 is very broad and encompasses mammalian (e.g., HeLa, BHK, and CHO), piscine, amphibian, and reptilian cells (Goorha and Granoff 1979). The in vivo host range is certainly far more restricted than the in vitro range and likely reflects the inability of the virus to replicate at temperatures above 32°C, and perhaps differences in the availability

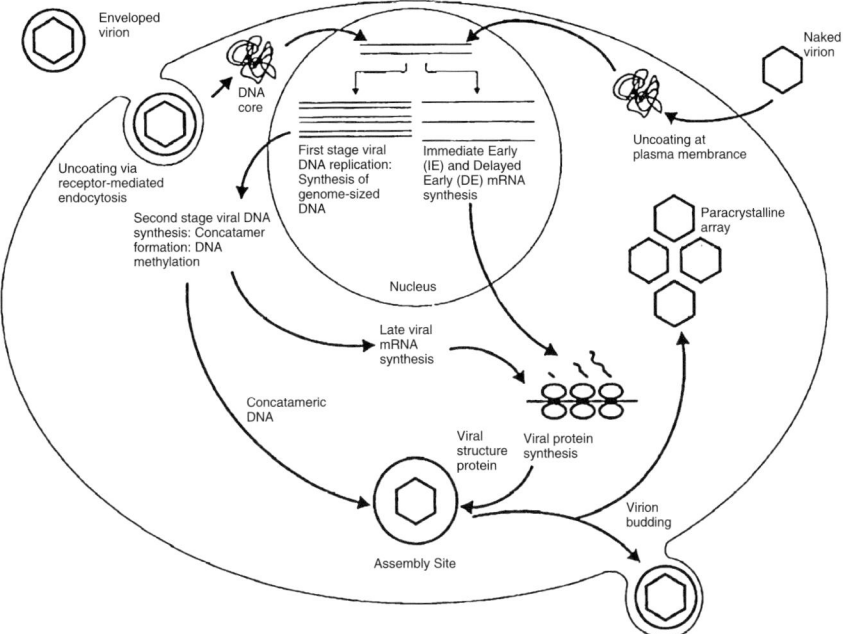

**Fig. 2** Iridovirid replication cycle. The life cycle of frog virus 3 (FV3) is illustrated. See text for details. (From Williams et al. 2005, with permission)

or expression of cellular receptors in whole animals, as well as differences in cellular physiology, immune response, etc. Transmission electron micrographic observations suggest that enveloped viruses enter cells via receptor-mediated endocytosis and are uncoated in lysosomes, whereas naked virions uncoat at the plasma membrane (Braunwald et al. 1985). In the latter case, viral DNA cores are released into the cytoplasm following virus–cell membrane interaction and are subsequently transported into the nucleus. Enveloped IIV particles may uncoat singly or in groups within vesicles (Younghusband and Lee 1969; Matheson and Lee 1981), whereas naked particles uncoat within the cytoplasm (Leutenegger 1967; Kelly and Tinsley 1974).

## Nuclear Events

### Early Gene Transcription

As with other large DNA viruses, iridovirid genes are expressed in a well-regulated temporal cascade involving the sequential expression of immediate-early (IE), delayed-early (DE), and late viral messages (Willis and Granoff 1978; D'Costa et al. 2001, 2004). However, unlike herpesviruses and poxviruses, where transcription is confined to nuclear or cytoplasmic sites, respectively, FV3 IE transcription takes place in the nucleus, whereas late viral message synthesis likely occurs in the cytoplasm (Goorha et al. 1978; Willis and Granoff 1978). Host RNA polymerase II (Pol II) is thought to be responsible for the synthesis of IE messages using input viral genomes as template (Goorha 1981). Support for the role of host Pol II is based on the observation that viral transcription is blocked by treatment of infected cells with α-amanitin, a potent inhibitor of Pol II, but is not blocked in α-amanitin-resistant cells. In marked contrast to herpesviruses, where transfection of naked viral DNA into permissive cells results in a productive infection, naked iridovirid DNA cannot be transcribed and therefore is not infectious. However, introduction of FV3 DNA into cells that were previously treated with UV-inactivated FV3 resulted in the successful transcription of IE viral mRNA and a productive virus infection. The simplest interpretation of this phenomenon is that IE transcription requires the presence of one or more virion-associated, transcriptional transactivators (VATT) in addition to host Pol II and that these are supplied by UV- (but not heat-) inactivated FV3 (Willis and Granoff 1985; Willis et al. 1990a, 1990b). This phenomenon is not unique to FV3 since nongenetic reactivation has also been observed with IIV-6 (Cerutti et al. 1989). At this time, neither the identity nor the precise mode of action of the VATT are known. It is not clear if VATT modifies the viral DNA template or if it interacts with Pol II to initiate transcription. Following synthesis of viral IE transcripts and their subsequent translation, one or more IE proteins (referred to as virus-induced transcriptional transactivators, VITT) are required for expression of DE and late viral transcripts. VITTs are thought to be required for transcription of the highly methylated viral genome, and a factor present in cells productively infected with FV3 has been shown to facilitate transcription of a methylated adenovirus promoter (Willis et al. 1989, 1990a). Collectively, IE and DE transcripts prepare the way for virion formation since IE transcripts are thought to encode regulatory proteins, whereas DE messages likely encode catalytic proteins, such as the viral DNA polymerase (Lua et al. 2005). While this view is generally correct, recent work involving an antisense morpholino oligonucleotide (asMO) targeted against an FV3 18-kDa IE protein suggests that not all IE proteins are key viral regulators. In this study, prior treatment of infected cells with the asMO reduced synthesis of the 18kDa protein by roughly 80% without affecting the expression of other viral proteins or reducing virus yields (Sample et al. 2007).

## First-Stage Viral DNA Replication

Following synthesis of the viral DNA polymerase and its subsequent translocation into the nucleus, viral DNA replication commences within the nucleus and results in the synthesis of genome-size to twice genome-size DNA molecules (Goorha 1982). As with other large DNA viruses, viral DNA synthesis is sensitive to inhibition by phosphonoacetic acid (PAA), an inhibitor of viral DNA polymerases, and aphidicolin, an inhibitor of eukaryotic DNA polymerase α (Chinchar and

Granoff 1984). Inhibition of viral DNA synthesis results in a marked inhibition of late viral gene expression, suggesting that late gene expression is dependent upon viral DNA synthesis (Chinchar and Granoff 1984, 1986). It is not clear whether the need for viral DNA synthesis reflects the requirement for newly synthesized viral templates or the requirement for late transcription to take place in the cytoplasm using a viral-encoded RNA polymerase. Genetic evidence suggests that at least one viral protein, probably the viral DNA polymerase, is required for first-stage viral DNA replication (Goorha et al. 1981; Chinchar and Granoff 1986). Newly synthesized viral DNA molecules are subsequently transported to the cytoplasm where they serve as templates for the formation of concatameric DNA, i.e., second-stage DNA synthesis.

## Cytoplasmic Events

### Viral DNA Methylation

Following transport of newly synthesized viral DNA to the cytoplasm, cytosine residues within CpG motifs are methylated by a virus-encoded DNA methyltransferase (DMTase; Willis et al. 1984; Kaur et al. 1995). Ultimately, 20%–25% of the cytosine residues are methylated, resulting in the highest level of genome methylation known among animal viruses. Inhibition of methylation by treatment of infected cells with 5-azacytidine lowered virus yields by 100-fold but did not reduce viral RNA or protein synthesis, suggesting that methylation is not necessary for viral gene expression (Goorha et al. 1984). However, treatment with 5-azacytidine reduced the size of replicating viral DNA (measured under denaturing conditions) indicating that undermethylated viral DNA was susceptible to DNAse-mediated degradation. These and other results suggest that the viral DMTase may be part of a restriction modification system, and that methylation protects viral DNA from endonucleolytic attack (Essani et al. 1987). While this interpretation is consistent with the above data, it is also possible that methylation of the viral genome may play a role in blocking the induction of pro-inflammatory cytokines following interaction of unmethylated viral DNA with Toll-like receptor 9 (TLR 9). In mammalian systems, interaction of unmethylated bacterial or viral DNA with TLR 9 results in the activation of NF-κB and the induction of type I interferons and inflammatory cytokines (Bauer et al. 2001; Boehme and Compton 2004). Whether unmethylated FV3 also triggers an inflammatory response remains to be seen.

### Second-Stage Viral DNA Replication: Concatamer Formation

In addition to methylation, viral DNA undergoes a second round of DNA synthesis, which results in the formation of large, concatameric structures that are more than ten times larger than genome-sized units (Goorha 1982). Genetic evidence suggests that a second viral protein, distinct from the viral DNA polymerase, is required for concatamer formation (Goorha and Dixit 1984). It is not known whether this protein plays a direct or indirect role in concatamer formation, e.g., by facilitating the transport of viral DNA from nucleus to the cytoplasm or by actively catalyzing the formation of concatamers. In addition, although concatameric viral DNA is likely present in cytoplasmic viral assembly sites (AS), it is not clear whether concatamers are constructed in preformed AS, or whether AS develop around sites of concatamer synthesis (Chinchar et al. 1984).

## Late Viral Gene Transcription

The identification of viral homologs of the two largest subunits of host Pol II (designated vPol-IIα and -IIβ) in all iridovirids sequenced to date strongly suggests that late in infection viral transcription is catalyzed by a virus-modified, or virus-encoded, DNA-dependent RNA polymerase (Williams et al. 2005; Eaton et al. 2007). Our current working model postulates that IE transcription takes place in the nucleus in reactions catalyzed by host Pol II, whereas late viral transcription occurs in the cytoplasm, perhaps within viral AS, and is catalyzed by vPol II. It is also possible that late in infection vPol II translocates to the nucleus and directs transcription of IE mRNAs. Experimental support for the role of vPol II in late viral RNA synthesis has recently been obtained. Inhibition of vPol II formation, using an asMO targeted to vPol IIα, resulted in a marked reduction in the synthesis of late viral gene products and viral titers, but did not affect early protein synthesis (Sample et al. 2007). These results are consistent with the hypothesis that vPol IIα is an integral part of the RNA polymerase and is responsible for late viral mRNA synthesis. However, it remains to be determined if the viral RNA polymerase is entirely virus-encoded or if it is a chimera composed of host and viral components.

## Viral Protein Synthesis

Late viral proteins are thought to encode structural elements of the virion and other virion-associated proteins. Aside from the MCP, which makes up the bulk of the viral capsid, the identities of other virion-associated proteins are largely unknown. Approximately 30 proteins are virion-associated, but whether they serve structural roles and are true capsid proteins, or whether they are virion-associated regulatory or catalytic proteins, is not known (Willis et al. 1985). Interestingly, complementation analysis of FV3 temperature-sensitive (ts) mutants indicates that at least 12 viral genes are involved in virion assembly (Chinchar and Granoff 1986). Defects in any one of these 12 genes had no apparent effect on the synthesis of viral protein, RNA or DNA, or assembly site formation, but reduced the yield of infectious virions by approximately 1,000-fold. Unfortunately, it is not known whether noninfectious virus particles are formed, or whether the specific ts defects block formation of recognizable virion structures. Early work suggested that late viral protein synthesis is translationally regulated and requires a viral-encoded function, perhaps analogous to a virus-encoded translational initiation factor (Raghow and Granoff 1983). While the biochemical data supporting the existence of a translational transactivator appear solid, recent studies have revealed that in vitro synthesized early and late viral transcripts were translated with roughly equal efficiency in cell-free protein synthesizing systems, suggesting that a virus-induced factor was not needed for the in vitro translation of late viral transcripts (Mao et al. 1996; J. Mao and V.G. Chinchar, unpublished observations). This discrepancy has yet to be resolved.

## Viral Assembly Sites and Virion Formation

Virion morphogenesis takes place in morphologically distinct areas of the cytoplasm termed viral assembly sites (AS). When visualized by electron microscopy (Fig. 3A), AS appear as clear regions within the cytoplasm that are devoid of large cellular organelles and are surrounded by mitochondria and intermediate filaments (Murti et al. 1988; Murti and Goorha 1989, 1990; Huang et al. 2007). While the position of the mitochondria may reflect their exclusion from the enlarging AS, intermediate filaments are thought to play an active role by anchoring the AS within the cytoplasm and excluding cellular elements that might interfere with virion morphogenesis (Murti et al. 1988). Alternatively, as suggested for hepatitis B virus, and African swine fever virus (ASFV), the clustering of mitochondria may reflect the need for

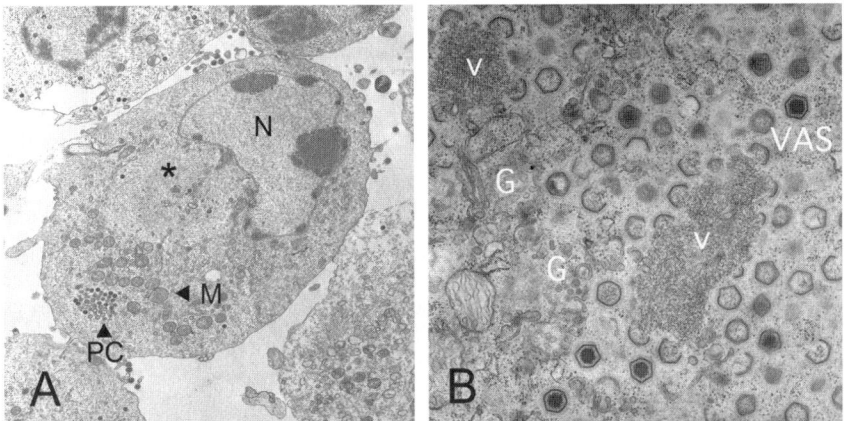

**Fig. 3** Transmission electron micrographs of iridovirid-infected cells. **a** A FV3-infected fathead minnow cell displaying a nucleus (*N*) showing chromatin condensation and margination, a well-formed, electrolucent viral assembly site (*) containing a few viral particles, mitochondria (*M*), and a paracrystalline array of newly assembled virions (*PC*) are shown. (From Sample et al. 2007, with permission). **b** An enlargement of a viral assembly site (*VAS*) within an LCDV-infected cell from the dermis of a Japanese flounder. Viroplasm (*v*), Golgi (*G*) are identified. The VAS contains both full and empty particles as well as several incomplete particles

high levels of ATP in virion assembly (Kim et al. 2007; Netherton et al., 2007). Feulgen staining and immunofluorescent staining with polyclonal mouse anti-FV3 serum or monoclonal antibodies targeted to specific viral proteins confirmed that AS contain DNA and viral proteins (Chinchar et al. 1984). Full and empty viral capsids, along with variable numbers of intermediate forms, are readily observed in iridovirid-infected cells (Fig. 3A,B). Elongated linear, bent, and circular structures (referred to as atypical elements) are sometimes seen, but it is not known if these are intermediates in virion morphogenesis or aberrant products of virus assembly (Huang et al. 2005; Sample et al. 2007). Transmission electron microscopy suggests that virion assembly occurs adjacent to electron dense follicular structures (Huang and Zhang 2007; Fig. 3B). Studies with African swine fever virus (ASFV), a large DNA virus with structural similarity to FV3, may shed light on iridovirid virion morphogenesis. Here virion formation is thought to begin by recruiting endoplasmic reticulum cisternae that give rise to viral membranes. These membranes develop into icosahedral structures through the progressive assembly of capsid proteins. Concomitantly, the core shell is formed under the inner envelope and viral DNA and nucleoproteins are packaged (Epifano et al. 2006). Currently, it is unclear whether iridovirid capsid proteins associate with cellular membranes, as is seen with ASFV, and serve as intermediates in iridovirid assembly. Furthermore, recent work with ASFV indicates that viral assembly sites resemble cellular aggresomes. Aggresomes are perinuclear structures, enclosed within a vimentin cage, that recruit cellular chaperones and proteases in response to high levels of misfolded or unassembled proteins (Heath et al. 2001). It is possible that aggresome formation, a cellular response for dealing with potentially toxic misfolded proteins, has been exploited by ASFV and iridovirids as a means of generating a cytoskeletal scaffold for concentrating structural proteins at the sites of virus assembly (Wileman 2006).

Virions are thought to package viral DNA by a "headful" mechanism in which a unit length of DNA plus an additional 5%–50% of the genome, depending upon the specific virus, is encapsidated. This method of packaging results in a DNA molecule that is circularly permuted and terminally redundant (Goorha and Murti 1982). Following their formation, virions accumulate in paracrystalline arrays within the cytoplasm or are transported to the plasma membrane where they acquire an envelop by budding. However, in cells infected by FV3, more than 90% of virions remain cell-associated.

Experiments using asMOs and short, interfering RNAs (siRNAs) targeted to the MCP illustrate the important role that the MCP plays in virion morphogenesis. Xie et al. (2005) showed that treatment of FHM cells with siRNAs targeted to the MCP transcript of TFV, a strain of FV3, resulted in marked reductions in CPE, virus yield, and the assembly of particles. We recently confirmed and extended these findings using an asMO targeted against the FV3 MCP. FHM cells pretreated with an anti-MCP asMO and infected 24 h later with FV3 showed an 80% reduction in MCP synthesis and a 90% drop in virus titer. Moreover, transmission electron microscopy showed that circular, bent, and tubular atypical elements were the predominant products detected within the AS of cells treated with an asMO targeted to the MCP (Sample et al. 2007). Interestingly, treatment of cells with an asMO targeted to the vPol-IIα transcript blocked the synthesis of the MCP protein and other late proteins, but did not result in the appearance of atypical elements, suggesting that their formation requires the presence of one or more yet to be identified late proteins along with the absence of the MCP.

## Viral Effects on Host Cell Function

### Inhibition of Host Cell Macromolecular Synthesis

Iridovirid infection has a dramatic inhibitory effect on cellular protein, RNA, and DNA synthesis (Raghow and Granoff 1979; Cerutti and Devauchelle 1980; Chinchar and Dholakia 1989). Interestingly, host macromolecular synthesis is blocked following productive infection, or treatment with heat-inactivated or UV-inactivated virus, indicating that the inhibitory component(s) is possibly a viral structural protein rather than a newly synthesized viral gene product. Alternatively, the interaction of virions with one or more cellular receptors might trigger shutoff through an unidentified signaling pathway. The molecular basis for translational shutoff and the selective synthesis of viral proteins is not known but likely reflects a combination of processes including the additive effects of (i) host message degradation (Chinchar and Yu 1992), (ii) the synthesis of large amounts of translationally efficient viral messages (Chinchar and Yu 1990), (iii) a global block in translation triggered by the PKR (a protein kinase activated by dsRNA) - mediated phosphorylation and subsequent inactivation of eukaryotic translational initiation factor 2α (eIF-2α) (Chinchar and Dholakia 1987), and (iv) the presence of a viral homolog of eIF-2α (designated vIF-2α) that is thought to act as a pseudo-substrate and prevent eIF-2α phosphorylation by binding PKR (Essbauer et al. 2001; Jancovich et al. 2003). The block to host DNA synthesis is thought to be due to the prior inhibition of cellular protein and RNA synthesis and not to a direct effect on cellular DNA replication, whereas nothing is known about the molecular basis for transcriptional inhibition.

By analogy to the vaccinia virus K3L protein, the ranavirus vIF-2α gene product is thought to maintain protein synthesis in virus-infected cells by binding PKR and blocking its interaction with eIF-2α (Beattie et al. 1991; Kawagishi-Kobayashi et al. 1997). However, in contrast to other ranaviruses, some strains of FV3 contain a truncated version of vIF-2α that likely resulted from an upstream deletion that apparently fused the first 11 amino acids from a small upstream ORF to the C-terminal 65 amino acids of vIF-2α. Contrary to expectations, FV3 strains bearing the truncated vIF-2α homolog efficiently synthesized their own proteins and rapidly shut off host cell translation (Chinchar and Dholakia 1989; Tan et al. 2004). This was not due to residual activity within the truncated protein since homology to K3L and eukaryotic eIF-2α resides in the N-terminal 100 amino acids that were lost. Since other ranaviruses contain full-length copies of vIF-2α, it is not clear whether vIF-2α was absent in the original FV3 isolate, or whether it has been lost following repeated passages in cultured cells over the last 40 years. Perhaps relevant to this situation is the observation that serial, low multiplicity passages of *Autographa californica* nucleopolyhedrovirus resulted in the selection of viable viruses with deletions or insertions in the DNA (Miller 1996). While the absence of a full-length version of vIF-2α in FV3 casts doubt on the role of this protein in maintaining viral protein synthesis late in infection, it is possible that FV3 may contain a second protein that acts upstream of vIF-2α. Such is the case in vaccinia virus, where the E3L protein binds dsRNA and blocks PKR activation (Langland and Jacobs 2002).

Apart from its role in protein synthesis, vIF-2α may also influence virulence. FV3 replicates efficiently in vitro, but is markedly less virulent in vivo than *Rana catesbeiana* virus Z (RCV-Z), a ranavirus that possesses a full-length vIF-2α. Thus, infection of bullfrog (*R. catesbeiana*) tadpoles with RCV-Z proved lethal, whereas infection with FV3 containing a truncated vIF-2α gene resulted in little or no mortality (Majji et al. 2006). Moreover, prior infection with FV3 protected tadpoles from subsequent lethal challenge with RCV-Z (Majji et al. 2006). Whether vIF-2α plays an additional role in viral biogenesis is not known. However, recent evidence that PKR is involved in NF-κB activation and the induction of pro-inflammatory cytokines (Tan and Katze 1999; Williams 1999) suggests that if vIF-2α is able to block this pathway by binding PKR, then the onset of an immune response might be delayed and virus yields enhanced. While this immune evasion pathway is unlikely to be important following high multiplicity, synchronous infection of cultured cells, it could have important consequences for infections in vivo that may be initiated with low levels of virus that require multiple rounds of replication prior to the development of clinical disease.

**Apoptosis**

Aside from the marked inhibition of cellular synthetic functions, iridovirid infection also results in the induction of apoptosis. As with CPE, apoptosis is triggered following either productive infection or treatment with heat-inactivated or UV-inactivated virus or a soluble virion extract (Essbauer and Ahne 2002; Chinchar et al. 2003; Hu et al. 2004; Imajoh et al. 2004; Paul et al. 2007; Huang et al. 2007a). In FV3-infected cells, apoptosis begins 6–8 h postinfection (p.i.) and is characterized by DNA fragmentation, chromatin condensation, apoptotic body formation, and the appearance of phosphotidylserine on the outer leaflet of the cellular membrane. Iridovirus-induced apoptosis is likely mediated by cellular caspases since Z-VAD-FMK, a pan-caspase inhibitor, blocks apoptosis (Chinchar et al. 2003; Paul et al. 2007). In addition, Huang et al. (2007a) reported mitochondrial fragmentation, the activation of caspases 3 and 9, and an increase in intracellular $Ca^{+2}$ in RGV-infected cells. Although it is not known whether apoptosis is a consequence of PKR activation, translational shutoff, or occurs independently of those events, Paul et al. (2007) demonstrated that the amount of soluble virion extract required to induce apoptosis was 1,000-fold lower than the dose required to shutdown protein synthesis. This result suggests that in the IIV-6 system translation shutoff may not be required to induce apoptosis. Furthermore, IIV-6 encodes a protein with homology to the baculovirus inhibitor of apoptosis protein (IAP), and it is likely that other iridovirids encode proteins that delay the onset of apoptosis. It may be advantageous for iridovirids to seek a balance between inducing and inhibiting apoptosis. In this view, if apoptosis occurs too soon, virus yields will be depressed, but if apoptosis is unduly delayed, necrotic cells may trigger a much more robust pro-inflammatory response than that induced by apoptotic cells.

## Viral Immune Evasion, Host Range, and Virulence Proteins

Based on analogy to poxviruses and herpesviruses, it is likely that iridovirids encode multiple proteins that modulate host immune responses (Alcami and Koszinski 2000; Tortorella et al. 2000; Johnston and McFadden 2003). To date, several proteins that potentially modulate cellular immune responses to infection have been identified, including the aforementioned vIF-2α, a CARD (caspase recruitment domain)-containing protein (designated vCARD), an hydroxysteroid dehydrogenase (vHSD), a viral homolog of the tumor necrosis factor receptor (vTNFR), the aforementioned IAP, and several ORFs encoding putative proteins containing immunoglobulin- or MHC-like domains (Jakob et al. 2001; Jancovich et al. 2003; Essbauer et al. 2004; Song et al. 2004; Tan et al. 2004). Although vIF-2α may play a role in maintaining viral protein synthesis in infected cells, recent data suggest that it might also be involved in blocking induction of an inflammatory response (Tan and Katze 1999; Williams 1999; Gil et al. 2004). Likewise, since DEATH domains, DEATH effector domains, CARDs, and pyrin motifs are thought to mediate inflammatory responses via protein–protein interactions, vCARD may block activation of pro-inflammatory or apoptotic responses, or, conversely, activate a pathway that is dependent on CARD–CARD interactions (Bouchier-Hayes and Martin 2002; Johnson and Gale 2006; Werts et al. 2006; Holm et al. 2007). Although its mechanism of action is different, the viral homolog of the TNFR may also block activation of proinflammatory genes by binding TNF and preventing signaling through the authentic TNF receptor (Essbauer et al. 2004). Similar to its vaccinia virus homolog, vHSD may play a role in modulating steroid biosynthesis and thereby impair immunity following in vivo infections (Moore and Smith 1992; Reading et al. 2003). Among ranaviruses, overexpression of the viral homolog of HSD suppressed the cytopathic effect, suggesting that it may play a role in vitro as well as in vivo (Sun et al. 2006). Inasmuch as the current catalog of potential iridovirid immune evasion proteins is short, the relatively large number of poxvirus proteins involved in immune evasion, virulence, and host range suggests that additional iridovirus proteins that affect these functions await discovery. Moreover, we anticipate that as more is revealed about the immune systems of iridovirid hosts, additional immune evasion, virulence, and host range modifiers will be identified by homology to their cellular counterparts.

In addition to putative immune evasion proteins, iridovirids and other large DNA viruses such as ASFV, poxviruses, and herpesviruses encode a number of proteins (dUTPase, ribonucleotide reductase, thymidylate synthase) with marked homology to cellular genes involved in dTTP synthesis (Oliveros et al. 1999; Langelier et al. 2002; Lembo et al. 2004; Glaser et al. 2006; Zhao et al. 2007; Zhang et al. 2007). Inclusion of these enzymes within the viral proteome is thought to ensure sufficient quantities of dNTPs for viral nucleic acid synthesis and to lessen the chances of misincorporating uracil into viral DNA (Oliveros et al. 1999; Zhang et al. 2007). Moreover, these enzymes likely play roles in determining host range. For example, Oliveros et al. (1999) found that dUTPase was not required for replication in dividing Vero cells, but was needed for ASFV replication in swine macrophages. Among ranaviruses, overexpression of dUTPase did not enhance the replication of RGV in dividing EPC cells, an established fish cell line (Zhao et al. 2007).

## Host Immune Response

Although there is little specific information about the host immune response to iridovirid infection, both humoral and cell-mediated immunity likely play roles in the prevention of, and recovery from, virus infection. For example, *Xenopus* mount effective B cell and T cell responses against FV3 infection (Morales and Robert 2007; Maniero et al. 2006), and antibodies targeted to other ranaviruses can be detected in infected frogs (Zupanovic et al. 1998a). Moreover, vaccination is effective in preventing disease due to RSIV infection (Caipang et al. 2006a, 2006b), and prior infection of bullfrog tadpoles with relatively avirulent FV3 protects against subsequent challenge with virulent RCV-Z (Majji et al. 2006). At the molecular level, ISKNV infection has been shown to induce in mandarin fish a variety of putative antiviral proteins, including homologs of a VHSV-induced protein, Gig2, viperin, Mx, CC chemokines, the immunoglobulin heavy chain etc. (He et al. 2006). As the immune systems of lower vertebrates become better understood, it is likely that their role in protecting fish, amphibians, and reptiles from iridovirid infections will become clearer and utilized to develop more effective vaccination strategies.

## Biology and Ecology of Iridovirid Infections

### FV3 and Other Amphibian Ranaviruses

FV3 was serendipitously isolated more than 40 years ago in the course of an attempt to develop cell lines that would support the growth of Lucke herpesvirus. Granoff et al. (1966) observed that cell monolayers prepared from the kidneys of normal and tumor-bearing frogs (*R. pipiens*) spontaneously underwent lysis. Analysis of the resulting infectious agents (designated frog virus-1, -2, -3, etc.) showed them to be large cytoplasmic DNA viruses with icosahedral symmetry.

While these and other similar viruses were likely different isolates of the same virus species, the isolate designated FV3 became the focus of further study because of its putative association with renal adenocarcinoma (Clark et al 1968; Granoff et al. 1966). Subsequent analysis indicated that FV3 played no part in tumor formation, but the virus continued to be studied and is now the best characterized iridovirid and the type species of the genus *Ranavirus* (Goorha and Granoff 1979).

The isolation of FV3 from ostensibly healthy anurans suggested that it was a pathogen of low virulence. This inference was confirmed early on by Tweedel and Granoff (1968) who showed that embryos and tadpoles were killed by injection with as little as 900 PFU of infectious virus, whereas adult frogs (*R. pipiens*) survived injection of $10^6$ PFU per animal[-]. These results were recently confirmed in a *Xenopus* model by Robert and his co-workers, who showed that tadpoles are relatively sensitive to infection by FV3, whereas adults mount an effective immune response and successfully resist infection unless their immune defenses are compromised by sublethal doses of γ-irradiation or depletion of CD8[+] T cells (Gantress et al. 2003; Robert et al. 2005). Moreover, FV3 displays a strong tropism for the proximal tubular epithelium of the kidney, but rarely disseminates beyond that organ in immunocompetent animals. Recently, Maniero et al. (2006) showed that *Xenopus* develop antibodies to FV3 that neutralize the virus in vitro and provide partial protection to susceptible larvae, whereas Morales and Robert (2007) found that protection also correlated with the expansion of CD8[+] cells. Collectively, these results suggest that both B cell (humoral) and T cell (cell-mediated) responses are critical for viral clearance and protection from disease.

Following the initial isolation of FV3 from North American frogs, numerous other isolations of FV3-like viruses were made in North America (Mao et al. 1999; Green et al. 2002; Greer et al. 2005; Miller et al. 2007), the United Kingdom (Cunningham et al. 1996), China (Zhang et al. 2001, 2006), Thailand (Kanchanakhan 1998), and South America (Zupanovic et al 1998b; Galli et al. 2006; Fox et al. 2006). In some cases, e.g., tadpole edema virus (TEV), Rana United Kingdom (RUK) virus, and Bufo United Kingdom (BUK) virus, viral protein profiles, RFLP analysis, and sequence analysis of the MCP support the notion that these are isolates of FV3 and not distinct viral species (Hyatt et al. 2000). However, in other cases, e.g., BIV and seven recent isolates from South America, RFLP profiles differed markedly from FV3 and are consistent with their classification as separate species or tentative species. Furthermore, unlike the early frog virus isolates that originated from ostensibly healthy adult animals, many of the recent isolates came from tadpoles or frogs that were patently diseased. Although previously considered relatively benign, it is now apparent that FV3-like viruses are capable of causing fatal disease in amphibian larvae and adults. In some cases, disease occurred among farmed frogs, suggesting that stress due to overcrowding may have triggered immune suppression and contributed to clinical disease (Kanchanakhan 1998; Zhang et al. 2001). These results are consistent with findings in the FV3/*Xenopus* model described above wherein the outcome of infection depended upon the degree of immunosuppression (Robert et al. 2005). Reports of ranavirus disease in other species, including reptiles (Chen et al. 1999; Hyatt et al. 2002; De Voe et al. 2004;

Marschang et al. 2005; Allender et al. 2006; Johnson et al. 2007) are becoming increasingly common, but the causative agents have yet to be fully characterized.

While it is clear from experimental transmission studies that larvae and immunocompromised adults are more susceptible to infection than healthy adult frogs, the ways in which FV3 is transmitted and maintained in wild populations are less clear. Recent studies by Pearman et al. (2004, 2005) and Harp and Petranka (2006) examined various aspects of the ecology of FV3 and FV3-like viruses. Pearman et al. (2004, 2005) found that FV3 was pathogenic for *Rana latastei* and that susceptibility to FV3 infection varied among geographically separated populations that differed in genetic diversity. In addition, they suggested that mortality depended on contact between infected and uninfected individuals and was influenced by the concentration of the virus, cannibalism, and necrophagy. Consistent with those results, Harp and Petranka (2006) found that ranavirus was transmitted to healthy *R. sylvatica* tadpoles by scavenging of infected carcasses and by exposure to water containing infected animals or carcasses.

**Ambystoma Tigrinum Virus**

Whereas outbreaks of FV3-like disease have been detected worldwide, clinical disease triggered by *Ambystoma tigrinum virus* (ATV) has been seen, with only one non-natural exception, in ambystoid salamanders in western North America. ATV was initially isolated from salamanders (*A. tigrinum stebbensi*) in southern Arizona, and related viral strains were subsequently isolated in Colorado, Utah, North Dakota, Saskatchewan, and Manitoba (Jancovich et al 1997; Bollinger et al. 1999; Docherty et al. 2003). Along its extensive north–south range, ATV isolates are similar, although isolates can be differentiated by the presence of point mutations within the MCP gene, RFLP profiles, and the presence or absence of one or more copies of a 16-bp repeat within the ORF 50R/51L intergenic region (Jancovich et al. 2005). The only isolate east of the Mississippi River was made in the axolotl colony at Indiana University following introduction of wild salamanders from Colorado and is genetically similar to ATV isolates from that state (Davidson et al. 2003).

In contrast to FV3 infections, where mortality is more common in larvae than adults, both larval and adult salamanders succumb to ATV infection, and mortality in affected ponds often exceeds 90% (Jancovich et al 1997; Bollinger et al. 1999). Temperature influences the extent of mortality and time to death as most salamanders infected at 26°C survive, whereas those infected at 18°C succumb to infection (Rojas et al. 2005). Clinical disease is accompanied by marked necrosis and hemorrhage within internal organs (spleen, liver, kidney, gastrointestinal tract), skin polyps, skin sloughing, and extrusion of a thick discharge from the vent. Populations rebound in affected ponds, but it is not clear whether this reflects the presence of a few surviving immune animals, or recolonization by naïve animals. Experimental attempts to determine the host range of ATV demonstrated that various salamander species (*Ambystoma graciale*, *A. californiense*, *Notophthalmus viridescens*) are susceptible to infection, but bullfrogs (*R. catesbeiana*) and fish (*Gambusia affinis*, *Lepomis cyanellus*, *Oncorhynchus mykiss*) are resistant to infection (Jancovich et al. 2001; Picco et al. 2007). Recently, ATV has been shown to infect wood frogs (*R. sylvatica*) and leopard frogs (*R. pipiens*) although with reduced efficiency and lower virulence, suggesting that its host range may be larger than originally thought (Schock et al. 2008). Despite this intriguing result, the role of anurans and possibly fish in the ecology of ATV infections is unknown. It remains to be determined whether sympatric fish and frog species serve as reservoirs for ATV following the die-off of susceptible salamander populations, or whether sublethally infected meta-morphs serve as an intraspecific reservoir and reintroduce the virus into uninfected larval populations (Brunner et al. 2004). Brunner et al. (2007) showed that ATV is effectively transmitted between animals directly by biting, cannibalism, necrophagy or mere physical contact, and indirectly through water and fomites. Moreover, larval salamanders are capable of transmitting virus to naïve animals soon after infection and their infectivity increases with time. Infection rates increase with viral dose and were affected by clutch identity and life stages. Surprisingly, metamorphs were less likely to be infected than larvae, but once infected were more likely to die (Brunner et al. 2005). ATV remains viable in moist soil for several days, but loses infectivity rapidly upon drying. The basis for the apparent sudden emergence of ATV in western North America is not known, but phylogenetic analysis and the marked virulence seen in infected larval and adult salamanders suggests that ATV recently arose from a fish ranavirus that jumped species and now infects salamanders (Jancovich et al. 2005). Moreover, human introduction of ATV through the bait trade has likely contributed to disease emergence (Storfer et al. 2007).

## EHNV and Other Australian Ranaviruses

The first report of a ranavirus infecting fish was from Australia. The virus, epizootic hematopoietic necrosis virus (EHNV), was isolated from wild redfin perch (*Perca fluviatilis*) (Langdon et al. 1986) and is unique to Australia. EHNV was the cause of spectacular epizootic mortalities in redfin perch and recurrent epizootics in farmed rainbow trout (*Oncorhynchus mykiss*) (Langdon and Humphrey 1987; Whittington et al. 1994). The host range of the virus now includes recrea-tional fish (Whittington et al. 1996) and the associated clinical disease (epizootic hematopoietic necrosis, EHN) is listed by the Office International Epizooties (OIE) as the cause of significant finfish losses at the zonal/regional level. Furthermore, EHNV has the potential for international spread via live animals or fomites (Office International des Epizooties 2006). Clinically similar systemic necrotizing syndromes have been reported in farmed catfish (*Ictalurus melas*) in France (Pozet et al. 1992), sheatfish (*Silurus glanis*) in Germany (Ahne et al. 1989), turbot (*Scophthalmus maximus*) in Denmark (Bloch and Larson 1993) and other fish species in Europe (Ariel et al. 1999; Tapiovaara et al. 1998), but they appear to be caused by viruses distinct from EHNV (Ahne et al. 1998).

EHNV causes marked mortality in redfin perch and may kill greater than 90% of infected fish (Pierce et al. 1991). Typically fingerling and juvenile fish are affected, but in areas in which EHNV has been newly introduced, adults are also susceptible (Langdon and Humphrey 1987; Langdon et al. 1986). Natural epizootics are frequently reported in summer when large numbers of presumably nonimmune young fish are present and school in warm shallow waters. EHN is also reported in rainbow trout where the mortality rates are lower and the disease is generally difficult to identify. Outbreaks of the disease appear to be related to poor husbandry (Whittington and Reddacliff 1995).

While numerous amphibian iridovirids have been described worldwide, only two have been identified within Australia. The first was Bohle iridovirus (BIV), which was isolated from a native Australian frog (*Limnodynastes ornatus*) in Northern Queensland (Speare and Smith 1992) and was also shown to be highly pathogenic for barramundi (*Lates calcarifer*), a commercially and recreationally important fish species, following experimental infection (Moody and Owens 1994). Surprisingly, BIV has not been isolated from natural epizootics involving either frogs or fish and has only been isolated from tadpoles taken from the wild and allowed to metamorphose in captivity. A PCR assay for diagnosis of BIV infection has been developed (Pallister et al. 2007) and the putative promoter regions of three genes, including the MCP gene, have been identified (Pallister et al. 2005). A second ranavirus, Wamena virus (WV), was isolated from illegally imported juvenile green pythons (*Chondropython viridis*) and may not be native to Australia (Hyatt et al. 2002). Whereas EHNV and BIV are recognized by the ICTV as distinct viral species within the genus *Ranavirus* (Chinchar et al. 2005), WV represents a tentative ranavirus species based on comparison of the MCP and thymidine kinase sequences, dot blot analysis, antigen capture ELISA, and RFLP profiles (Hyatt et al. 2002; Coupar et al. 2005; Hyatt and Whittington 2005).

Experimental infections of cane toad (*Bufo marinus*) tadpoles with BIV led to mortality rates of up to 100% (Hyatt et al. 1998). Clinical signs following experimental infection of tadpoles with BIV include wasting and behavioral changes, i.e., tadpoles frequently lay on the bottom of the tank, often on their sides with their tails curved. Upon stimulation, they swam weakly and erratically in a forward whirling movement. In general, deaths begin about day 5, and peak at 7–8 days p.i. Following both EHNV and BIV infection, death is most likely due to the degeneration of hematopoietic cells and damage to the vascular endothelium. Examination of infected tadpoles revealed a systemic infection primarily affecting the major hematopoietic organs, i.e., the kidney, liver, and spleen (Hyatt et al. 1998). In the kidneys, acute hematopoietic necrosis is present, varying from individual infected cells to extensive areas of necrosis (Fig. 4A,B). Infected cells contain characteristic

---------------------------------------------------------------→

**Fig. 4** Histopathological changes within BIV-infected cells. **a** Light micrograph of a kidney from a *B. marinus* tadpole infected with BIV showing extensive necrosis (*n*) particularly within interstitial tissue. Glomerulus (*g*); tubules (*t*); interstitial tissue (*it*), which is the site of hematopoietic cells; melanin pigment (*p*); muscle of the base tail (*m*). Hematoxylin-eosin stain, ×200. **b** A comparable section has been labeled using the peroxidase-antiperoxidase (PAP) method with a primary antibody to EHNV, AEC as the chromogen, and counterstained with hematoxylin. PAP staining, *white arrow*. ×250. **c** Transmission electron micrograph from a section of the same kidney. Kidney tubule (*t*); interstitial cells (*ic*); assembly sites (*as*). Viruses (*v*) are associated with hematopoietic cells of the interstitial tissue. Bar represents 1 µm

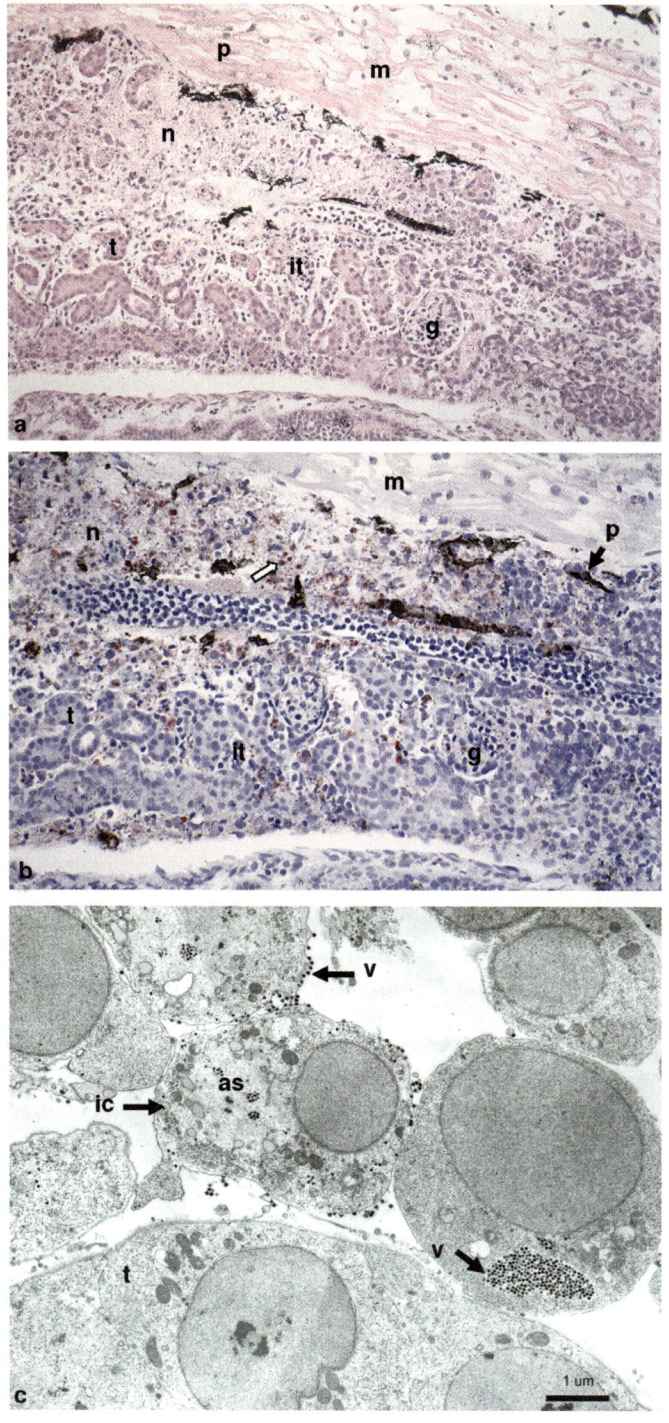

basophilic intracytoplasmic inclusions, which, upon electron microscopic examination, were identified as virus assembly sites (Fig. 4C). Within the liver, multifocal hepatocellular necrosis is apparent, whereas diffuse acute necrosis is seen in the spleen. Focal destruction of the pancreas is frequently observed, as is the degeneration of undifferentiated mesenchymal tissue in the developing limb buds. Another consistent and conspicuous change in infected rainbow trout and redfin perch is the presence of degenerate hematopoietic cells, basophilic debris, and fibrous material within blood vessels (Reddacliff and Whittington 1996). Degeneration of the vascular endothelium is observed in the liver, spleen, kidney, gill and heart, suggesting that the virus is endotheliotropic. The clinical signs and histopathology described above differ from that seen following infection with other ranaviruses (Cunningham et al. 1996; Jancovich et al. 1997; Bollinger et al. 1999) in that the Australian viruses do not cause major skin lesions or megalocytosis. Therefore, it appears that ranavirus pathology varies depending upon the virus species and host animal infected. Recently recombinant BIV has been constructed using homologous recombination to replace the viral homolog of eIF-2α with a cassette containing the neomycin resistance gene and the globin gene from adult cane toads (Pallister et al. 2007). Not only does this illustrate the feasibility of constructing iridovirid knock out mutants, but the expression of adult globin suggests that recombinant BIV may be useful in the biological control of cane toads.

## RSIV and Other Megalocytiviruses from Asia

*Megalocytivirus* is a newly established genus within the family *Iridoviridae*. Numerous megalocytivirus isolates have been described, but it has not yet been determined whether these represent distinct viral species or merely strains or isolates of a single species. Sequence analysis of the highly conserved MCP suggests that there may be two clusters within the genus (Do et al. 2005b; Wang et al. 2007). Alternatively, since the isolates come from different fish species, it is not clear if they represent host-specific species or a single species with a broad host range. The complete genomic DNA sequences of four megalocytiviruses has been determined: infectious spleen and kidney necrosis virus (ISKNV, He et al. 2001), red sea bream iridovirus (RSIV, Kurita et al. 2002), rock bream iridovirus (RBIV, Do et al. 2004), and orange spotted grouper iridovirus (OSGIV, Lu et al. 2005). ISKNV was chosen as the type species of the genus because it was the first megalocytivirus whose complete sequence was published. However, extensive work has been carried out in Japan on RSIV and provides much of what we know about the biology of this genus.

### Host Range

Epizootics caused by megalocytiviruses first occurred among red sea bream in 1990 (Nakajima and Maeno 1998; Nakajima and Kunita 2005). Subsequently, outbreaks were noted in brown spotted groupers (*Epinephelus tauvina*) in Singapore in 1992 (Chua et al. 1994) and among Malabar groupers (*E. malabaricus* ) and sea bass (*Lateolabrax* spp.) in southern Thailand in 1993 (Danayadol et al. 1996). Based on the presence of icosahedral particles within the cytoplasm and characteristic enlarged cells (designated inclusion body-bearing cells), megalocytiviruses have been found in ornamental tropical fishes such as dwarf gourami (*Cosa lalia*) that were imported into Australia from Singapore in 1988 (Anderson et al. 1993) and in orange chromide cichlids (*Etroplus maculatus* ) imported to Canada from Singapore in 1989 (Armstrong et al. 1989). Moreover, a similar disease was also observed in ornamental tropical fish such as angelfish (*Pterophyllum scalare*) in the UK (Rodger et al. 1997), and gouramis and swordtails (*Xiphophorus hellerii*) in Israel (Paperna et al. 2001), which were bred in those countries after being imported from Singapore. A megalocytivirus was also isolated from diseased African lamp-eyes (*Aplocheilichthys normani*) that were cultured in Indonesia and imported into Japan (Sudthongkong et al. 2001). Sequence analysis showed that the isolates designated GSDIV (grouper sleepy disease iridovirus), SBIV (sea bass iridovirus), DGIV (dwarf gourami iridovirus), RSIV, TGIV (Taiwan grouper iridovirus), and ALIV (African lampeye iridovirus) were essentially genetically identical and likely represent strains or isolates of the same viral species (Sudthongkong et al. 2002). Collectively, these reports suggest that megalocytiviruses originated in Southeast Asia and are pathogenic to both marine and freshwater fishes. The appearance of clinical disease in farmed and imported fish might reflect the effects of crowding, or transport- and handling-induced stresses on the immune system that exacerbate preexisting subclinical infections. Moreover, the age of the fish and water temperature appear to be important factors in determining the outcome of RSIV infection (Choi et al. 2006). Because grouper species are of high economic value, fish farming has likely resulted in virus dispersal and an increase in the prevalence of the disease in Southeast Asian countries such as Singapore, Thailand, Indonesia, Malaysia, and Taiwan. Mass mortalities of farmed grouper species and ornamental freshwater fish have resulted in large economic losses in those countries. The recent identification of markedly similar megalocytiviruses in Murray cod and dwarf gouramis, coupled with the high prevalence of megalocytivirus infections in retail aquarium shops in Sydney, Australia, suggests that the global trade in ornamental fish may also facilitate the spread of this virus to new host species in geographically distant regions (Go et al. 2006; Go and Whittington 2006).

## Isolation and Characterization of RSIV

In the summer of 1990, mass mortalities occurred among farmed red sea bream in Japan when the water temperature rose above 25°C. An iridovirus-like agent was detected in diseased fish and designated red sea bream iridovirus after the host species from which it was isolated (Inouye et al. 1992). RSIV is the most extensively studied of the megalocytiviruses and has been shown to infect more than 30 species of farmed marine fish including sea bass, amberjack, yellowtail

(*Seriola quinqueradiata*), striped jack (*Pseudocaranx dentex*), and rock bream (*Oplegnathus fasciatus*). It is likely that the importation of infected seedlings captured in the South China Sea was responsible for the introduction of RSIV into Japan. In addition, epizootics of RSIV-induced disease occurred in farmed red sea bream, rock bream, Japanese flounder (*Paralichthys olivaceus*), and turbot (*Scophthalmus maximus*) in Korea (Jung et al. 2000, Do et al. 2005a, 2005b), and in mandarin fish (*Siniperca chuatsi*) reared in offshore pens in the South China Sea (He et al. 2000) as well as in stocks of amberjack seedlings from Hainan Island. Although most studies to date have focused solely on identifying the viral agent responsible for disease, and developing accurate diagnostic tools, the transcriptional program of RSIV has recently been elucidated using DNA microarray technology (Lua et al. 2005, 2007). Given the economic importance of this viral agent, additional molecular studies should be forthcoming.

## Clinical Signs: Gross and Histopathological Findings

Clinically, diseased fish show a darkening of body color and become lethargic immediately before dying. The latter symptom is so prominent that the common name for the infection is sleepy disease. Splenomegaly is marked in diseased fish, and cell enlargement is evident (Fig. 5A). In some cases, necrosis of splenocytes and hematopoietic cells is observed. It is thought that the enlarged cells observed in the spleen, kidney, liver, gills, and heart of infected fish are cells of the macrophage/monocyte lineage (T. Miyazaki, unpublished observations). From a diagnostic viewpoint, these enlarged cells, which are termed inclusion body-bearing cells, are pathognomonic for megalocytivirus infections.

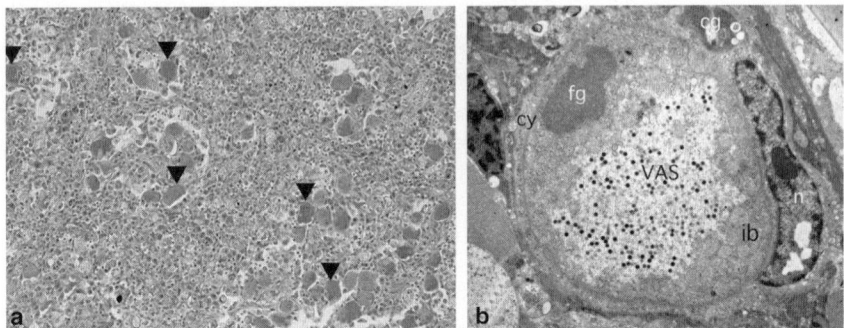

**Fig. 5** Histopathological changes within megalocytivirus-infected cells. **a** Spleen cells from red sea bream infected with RSIV were stained by Mayer's hematoxylin and eosin and examined by light microscopy. Characteristic of RSIV infection is the presence of large, inclusion body-bearing cells (indicated by *arrowheads*) within the spleen. **b** A mature inclusion body-bearing cell with a large, centrally located inclusion body containing masses of fine granules (*fg*) and coarse granules (*cg*) and a well-developed viral assembly site (*VAS*) containing numerous viral particles. The nucleus (*n*) and cytoplasm (*cy*) have been displaced to the periphery by the enlarging VAS. (**b** from Mahardika et al. 2004, with permission)

As with other iridovirids, inclusion bodies contain viral DNA and serve as the sites of virion formation. However, in contrast to ranavirus-infected cells, a membrane surrounds the inclusion body found within RSIV-infected spleen cells (compare Figs. 3A and 5B). Moreover, the inclusion body occupies most of the volume of the cell and displaces the cytoplasm and nucleus to the periphery. The membrane-bound inclusion contains not only the viral assembly site, but also mitochondria, rough and smooth endoplasmic reticulum, and two types of electron-dense matrices, one composed of fine granules and another composed of coarse or rough granules (Mahardika et al. 2004). It has been suggested that fine granules contain viral DNA and that virions form in close association with this structure. Rough granules are thought to represent viral DNA that has recently been transported from the nucleus to the cytoplasm (Mahardika et al. 2004). Although, membrane-bound inclusions were initially seen only in cells from infected fish, recent studies examining RSIV-infected grunt fin cells demonstrated the presence of inclusion bodies in vitro (K. Mahardika and T. Miyazaki, unpublished observations). However, it remains to be determined if assembly site formation among megalocytiviruses is a fundamentally different process from that seen among other iridovirids or whether the generation of inclusion body-bearing cells represents the engulfment of infected cells by neighboring cells or macrophages (Overholtzer et al., 2007).

**Vaccine Development**

Given the commercial value of the fish species affected, vaccine development has been a priority. An injectable vaccine using formalin-inactivated RSIV is efficacious in protecting red sea bream, yellowtail, amberjack, kelp grouper, and other species from RSIV infection and has been commercialized (Nakajima et al. 1999, 2002; Caipang et al. 2006a). Moreover, a DNA vaccine and an oral vaccine using liposome-entrapped RSIV antigens are also effective (Caipang et al. 2006b; T. Miyazaki, unpublished observations). Because oral vaccination is more practical for large-scale administration to susceptible fish, commercialization of an oral RSIV vaccine will be a significant breakthrough in the protection of farmed fish against RSIV infection.

## *Invertebrate Iridescent Viruses*

Invertebrate iridescent viruses (IIVs) infect mostly arthropods, particularly the immature stages of insects inhabiting damp or aquatic environments. The IIVs are currently classified into two genera, *Iridovirus* and *Chloriridovirus*, based on particle size and genetic characteristics. Members of the genus *Iridovirus* have a particle diameter of approximately 120–130 nm in ultrathin section and have been isolated from many orders of insects, terrestrial isopods, and shrimp (Williams 1998; Tang et al. 2007). There are also a number of reports of iridovirus infections of marine and freshwater invertebrates, including bivalve, gastropod and cephalopod mollusks, a daphnid, and an annelid worm, although the relationship between these viruses and those that infect insects and isopods is uncertain (for a recent example, see Gregory et al. 2006). In contrast, members of the genus *Chloriridovirus* are larger (~180 nm in ultrathin section) and infect the aquatic stages of mosquitoes and midges. The first IIV was isolated from soil-dwelling cranefly larvae (*Tipula paludosa*) in England in 1954, and subsequently IIV infections have been reported from every continent except Antarctica.

## IIV Species Relationships

IIVs have received little attention over the past several decades and there is a paucity of genomic sequence information and other comparative studies that form the basis for defining species within these genera. Currently, the genus *Iridovirus* comprises two species, *Invertebrate iridescent virus 1* (IIV-1) and *Invertebrate iridescent virus 6* (IIV-6), and 11 tentative species whose status can only be resolved as additional information becomes available. Sequence comparisons of the MCP, DNA hybridization studies, and serological evidence indicate that isolates in the genus *Iridovirus* can be divided into three groups, or complexes (Williams and Cory 1994; Webby and Kalmakoff 1998). The largest group comprises IIV-1, -2, -9, -16, -22, -23, -24, -29, -30, *Anticarsia gemmatalis* IV, and an undescribed isolate from a weevil. The second group comprises IIV-6 and related strains of this virus (Gryllus IV, IIV-21, IIV-28). The third group comprises IIV-31 from terrestrial isopods and Pj-IV from the Japanese beetle (*Popillia japonica*). Recognition of these complexes should prove valuable for the future classification of novel IIV isolates.

The taxonomic situation is even more unsatisfactory for the genus *Chloriridovirus*, which, despite numerous records of infections in many species of mosquitoes and midges of medical and veterinary importance, consists of a single species, *Invertebrate iridescent virus 3* (IIV-3) from the saltmarsh mosquito, *Ochlerotatus* (*Aedes*) *taeniorhynchus*. DNA hybridization and sequence information based on the MCP and DNA polymerase support the classification of this virus in a separate genus (Williams and Cory 1994; Stasiak et al. 2000). However, recent analysis of the complete genome of IIV-3 indicates that some invertebrate iridoviruses that had previously been classified along with IIV-1 within the genus *Iridovirus* now cluster closer to IIV-3 than to IIV-6 (Delhon et al. 2006).

## Genomic Characteristics

Unlike virtually all vertebrate iridovirids, the IIV genome is not methylated, or methylated at a very low level that likely reflects the activity of cellular DNA methyltransferases. To date, only two IIV genomes have been sequenced completely: IIV-3 and IIV-6 (Jakob et al. 2001; Delhon et al. 2006). The genome of

IIV-6 is 212 kbp (unique portion) with a G+C content of 28.6% (Jakob et al. 2001). Although the original description of the IIV-6 genome predicted 468 ORFs, this value is likely an overestimate since it includes both overlapping and nonoverlapping ORFs. If only nonoverlapping reading frames are considered, then IIV-6 likely encodes approximately 200 ORFs, a figure consistent with the potential genetic content of vertebrate iridoviruses (Williams et al. 2005; Eaton et al. 2007; Tsai et al. 2007). Clearly, the only way to resolve the authenticity of potential overlapping reading frames is to identify viral transcripts by Northern blotting or RT-PCR. Using that approach, D'Costa et al. (2004) identified 137 IIV-6 transcripts, a value that is likely an underestimate due to the presence of low copy number transcripts. The genome of IIV-3 (genus *Chloriridovirus*) is 191 kbp (unique portion) with a G+C content of 48% and encodes 126–143 putative nonoverlapping ORFs (Delhon et al. 2006; Eaton et al. 2007; Tsai et al. 2007). Despite the aforementioned clustering of IIV sequences, IIV-3 and IIV-6 are considered members of separate viral genera based on the lack of colinearity between their genomes, the low sequence identity seen between homologous genes, virion size, and host range.

In addition to the core iridovirid genes involved in replication and nucleotide metabolism, other genes of interest in IIV-6 include an inhibitor of apoptosis (*iap*), an NAD$^+$-dependent DNA ligase, and putative homologs of peptides with antibacterial and antifungal properties (Jakob et al. 2001; Tanaka et al. 2003). Two adjacent ORFs have been detected in IIV-6 with homology to eukaryotic poly(ADP-ribosyl)transferase (pART), an enzyme that plays an important role in genome repair and maintenance (Otto et al. 2005). Between one and three copies of homologs to *bro* genes (baculovirus repeated ORFs) of lepidopteran ascoviruses have been detected in IIV-6 and IIV-31 (Bideshi et al. 2003). *Bro* genes are a multigene family of unknown function that may influence host DNA replication or transcription by regulating host chromatin structure (Zemskov et al. 2000). They are present in baculoviruses, entomopoxviruses, phycodnaviruses, and a number of bacteriophages and bacterial transposons. Interestingly, the large subunit of the IIV-6 ribonucleotide reductase contains an intein, a form of selfish genetic element that removes itself from the protein by posttranslational autocatalytic splicing (Pietrokovski 1998). Thirty-three genes are unique to IIV-3 and include a putative S/T protein kinase, a mutT-like protein, and RNA Pol II subunits with similarities to those of fungi (Delhon et al. 2006).

As with other iridovirids, IIV genomes contain extensive regions of repetitive DNA that account for 20% (IIV-3) to over 25% (IIV-9) of the genome (Fisher et al. 1988; McMillan et al. 1990; Delhon et al. 2006). The coding function of these regions is unknown, although late transcripts were detected in the IIV-9 repeat region. The pattern of repetitive DNA in the genome of IIV-6 is complex and involves boxes of tandem repeats and others with a number of different interdigitated repeat sequences of variable size and homology. Promoter regions for the MCP and DNA polymerase genes of IIV-6 have been located 29–53 and 6–27 nucleotides upstream of their respective transcriptional start sites (Nalcacioglu et al. 2003).

## Structural Characteristics

Structural studies of IIV are far more advanced than those of vertebrate iridoviruses and have revealed much about virion architecture. Building on early electron microscope studies (Wrigley 1969, 1970; Stoltz 1971, 1973), high-resolution cryo-electron microscopy and three-dimensional image reconstructions revealed IIV-6 particles in quasi-crystalline hexagonal arrays with an interparticle distance of 40–60 nm (Yan et al. 2000). Particle diameter was calculated at 162 nm along the two- and threefold axes of symmetry and 185 nm along the fivefold axis; considerably larger than the diameter observed in ultrathin sections or by negative staining. The outer capsid comprises a pseudohexagonal array of trimeric capsomers, 8 nm in diameter and 7.5 nm high. A thin fiber projects radially from the surface of each capsomer and likely regulates interparticle distance, a key characteristic for the iridescence of infected hosts. Such fibers are also seen in some vertebrate iridoviruses, such as LCDV-1 (Zwillenberg and Wolf 1968), but are not seen in enveloped virions released by budding. The major capsid protein of 51.4 kDa represents roughly 40% of the particle weight and forms hexavalent capsomers comprising an external noncovalent trimer and an internal trimer linked by disulfide bonds. The capsomers are arranged into trisymmetron and pentasymmetron facets. Each particle comprises 1,460 capsomers and an additional 12 pentavalent capsomers located singly at the center of each pentasymmetron. The triangulation number (T) is 147. Large IIVs that infect mosquitoes and midges have larger facets, giving a likely 1,560 subunits per particle (Stoltz 1971, 1973). A lipid bilayer, 4 nm thick, surrounds a highly hydrated DNA core that is arranged in a long coiled filament 10 nm in diameter. The lipid component is intimately associated with an additional inner shell beneath the fused layer of the capsid. Core and capsid polypeptides appear to be connected by intermembrane proteins passing through the lipid layer. Sensitivity to lipid solvents in IIVs varies with the assay system used (Martínez et al. 2003). Most of the polypeptide diversity of IIVs appears to be associated with the core and lipid membrane (Cerutti and Devauchelle 1990).

## Host Range

The host range of IIVs depends on the route of infection. Most IIVs cause lethal infections in a broad range of insects and other arthropod species following injection of the virus. In contrast, host range following oral administration tends to be far more restricted. The best studied example is IIV-6, which, in laboratory studies, infects species from all major insect orders and a number of other arthropods, including isopods and a centipede (Ohba 1975; Ohba and Aizawa 1979). Others, such as IIV-3, IIV-16, or IIV-24, naturally infect just one or two closely related host species. IIVs are clearly capable of replication in host species that do not develop signs of disease (Ohba 1975; Ward and Kalmakoff 1991), but systematic studies are required to determine the range of species susceptible to such asymptomatic infections. Virus propagation in vitro is readily achieved in dipteran and lepidopteran cell lines, and in recently developed cell lines from Homoptera and Coleoptera. Almost all IIVs, with the notable exceptions of IIV-3, IIV-16, and IIV-24, can be grown in massive quantities in the standard laboratory host, *Galleria mellonella* (Lepidoptera: Pyralidae). The in vivo host range of invertebrate iridoviruses does not extend to vertebrates (Kelly and Robertson 1973; Ohba and Aizawa 1982). Although an IIV-6-like virus has been reported from reptiles (Just et al. 2001), the need for caution regarding this report has been emphasized previously (Williams et al. 2005). Resolution of this controversy will require additional study since members of some viral families (e.g., *Flaviviridae*) are able to replicate in both invertebrates and endothermic vertebrates.

## IIVs Cause Lethal and Sublethal Disease

IIVs replicate extensively in most host tissues, especially the epidermis, muscle, fat body, nerves, hemocytes, and areas of the gut. Virus particles assemble into paracrystalline arrays within the cell cytoplasm and cause the iridescence that is characteristic of patent infections. Light reflected from the surface of close packed particles causes interference with incident light, a phenomenon known as Bragg diffraction. Purified pellets of IIV particles are also iridescent. Patently infected holometabolous insects, i.e., species that undergo complete metamorphosis, often die in the larval stage, but those that survive to pupate frequently show marked deformities and usually die in the pupal stage. IIVs can also cause patent infection in the adult stages of isopods, bees, and crickets. Some isolates have unusually long external fibrils attached to the capsid that increase the interparticle distance; these isolates do not display iridescence at visible wavelengths (Stoltz et al. 1968).

It is now apparent that iridescence is not a reliable indicator of IIV infection given that many hosts can be infected sublethally and show no obvious signs of disease. The density of IIV particles in sublethally infected insects is far lower than seen in the cells of patently infected individuals (Tonka and Weiser 2000). Such covert infections have been reported in natural populations of blackflies, *Simulium variegatum* (Williams 1993), and mayflies, *Ecdyonurus torrentis* (Tonka and Weiser 2000), and in laboratory populations of a mosquito, *Aedes aegypti* (Marina et al. 1999), and a moth, *Galleria mellonella* (Constantino et al. 2001). Covert infections have been detected by PCR amplification of the MCP gene, electron microscope observations, and insect bioassay techniques. Covert infection of *Ae. aegypti* by IIV-6 results in an increase in larval development time and reductions in adult body size and longevity. The reproductive capacity of covertly infected mosquitoes is reduced by 22%–50% compared to healthy mosquitoes (Marina et al. 2003a, 2003b, 2003c).

## Ecology

The incidence of patent disease in host populations is typically very low, although occasional epizootics have been observed in lepidopteran species (Sikorowski and Tyson 1984; Sieburth and Carner 1987), crickets (Fowler 1989), craneflies (Ricou 1975), mosquitoes (Fedorova 1986), and blackfly larvae (Hernández et al. 2000). Studies on blackflies and Lepidoptera have indicated that there exists considerable genotypic variation in IIV populations; individual insects collected at the same place and time differed in their respective RFLP profiles and in the size of the restriction fragments that hybridized to an MCP gene probe in Southern blots (Williams and Cory 1994; Williams, unpublished data).

There is clear evidence of seasonality in the prevalence of IIV infections that correlates with fluctuations in precipitation and soil humidity and/or host densities (Ricou 1975; Grosholz 1993; Hernández et al. 2000). The persistence of IIV-6 in soil depends very much on moisture content (Reyes et al. 2004) and the presence of clay and iron minerals (Christian et al. 2006). In contrast, IIV persistence in water is markedly affected by solar ultraviolet radiation; a rapid loss of activity is observed following exposure to direct sunlight (Hernández et al. 2005).

The route of infection is uncertain for most IIV-host systems. Cannibalism or predation of infected individuals appears to be the principal mechanism of transmission in populations of mosquitoes (Linley and Nielsen 1968a, 1968b), isopods (Federici 1980; Grosholz 1992), tipulids (Carter 1973a, 1973b), crickets (Fowler 1989), and cannibalistic Lepidoptera (Williams and Hernández 2006). IIV particles can be transmitted from infected to susceptible insects by hymenopteran parasitoids and entomopathogenic nematodes (Mullens et al. 1999; Lopez et al. 2002). IIV-3 survives in populations of *O. taeniorhynchus* by alternating cycles of horizontal transmission among cannibalistic larvae and vertical transmission from adult female mosquitoes that acquire a sublethal infection shortly before pupating (Linley and Nielsen 1968a, 1968b). Horizontal transmission is more prevalent in high-density populations of some species wherein the frequency of aggressive interactions between conspecifics and the probability of wounding is greater than at low densities (Grosholz 1993; Marina et al. 2005). Physical abrasion also results in an increased probability of infection (Carter 1973b; Undeen and Fukuda 1994).

## Economic Importance

Theoretical studies suggest that the impact of sublethal disease on the population dynamics of certain species may be considerable and merits further investigation, particularly in the case of medically important insect vectors such as mosquitoes (Boots et al. 2003; Marina et al. 2003a). However, IIVs are not presently considered to be useful agents for the biological control of insect pests because of the low prevalence of lethal disease, their potentially broad host range, and their poor ability to be transmitted by ingestion. IIV infections are suspected to periodically devastate populations of the mopane worm, which is an economically important species of caterpillar harvested for human consumption in southern African countries. Oyster populations also suffer serious disease, although the causative agents have been poorly characterized (Elston 1997). Recently, the use of IIVs in industrial applications has aroused some interest, specifically their role as bioscaffolds in the construction of metal nanostructures with unique optical and dielectric properties for use in nanotechnological applications (Juhl et al. 2004; Radloff et al. 2005a,b).

## LCDV and White Sturgeon Iridovirus

Although LCDV was not the first iridovirid to be isolated (that distinction goes to IIV-1), the characteristic signs of LCDV infection meant that lymphocystis disease was the first iridovirid disease to be described. Fish infected with what we now know to be LCDV were identified more than a century ago based on the distinctive appearance of diseased animals (reviewed in Weissenberg 1965; Wolf 1988; Bremont and Bernard 1995; Chinchar 2000). Infection with LCDV induces wart-like lesions generally on the skin, but also on internal organs, in numerous species of freshwater and marine fish (Dukes and Lawler 1975). Unlike papillomavirus-induced warts that result from an increased number of infected cells (hyperplasia), LCDV-mediated warts represent the enlargement of single infected cells. Lymphocystis cells are commonly 100 μm or more in diameter and can coalesce to form large masses several millimeters in diameter. Infected cells possess a thick hyaline capsule, a central enlarged nucleus, and prominent basophilic DNA inclusions (Wolf 1988). Low numbers of LCDV-induced warts are not in themselves life-threatening, but in high numbers they may impair respiration, feeding, and movement, making infected fish more susceptible to predation (http://www.maine. gov/ifw/fishing/fishlab/vol4issue10.htm). Moreover, external lesions reduce the commercial value of both food fish and aquarium stock.

Despite its early identification, LCDV has not been as extensively studied at the molecular level as other iridovirids because of its inability to be readily propagated in cell culture. Recently, the complete genomic sequences of what are likely two different viral species, one from plaice and flounder isolated from the Atlantic Ocean (LCDV-1, Tidona and Darai 1997) and a second from olive flounder in China (LCDV-C, Zhang et al. 2004) have been described. Additional putative species/isolates from infected dabs (LCDV-2, Wolf 1988), rockfish (Kitamura et al. 2006; Kim et al. 2007), and largemouth bass (Hanson et al. 2006) have been identified but not completely sequenced. Aside from the standard array of replicative genes (Eaton et al. 2007), LCDV contains as least two novel putative immune evasion genes, a protein with homology to members of the tumor necrosis factor receptor family (Essbauer et al. 2004) and another with similarity to a G protein-coupled receptor (Huang et al. 2007b).

Although diagnosis of LCDV infections is often made based on clinical and histological signs (i.e., wart-like external lesions), serological and molecular techniques have been developed to enhance the rapidity and specificity of diagnosis (Cano et al. 2006). No vaccines are currently available for preventing LCDV-induced disease, and there is no specific treatment for an ongoing infection. A DNA vaccine has been developed, but its efficacy is unknown (Zheng et al. 2006). Little is known about immunity to LCDV, but there is the suggestion that by replicating primarily in the skin, LCDV is shielded from an antiviral response until late in infection. Although fish can be reinfected with LCDV, subsequent infections are not as extensive, suggesting that immunity plays a role in ameliorating infection. Transmission of LCDV is thought to occur by direct contact between infected and uninfected fish and is likely increased by high population densities. In surviving fish, LCDV lesions resolve spontaneously. The recent identification of a single major genetic locus controlling susceptibility to LCDV infection in Japanese flounder opens the way to selective breeding programs designed to develop flounder populations that are highly resistant to lymphocystis disease (Fuji et al. 2006).

WSIV is currently an unclassified member of the family *Iridoviridae* and causes lethal infections in farm-reared juvenile white sturgeon (Georgiadis et al. 2001; Hedrick et al. 1992b; Watson et al. 1998). Several procedures commonly encountered in the culture of white sturgeon, such as rearing at high densities, contribute to patent disease because of infection originating from asymptomatic individuals (Georgiadis et al. 2001, 2002; Drennnan et al. 2005). WSIV has an affinity for epithelial cells and infects the gills, skin, olfactory organ, barbells, and esophagus. Electron microscopy identified large (~270 nm in diameter) icosahedral virions within infected cells and tentatively classified this agent as an iridovirus (Hedrick et al. 1990). Preliminary sequence analysis showed little similarity to other known viruses with the exception of partial sequence similarity to the MCP of mimivirus and members of the *Iridoviridae* and *Phycodnaviridae* families. A PCR assay has been developed to detect the presence of the WSIV MCP gene (Kwak et al. 2006) and has proved effective in the diagnosis of asymptomatically infected individuals (Drennan et al. 2007).

## Conclusions and Future Directions

The fundamentals of the iridovirid replication strategy were elucidated by Granoff, Willis, Goorha, and their co-workers more than 20 years ago. However, despite this commendable start, much remains to be done in terms of determining the precise roles that specific viral proteins play in the virus life cycle. Moreover, we now appreciate the fact that viral proteins not only control viral biogenesis, but also likely regulate evasion of the immune system, host range, and virulence. Recent studies indicate that gene silencing using siRNAs and targeted knockdown techniques based on asMOs offer the opportunity to block expression of specific viral genes and infer their function from the resulting changes in phenotype. In addition, it may soon be possible to knock out specific viral genes via homologous recombination and assess the role of these deletion mutants both in vitro and in vivo. From the immunological point of view, the FV3/*Xenopus* model developed by Robert and his colleagues offers an excellent approach for examining antiviral responses in a well-characterized amphibian system. Moreover, the growing awareness of the impact of iridovirid-mediated mortality in various fish and amphibian species suggests that ecologists, ichthyologists, fish and frog farmers, virologists, and immunologists need to pool their efforts to understand the biological and ecological basis underlying the die-off phenomena. Efforts along these lines are currently underway (Grant et al. 2005; Inendino et al. 2005; Pearman and Garner 2005; http://lifesciences.asu.edu/irceb/amphibians). With the growing global dependence on aquaculture as a human food source and increasing concerns about the decline in natural amphibian populations, it is critical that we understand the biology of one of the major families of viral pathogens infecting cold-blooded vertebrates. In addition, it is time that the life cycle of invertebrate iridoviruses be subjected to the same molecular dissection that has been applied to their vertebrate counterparts. Specifically, molecular tools should be employed to elucidate the factors that modulate the virulence of IIV infections in insects, to determine the relationship between covert and patent infections, and to elucidate strategies of vertical transmission. Additional sequence information is also required to resolve the large number of tentative IIV species in the *Iridovirus* genus, and to delineate the diversity present in the *Chloriridovirus* genus that currently comprises just a single member.

Key areas of research in the coming years:

- Identification of the molecular composition of the viral RNA polymerase and determination of its precise role in viral transcription
- Identification of the cellular and molecular elements of antiviral immunity in fish and amphibians infected with ranaviruses
- Identification of the temporal class and promoter elements of key viral genes
- Elucidation of the function of key viral genes controlling replication, immune evasion, virulence, and host range through the use of asMO, siRNA, and gene knockout experiments
- Development of effective vaccines to protect captive and commercially important species from iridovirus-induced disease
- Clarification of the taxonomic relationships among these viruses, specifically the quantification of intra- and interspecific genotypic variation and the designation of genetic, biological, and ecological species-defining criteria that can be uniformly applied to the members of each genus
- Delineation of the factors that determine virulence and vertical transmission among invertebrate iridoviruses

**Acknowledgements** This work was partially supported by grants from the National Science Foundation, United States (Award No. DEB 02-13851).

# References

Ahne W, Schlotfeldt HJ, Thomsen I (1989) Fish viruses: isolation of an icosahedral cytoplasmic deoxyribovirus from sheatfish (*Silurus glanis*). J Vet Med B 36:333–336

Ahne W, Bearzotti M, Bremont M, Essbauer S (1998) Comparison of European systemic piscine and amphibian iridoviruses with epizootic haematopoietic necrosis virus and frog virus 3. J Vet Med B 45:373–383

Alcami A, Koszinowski UH (2000) Viral mechanisms of immune evasion. Mol Med Today 6:365–372

Allen MJ, Schroeder DC, Holden MTG, Wilson WH (2006) Evolutionary history of the *Coccolithoviridae*. Mol Biol Evol 23:86–92

Allender MC, Fry MM, Irizarry AR, Craig L, Johnson AJ, Jones M (2006) Intracytoplasmic inclusions in circulating leukocytes from an Eastern Box turtle (*Terrapene carolina carolina*) with an iridoviral infection. J Wildlife Dis 42:677–684

Anderson IG, Prior HC, Rodwell BJ, Harris GO (1993) Iridovirus-like virions in imported dwarf gourami (*Colisa lalia*) with systemic amoebiasis. Aust Vet J 70:66–67

Armstrong RD, Ferguson HW (1989) Systemic viral disease of the chromid cichlid *Etroplus maculatus*. Dis Aquat Org 7:155–157

Ariel E, Tapiovaara H, Olesen NJ (1999) Comparison of pike-perch (*Stizostedion lucioperca*), cod (*Gadus morhua*) and turbot (*Scophthslmus maximus*) iridovirus isolates with reference to other piscine and amphibian iridovirus isolates. European Association of Fish Pathologists, VIII. International Conference on Diseases of Fish and Shellfish, Rhodes, Greece, 20–24 September

Bauer S, Kirschning CJ, Hacker H, Redecke V, Hausmann S, Akira S, Wagner H, Lipford GB (2001) Human TLR9 confers responsiveness to bacterial DNA via species-specific CpG motif recognition. Proc Natl Acad Sci U S A 98:9237–9242

Beattie E, Tartaglia J, Paoletti E (1991) Vaccinia virus-encoded eIF-2 alpha homolog abrogates the antiviral effect of interferon. Virology 183:419–422

Bideshi DK, Renault S, Stasiak K, Federici BA, Bigot Y (2003) Phylogenetic analysis and possible function of *bro*-like genes, a multigene family widespread among large double-stranded DNA viruses of invertebrates and bacteria. J Gen Virol 84:2531–2544

Bloch B, Larsen JL (1993) An iridovirus-like agent associated with systemic infection in cultured turbot *Scophthalmus maximus* fry in Denmark. Dis Aquat Org 15:235–240

Boehme KW, Compton T (2004) Innate sensing of viruses by Toll-like receptors. J Virol 78:7867–7873

Bollinger TK, Mao J, Schock D, Brigham RM, Chinchar VG (1999) Pathology, isolation and molecular characterization of an iridovirus from tiger salamanders in Saskatchewan. J Wildlife Dis 35:413–429

Boots M, Greenman J, Ross D, Norman R, Hails R, Sait S (2003) The population dynamical implications of covert infections in host-microparasite interactions. J Anim Ecol 72:1064–1072

Bouchier-Hayes L, Martin SJ (2002) CARD games in apoptosis and immunity. EMBO Rep 3:616–621

Braunwald J, Tripier F, Kirn A (1979) Comparison of the properties of enveloped and naked frog virus 3 (FV3) particles. J Gen Virol 45:673–682

Braunwald J, Nonnenmacher H, Tripier-Darcy F (1985) Ultrastructural and biochemical study of frog virus 3 uptake by BHK-21 cells. J Gen Virol 66:283–293

Bremont J, Bernard M (1995) Molecular biology of fish viruses: a review. Vet Res 26:341–351

Brunner JL, Schock DM, Collins JPl, Davidson EW (2004) The role of an intraspecific reservoir in the persistence of a lethal ranavirus. Ecology 85:560–566

Brunner JL, Richards K, Collins JP (2005) Dose and host characteristics influence virulence of ranavirus infections. Oecolgia 144:399–406

Brunner JL, Schock DM, Collins JP (2007) Transmission dynamics of the amphibian ranavirus *Ambystoma tigrinum virus*. Dis Aquat Org 77:87–95

Caipang CM, Hirono I, Aoki T (2006a) Immunogenicity, retention, and protective effects of the protein derivatives of formalin-inactivated red seabream iridovirus (RSIV) vaccine in red seabream, *Pagrus major*. Fish Shellfish Immunol 20:597–609

Caipang CM, Takano T, Hirono I, Aoki T (2006b) Genetic vaccines protect red seabream, *Pagrus major*, upon challenge with red seabream iridovirus (RSIV). Fish Shellfish Immunol 22:130–138

Cano I, Ferro P, Alonso MC, Bergmann SM, Romer-Oberdorfer A, Garcia-Rosado E, Castro D, Borrego JJ (2006) Development of molecular techniques for detection of lymphocystis disease virus in different marine fish species. J Appl Microbiol 102:32–40

Carter JB (1973a) The mode of transmission of *Tipula* iridescent virus. I. Source of infection. J Invert Pathol 21:123–130

Carter JB (1973b) The mode of transmission of *Tipula* iridescent virus. II. Route of infection. J Invertebr Pathol 21:136–143

Cerutti M, Devauchelle G (1980) Inhibition of macromolecular synthesis in cells infected with an invertebrate virus (iridovirus type 6 or CIV). Arch Virol 63:297–303

Cerutti M, Devauchelle G (1990) Protein composition of Chilo iridescent virus. In: Darai G (ed) Molecular biology of iridoviruses, Kluwer, Boston, pp 81–112

Cerutti M, Cerutti P, Devuachelle G (1989) Infectivity of vesicles prepared from Chilo iridescent virus inner membrane: evidence for recombination between associated DNA fragments. Virus Res 12:299–314

Chen ZX, Zheng JC, Jiang YL (1999) A new iridovirus isolated from soft-shelled turtle. Virus Res 63:147–151

Chinchar VG (2000) Ecology of viruses of cold-blooded vertebrates. In: Hurst CJ (ed) Virus ecology, Academic, New York, pp 413–445

Chinchar VG (2002) Ranaviruses (family *Iridoviridae*): emerging cold-blooded killers. Arch Virol 147:447–470

Chinchar VG, Dholakia JN (1989) Frog virus 3-induced translational shut-off: activation of an eIF-2 kinase in virus-infected cells. Virus Res 14:207–224

Chinchar VG, Granoff A (1984) Isolation and characterization of a frog virus 3 variant resistant to phosphonoacetate: genetic evidence for a virus-specific DNA polymerase. Virology 138:357–361

Chinchar VG, Granoff A (1986) Temperature-sensitive mutants of frog virus 3: biochemical and genetic characterization. J Virol 58:192–202

Chinchar VG, Mao J (2000) Molecular diagnosis of iridovirus infections in cold-blooded animals. Sem Avian Exotic Pet Med 9:27–35

Chinchar VG, Yu W (1990) Frog virus 3-mediated translational shut-off: frog virus 3 messages are translationally more efficient than host and heterologous viral messages under conditions of increased translational stress. Virus Res 16:163–174

Chinchar VG, Yu W (1992) Metabolism of host and viral mRNAs in frog virus 3-infected cells. Virology 186:435–443

Chinchar VG, Goorha R, Granoff A (1984) Early proteins are required for the formation of frog virus 3 assembly sites. Virology 135:148–156

Chinchar VG, Bryan L, Wang J, Long S, Chinchar GD (2003) Induction of apoptosis in frog virus 3-infected cells. Virology 306:303–312

Chinchar VG, Essbauer S, He JG, Hyatt A, Miyazaki T, Seligy V, Williams T (2005) *Iridoviridae.* In: Fauquet CM, Mayo MA, Maniloff J, Desselberger U, Ball LA (eds) Virus taxonomy: 8th report of the International Committee on the Taxonomy of Viruses, Elsevier, London, pp 163–175

Choi S-K, Kwon S-R, Nam Y-K, Kim S-K, Kim K-H (2006) Organ distribution of red sea bream iridovirus (RSIV) DNA in asymptomatic yearling and fingerling rock bream (*Oplegnathus fasciatus*) and effects of water temperature on transition of RSIV into acute phase. Aquaculture 256:23–26

Chua HC, Ng, ML, Woo JJ, Wee JY (1994) Investigation of outbreaks of a novel disease, "Sleepy Grouper Disease," affecting the brown-spotted grouper, *Epinephelus tauvina* Forskal. J Fish Dis 17:417–427

Christian P, Richards AR, Williams T (2006) Differential adsorption of occluded and non-occluded insect pathogenic viruses to soil forming minerals. Appl Environ Microbiol 72:4648–4652

Clark HF, Brennan JC, Zeigel RF, Karzon DT (1968) Isolation and characterization of viruses from the kidneys of *Rana pipiens* with renal adenocarcinoma before and after passage in the red eft (*Triturus viridescens*). J Virol 2:629–640

Constantino M, Christian P, Marina CF, Williams T (2001) A comparison of techniques for detecting *Invertebrate iridescent virus 6.* J Virol Meth 98:109–118

Coupar BEH, Goldie SG, Hyatt AD, Pallister JA (2005) Identification of a Bohle iridovirus thymidine kinase gene and demonstration of activity using vaccinia virus. Arch Virol 150:1797–1812

Cunningham AA, Langton TES, Bennett PM, Lewin JF, Drury SEV, Gough RE, MacGregor SK (1996) Pathological and microbiological findings from incidents of unusual mortality of the common frog *Rana temporaria.* Phil Trans R Soc Lond B 351:1539–1557

Danayadol Y, Direkbusarakom S, Boonyaratpalin S, Miyazaki T, Miyata M (1996) An outbreak of iridovirus-like infection in brown-spotted grouper (*Epinephelus malabaracus*) cultured in Thailand. Aquat Anim Health Res Inst Newsletter 5:6

Davidson EW, Jancovich JK, Borland S, Newberry M, Gresens J (2003) Dermal lesions, hemorrhage, and limb swelling in laboratory axolotls. Lab Animal 32:23–24

D'Costa SM, Yao H, Bilimoria SL (2001) Transcription and temporal cascade in Chilo iridescent virus infected cells. Arch Virol 146:2165–2178

D'Costa SM, Yao HJ, Bilimoria SL (2004) Transcriptional mapping in Chilo iridescent virus infections. Arch Virol 149:723–742

Delhon G, Tulman ER, Afonso CL, Lu Z, Becnel JJ, Moser BA, Kutish GF, Rock DL (2006) Genome of invertebrate iridescent virus type 3 (mosquito iridescent virus) J Virol 80:8439–8449

Delius H, Darai G, Flügel RM (1984) DNA analysis of insect iridescent virus 6: evidence for circular permutation and terminal redundancy. J Virol 49:609–614

De Voe R, Geissler K, Elmore S, Rotstein D, Lewbart G, Guy J (2004) Ranavirus-associated morbidity and mortality in a group of captive Eastern box turtles (*Terrapene carolina carolina*). J Zoo Wildlife Med 35:534–543

Do JW, Moon C H, Kim HJ, Ko MS, Kim SB, Son JH, Kim JS, An EJ, Kim MK, Lee SK, Han MS, Cha SJ, Park MS, Park MA, Kim YC, Kim JW, Park JW (2004) Complete genomic DNA sequence of rock bream iridovirus. Virology 325:351–363

Do JW, Cha SJ, Kim YC, An EJ, Lee NS, Choi HJ, Lee CH, Park MS, Kim JW, Kim YC, Park JW (2005a) Phylogenetic analysis of the major capsid protein gene of iridovirus isolates from cultured flounder *Paralichthys olivaceus* in Korea. Dis Aquat Org 64:193–200

Do JW, Cha SJ, Kim JS, An EJ, Park MS, Kim JW, Lim YC, Park MA, Park JW (2005b) Sequence variation in the gene encoding the major capsid protein of Korean fish iridovirus. Arch Virol 150:351–359

Docherty DE, Meteyer CU, Wang J, Mao J, Case ST, Chinchar VG (2003) Diagnostic and molecular evaluation of three iridovirus-associated salamander mortality events. J Wildlife Dis 39:556–566

Drennan JD, Ireland S, LaPatra SE, Grabowski L, Carrothers TK, Cain KD (2005) High-density rearing of white sturgeon *Acipenser transmontanus* (Richardson) induces white sturgeon iridovirus disease among asymptomatic carriers. Aquacult Res 36:824–827

Drennan JD, LaPatra SE, Samson CA, Ireland S, Eversman KF, Cain KD (2007) Evaluation of lethal and non-lethal sampling methods for the detection of white sturgeon iridovirus infection in white sturgeon, *Acipenser transmontanus* (Richardson). J Fish Dis 30:367–379

Dukes TW, Lawler AR (1975) The ocular lesions of naturally occurring lymphocystis in fish. Can J Comp Med 39:406–410

Eaton HE, Metcalf J, Penny E, Tcherepanov V, Upton C, Brunetti CR (2007) Comparative genomic analysis of the family *Iridoviridae*: Re-annotating and defining the core set of iridovirus genes. Virol J 4:11

Elston R (1997) Bivalve mollusk viruses. World J Microbiol Biotechnol 13:393–403

Epifano C, Krijnse-Locker J, Salas ML, Rodriguez JM, Salas J (2006) The African swine fever virus nonstructural protein pB602L is required for formation of the icosahedral capsid of the virus particle. J Virol 80:12260–12270

Essani K, Goorha R, Granoff A (1987) Mutation in a DNA binding protein reveals an association between DNA methyltransferase activity and a 26,000 Da polypeptide in FV3-infected cells. Virology 161:211–217

Essbauer S, Ahne W (2002) The epizootic haematopoietic necrosis virus (*Iridoviridae*) induces apoptosis in vitro. J Vet Med B 49:25–30

Essbauer S, Bremont M, Ahne W (2001) Comparison of the eIF-2 alpha homologous proteins of seven ranaviruses (*Iridoviridae*). Virus Genes 23:347–359

Essbauer S, Fischer U, Bergmann S, Ahne W (2004) Investigations on ORF 167L of lymphocystis disease virus (*Iridoviridae*). Virus Genes 28:19–39

Federici BA (1980) Isolation of an iridovirus from two terrestrial isopods, the pill bug, *Armadillidium vulgare* and the sow bug, *Porcellio dilatatus*. J Invertebr Pathol 36:373–381

Fedorova VG (1986) On finding larvae of *Culex territans* Walk. and Dixidae infected with iridovirus in the forest zone of Novgorrod region (in Russian). Med Parazitol Mosk 3:86–87

Fischer M, Schnitzler P, Delius H, Darai G (1988) Identification and characterization of the repetitive DNA element in the genome of insect iridescent virus type 6. Virology 167:485–496

Fox SF, Greer AL, Torres-Cervantes R, Collins JP (2006) First case of ranavirus-associated morbidity and mortality in natural populations of the South American frog *Atelognathus patagonicus.* Dis Aquat Org 72:87–92

Fowler HG (1989) An epizootic iridovirus of Orthoptera (Gryllotalpidae: *Scaptericus borellii*) and its pathogenicity to termites (Isoptera: Cryptotermes). Rev Microbiol 20:115–120

Fuji K, Kobayashi K, Hasegawa O, Moura Coimbra MR, Sakamoto T, Okamoto N (2006) Identification of a single major genetic locus controlling the resistance to lymphocystis disease in Japanese flounder (*Paralichthys olivaceus*). Aquaculture 254:203–210

Galli L, Pereira A, Márquez A, Mazzoni R (2006) Ranavirus detection by PCR in cultured tadpoles (*Rana catesbeiana* Shaw 1802) from South America. Aquaculture 257:78–82

Gantress J, Bell A, Maniero G, Cohen N, Robert J (2003) *Xenopus*, a model to study immune responses to iridovirus. Virology 311:254–262

Gendrault J-L, Steffan A-M, Bingen A, Kirn A (1981) Penetration and uncoating of frog virus 3 in cultured rat Kupffer cells. Virology 112:375–384

Georgiadis MP, Hedrick RP, Johnson WO, Yun S, Gardner IA (2000) Risk factors for outbreaks of disease attributable to white sturgeon iridovirus and white sturgeon herpesvirus-2 at a commercial sturgeon farm. Am J Vet Res 61:1232–1240

Georgiadis MP, Hedrick RP, Carpenter TE, Gardner IA (2001) Factors influencing transmission, onset and severity of outbreaks due to white sturgeon iridovirus in a commercial hatchery. Aquaculture 194:21–35

Gil J, Garcia MA, Gomez-Puertas P, Guerra S, Rullas J, Nakano H, Alcami J, Esteban M (2004) TRAF family proteins link PKR with NF-κB activation. Mol Cell Biol 24:4502–4512

Glaser R, Litsky ML, Padgett DA, Baiocchi RA, Yang EV, Chen M, Yeh PE, Green-Church KB, Caligiuri MA, Williams MV (2006) EBV-encoded dUTPase induces immune dysregulation: implications for the pathophysiology of EBV-associated disease. Virology 346:205–218

Go J, Whittington R (2006) Experimental transmission and virulence of a megalocytivirus (Family *Iridoviridae*) of dwarf gourami (*Colisa lalia*) from Asia in Murray cod (*Maccullochella peelii peelii*) in Australia. Aquaculture 258:140–149

Go J, Lancaster M, Deece K, Dhungyel O, Whittington R (2006) The molecular epidemiology of iridovirus in Murray cod (*Muccullochella peeli peelii*) and dwarf gourami (*Colisa lalia*) from distant biogeographical regions suggests a link between trade in ornamental fish and emerging iridoviral diseases. Mol Cell Probes 20:212–222

Goorha R (1981) Frog virus 3 requires RNA polymerase II for its replication. J Virol 37:496–499

Goorha R (1982) Frog virus 3 DNA replication occurs in two stages. J Virol 43:519–528

Goorha R, Dixit P (1984) A temperature-sensitive mutant of frog virus 3 is defective in second stage DNA replication. Virology 136:186–195

Goorha R, Granoff A (1979) Icosahedral cytoplasmic deoxyriboviruses. In: Fraenkel-Conrat H, Wagner RR (eds) Comprehensive virology. Plenum, New York, pp 347–399

Goorha R, Murti KG (1982) The genome of frog virus 3, an animal DNA virus, is circularly permuted and terminally redundant. Proc Natl Acad Sci U S A 79:248–262

Goorha R, Murti G, Granoff A, Tirey R (1978) Macromolecular synthesis in cells infected by frog virus 3: VIII. The nucleus is a site of frog virus 3 DNA and RNA synthesis. Virology 84:32–50

Goorha R, Willis DB, Granoff A, Naegele RF (1981) Characterization of a temperature-sensitive mutant of frog virus 3 defective in DNA replication. Virology 112:40–48

Goorha R, Granoff A, Willis DB, Murti KG (1984) The role of DNA methylation in virus replication: inhibition of frog virus 3 replication by 5-azacytidine. Virology 138:94–102

Granoff A, Came PE, Breeze DC (1966) Viruses and renal carcinoma of *Rana pipiens*: I. The isolation and properties of virus from normal and tumor tissues. Virology 29:133–148

Grant EC, Inendino KR, Love WJ, Philipp DP, Goldberg TL (2005) Effects of practices related to catch-and-release angling on mortality and viral transmission in juvenile largemouth bass infected with largemouth bass virus. J Aquat Anim Health 17:315–322

Green DE, Converse KA, Schrader AK (2002) Epizootiology of sixty-four amphibian morbidity and mortality events in the U S A 1996–2001. Ann N Y Acad Sci 969:323–339

Greer AL, Berrill M, Wilson PJ (2005) Five amphibian mortality events associated with ranavirus infection in south central Ontario, Canada. Dis Aquat Org 67:9–14

Gregory CR, Latimer KS, Pennick KE, Benson K, Moore T (2006) Novel iridovirus in a nautilus (*Nautilus* spp.). J Vet Diagn Invest 18:208–211

Grosholz ED (1992) Interactions of intraspecific, interspecific and apparent competition with host-pathogen population dynamics. Ecology 73:507–514

Grosholz ED (1993) The influence of habitat heterogeneity on host-pathogen population dynamics. Oecologia 96:347–353

Hanson LA, Rudis MR, Vasquez-Lee M, Montgomery RD (2006) A broadly applicable method to characterize large DNA viruses and adenoviruses based on the DNA polymerase gene. Virol J 3:28

He JG, Wang SP, Zeng K, Huang ZJ, Chan SM (2000) Systemic disease caused by an iridovirus-like agent in cultured mandarinfish, *Siniperca chuatsi* (Basilewsky), in China. J Fish Dis 23:219–222

He JG, Deng M, Weng SP, Li Z, Zhou SY, Long QX, Wang XZ, Chan SM (2001) Complete genome analysis of the mandarin fish infectious spleen and kidney necrosis iridovirus. Virology 291:126–139

He JG, Lu L, Deng M, He HH, Weng SP, Wang XH, Zhou SY, Long QX, Wang XZ, Chan SM (2002) Sequence analysis of the complete genome of an iridovirus isolated from the tiger frog. Virology 292:185–197

He W, Yin ZX, Li Y, Huo WL, Guan HJ, Weng SP, Chan SM, Je JG (2006) Differential gene expression profile in spleen of mandarin fish *Siniperca chuasti* infected with ISKNV, derived from suppression subtractive hybridization. Dis Aquat Org 73:113–122

Heath CM, Windsor M, Wileman T (2001) Aggresomes resemble sites specialized for virus assembly. J Cell Biol 153:449–456

Hedrick RP, Groff JM, McDowell T, Wingfield WH (1990) An iridovirus infection of the integument of white sturgeon (*Acipenser transmontanus*). Dis Aquat Org 8:39–44

Hedrick RP, McDowell TS, Ahne W, Torhy C, de Kinkelin P (1992a) Properties of three iridovirus-like agents associated with systemic infections of fish. Dis Aquat Org 13:203–209

Hedrick RP, McDowell TS, Groff JM, Yun S, Wingfield WH (1992b) Isolation and properties of an iridovirus-like agent from white sturgeon *Acipenser transmontanus*. Dis Aquat Org 12:75–81

Hernández O, Maldonado G, Williams T (2000) An epizootic of patent iridescent virus disease in multiple species of blackflies in Chiapas, Mexico. Med Vet Entomol 14:458–462

Hernández O, Marina CF, Valle J, Williams T (2005) Persistence of invertebrate iridescent viruses in artificial tropical aquatic environments. Arch Virol 150:2357–2363

Holm GH, Zurney J, Tumilasci V, Leveille S, Danthi P, Hiscott, Sherry B, Dermody TS (2007) RIG-I and IPS-1 augment proapoptotic responses following mammalian reovirus infection via IRF-3. J Biol Chem 282:21953–21956

Hu GB, Cong RS, Fan TJ, Mei XG (2004) Induction of apoptosis in a flounder gill cell line by lymphocystis disease virus infection. J Fish Dis 27:657–662

Huang X, Zhang Q (2007) Improvement and observation of immunoelectron microscopic method for the localization of frog *Rana grylio* virus (RGV) in infected fish cells. Micron 38:599–606

Huang XH, Huang YH, Yuan XP, Zhang QY (2005) Electron microscopic examination of the viromatrix of *Rana grylio* virus in a fish cell line. J Virol Methods 133:117–123

Huang YH, Huang XH, Gui JF, Zhang QY (2007a) Mitochondrion-mediated apoptosis induced by Rana grylio virus infection in fish cells. Apoptosis 12:1569–1577

Huang YH, Huang XH, Zhang J, Gui JF, Zhang QY (2007b) Subcellular localization and characterization of G protein-coupled receptor homolog from lymphocystis disease virus isolated in China. Viral Immunol 20:150–159

Hyatt AD, Whittington RJ (2005) Ranaviruses of fish, amphibians and reptiles: diversity and the requirement for revised taxonomy. Dis Asian Aquacult 5:155–170

Hyatt AD, Parkes H, Zupanovic Z (1998) Identification, characterization and assessment of Venezuelan viruses for potential use as biological control agents against the cane toad (*Bufo marinus*) in Australia. Report from the Australian Animal Health Laboratory, CSIRO, Geelong, Australia

Hyatt AD, Gould AR, Zupanovic Z, Cunningham AA, Hengstberger S, Whittington RJ, Kattenbelt J, Coupar BE (2000) Comparative studies of piscine and amphibian iridoviruses. Arch Virol 145:301–331

Hyatt AD, Williamson M, Coupar BE, Middleton D, Hengstberger SG, Gould AR, Selleck P, Wise TG, Kattenbelt J, Cunningham AA, Lee J (2002) First identification of a ranavirus from green pythons (*Chondropython viridis*). J Wildlife Dis 38:239–252

Imajoh M, Sugiura H, Oshima S (2004) Morphological changes contribute to apoptotic cell death and are affected by caspase-3 and caspase-6 inhibitors during red sea bream iridovirus permissive replication. Virology 322:220–230

Inendino KR, Grant EC, Philipp DP, Goldberg TL (2005) Effects of factors related to water quality and population density on the sensitivity of juvenile largemouth bass to mortality induced by viral infection. J Aquat Anim Health 17:304–314

Inouye K, Yamano K, Maeno Y, Nakajima K, Matsuoka M, Wada Y, Sorimahi M (1992) Iridovirus infection of cultured red sea bream, *Pagrus major* (in Japanese). Fish Pathol 27:19–27

Iyer LM, Aravind L, Koonin EV (2001) Common origin of four diverse families of large eukaryotic DNA viruses. J Virol 75:11720–11734

Jakob NJ, Muller K, Bahr U, Darai G (2001) Analysis of the first complete DNA sequence of an invertebrate iridovirus: coding strategy of the genome of *Chilo* iridescent virus. Virology 286:182–196

Jancovich JK, Davidson EW, Morado JF, Jacobs BL, Collins JP (1997) Isolation of a lethal virus from the endangered tiger salamander *Ambystoma tigrinum stebbinsi*. Dis Aquat Org 31:161–167

Jancovich JK, Davidson EW, Seiler A, Jacobs BL, Collins JP (2001) Transmission of the *Ambystoma tigrinum* virus to alternative hosts. Dis Aquat Org 46:159–163

Jancovich JK, Mao J, Chinchar VG, Wyatt C, Case ST, Kumar S, Valente G, Subramanian S, Davidson EW, Collins JP, Jacobs BL (2003) Genomic sequence of a ranavirus (family *Iridoviridae*) associated with salamander mortalities in North America. Virology 316:90–103

Jancovich JK, Davidson EW, Parameswaran N, Mao J, Chinchar VG, Collins JP, Jacobs BL, Storfer A (2005) Evidence for emergence of an amphibian iridoviral disease because of human-enhanced spread. Mol Ecol 14:213–224

Johnson AJ, Pessier AP, Jacobson ER (2007) Experimental transmission and induction of ranaviral disease in Western ornate box turtles (*Terrapene ornata ornata*) and Red-eared sliders (*Trachemys scripta elegans*). Vet Pathol 44:285–297

Johnson CL, Gale M Jr (2006) CARD games between virus and host get a new player. Trends Immunol 27:1–4

Johnston JB, McFadden G (2003) Poxvirus immunomodulatory strategies: current perspectives. J Virol 77:6093–6100

Juhl S, Ha YH, Chan E, Ward V, Smith A, Dokland T, Thomas EL, Vaia R (2004) Bioharvesting: optical characteristics of *Wiseana* iridovirus assemblies. Polym Mat Sci Eng 90:317–318

Jung SJ, Oh MJ (2000) Iridovirus-like infection associated with high mortalities of striped beakperch, *Oplegnathus fasciatus* (Temminck et Schlegel), in southern coastal areas of the Korean peninsula. J Fish Dis 23:223–226

Just F, Essbauer S, Ahne W, Blahak S (2001) Occurrence of an invertebrate iridescent-like virus (*Iridoviridae*) in reptiles. J Vet Med B 48:685–694

Kanchanakhan S (1998) An ulcerative disease of the cultured tiger frog, *Rana tigrina*, in Thailand: Virological examination. AAHRI Newsletter 7:1–2

Kaur K, Rohozinski J, Goorha R (1995) Identification and characterization of the frog virus 3 DNA methyltransferase. J Gen Virol 76:1937–1943

Kawagishi-Kobayashi M, Silverman JB, Ung TL, Dever TE (1997) Regulation of the protein kinase PKR by the vaccinia virus pseudosubstrate inhibitor K3L is dependent on residues conserved between the K3L protein and the PKR substrate eIF2 alpha. Mol Cell Biol 17:4146–4158

Kelly DC, Robertson J (1973) Icosahedral cytoplasmic deoxyriboviruses. J Gen Virol 21:17–41

Kelly DC, Tinsley TW (1974) Iridescent virus replication: a microscope study of *Aedes aegypti* and *Antherea eucalypti* cells in culture infected with iridescent virus types 2 and 6. Microbios 9:75–93

Kim S, Kim HY, Lee S, Kim SW, Sohn S, Kim K, Cho H (2007) Hepatitis B virus X protein induces perinuclear mitochondrial clustering in microtubule- and dynein-dependent manners. J Virol 81:1714–1726

Kim TJ, Lee JI (2007) Sequence variation in the genes encoding the major capsid protein, the ATPase, and the RNA polymerase 2 (domain 6) of lymphocystis disease virus isolated from Schlegel's black rockfish, *Sebastes schlegelii* Hilgendorf. J Fish Dis 30:501–504

Kitamura SI, Jung SJ, Kim WS, Nishizawa T, Yoshimizu M, Oh MJ (2006) A new genotype of lymphocystis virus, LCDV-RF, from lymphocystis diseased fish. Arch Virol 151:607–615

Kumar S, Tamura K, Nei M (2004) MEGA3: Integrated software for Molecular Evolutionary Genetics Analysis and sequence alignment. Brief Bioinform 5:150–163

Kurita J, Nakajima K, Hirono I, Aoki T (2002) Complete genome sequencing of red seabream iridovirus (RSIV). Fisheries Sci 68:1113–1115

Kwak KT, Gardner IA, Farver TB, Hedrick RP (2006) Rapid detection of white sturgeon iridovirus (WSIV) using a polymerase chain reaction (PCR) assay. Aquaculture 254:92–101

Langdon JS, Humphrey JD (1987) Epizootic haematopoietic necrosis, a new viral disease in redfin perch, *Perca fluviatilis* L., in Australia. J Fish Dis 10:289–297

Langdon JS, Humphrey JD, Williams LM, Hyatt AD, Westbury H (1986) First virus isolation from Australian fish: an iridovirus-like pathogen from redfin perch, *Perca fluviatilis* L. J Fish Dis 9:263–268

Langelier Y, Bergeron S, Chabaud S, Lippens J, Guilbault C, Sasseville AMJ, Denis S, Mosser DD, Massie B (2002) The R1 subunit of herpes simplex virus ribonucleotide reductase protects cells against apoptosis at, or upstream of, caspase-8 activation. J Gen Virol 83:2779–2789

Langland JO, Jacobs BL (2002) The role of the PKR-inhibitory genes E3L and K3L, in determining vaccinia virus host range. Virology 299:133–141

Lembo D, Donalisio M, Hofer A, Cornagliea M, Brune W, Koszinowski U, Thelander L, Landolfo S (2004) The ribonucleotide reductase R1 homolog of murine cytomegalovirus is not a functional enzyme subunit but is required for pathogenesis. J Virol 78:4278–4288

Leutenegger R (1967) Early events of *Sericesthis* iridescent virus infection in hemocytes of *Galleria mellonella* (L.). Virology 32:109–116

Linley JR, Nielsen HT (1968a) Transmission of a mosquito iridescent virus in *Aedes taeniorhynchus*. I. Laboratory experiments. J Invertebr Pathol 12:7–16

Linley JR, Nielsen HT (1968b) Transmission of a mosquito iridescent virus in *Aedes taeniorhynchus*. II. Experiments related to transmission in nature. J Invertebr Pathol 12:17–24

López M, Rojas JC, Vandame R, Williams T (2002) Parasitoid-mediated transmission of an iridescent virus. J Invertebr Pathol 80:160–170

Lu L, Zhou SY, Chen C, Weng SP, Chan SM, He JG (2005) Complete genome sequence analysis of an iridovirus isolated from the orange spotted grouper *Epinephelus coiodes*. Virology 339:81–106

Lua DT, Yasuike M, Hirono I, Aoki T (2005) Transcription program of red seabream iridovirus as revealed by DNA microarrays. J Virol 79:15151–15164

Lua DT, Yasuike M, Hirono I, Kondo H, Aoki T (2007) Transcriptional profile of red seabream iridovirus in a fish model as revealed by DNA microarrays. Virus Genes 35:449–461

Mahardika K, Zafran, Yamamoto A, Myazaki T (2004) Susceptibility of juvenile humpback grouper (*Cromileptes altivelis*) to grouper sleepy disease iridovirus (GSDIV). Dis Aquat Org 59:1–9

Majji S, LaPatra S, Long SM, Bryan L, Sample R, Sinning A, Chinchar VG (2006) *Rana catesbeiana* virus Z (RCV-Z): a novel pathogenic ranavirus. Dis Aquat Org 73:1–11

Maniero GD, Morales H, Gantress J, Robert J (2006) Generation of a long-lasting, protective, and neutralizing antibody response to the ranavirus FV3 by the frog *Xenopus*. Dev Comp Immunol 30:649–657

Mao J, Tham TN, Gentry GA, Aubertin A, Chinchar VG (1996) Cloning, sequence analysis, and expression of the major capsid protein of the iridovirus frog virus 3. Virology 216:431–436

Mao JH, Hedrick RP, Chinchar VG (1997) Molecular characterization, sequence analysis, and taxonomic position of newly isolated fish iridoviruses. Virology 229:212–220

Mao J, Green DE, Fellers G, Chinchar VG (1999) Molecular characterization of iridoviruses isolated from sympatric amphibians and fish. Virus Res 63:45–62

Marschang RE, Braun S, Becher P (2005) Isolation of a ranavirus from a gecko (*Uroplatus fimbriatus*). J Zoo Wildlife Med 36:295–300

Marina CF, Arredondo-Jiménez J, Castillo A, Williams T (1999) Sublethal effects of iridovirus disease in a mosquito. Oecologia 119:383–388

Marina CF, Arredondo-Jiménez JI, Ibarra JE, Fernández-Salas I, Williams T (2003a) Effects of an optical brightener and an abrasive on iridescent virus infection and development of *Aedes aegypti*. Entomol Exp Appl 109:155–161

Marina CF, Ibarra JE, Arredondo-Jiménez JI, Fernández-Salas I, Liedo P, Williams T (2003b) Adverse effects of covert iridovirus infection on life history and demographic parameters of *Aedes aegypti*. Entomol Exp Appl 106:53–61

Marina CF, Ibarra JE, Arredondo-Jiménez JI, Fernández-Salas I, Valle J, Williams T (2003c) Sublethal iridovirus disease of the mosquito *Aedes aegypti* is due to viral replication not cytotoxicity. Med Vet Entomol 17:187–194

Marina CF, Fernández-Salas I, Ibarra JE, Arredondo-Jiménez JI, Valle J, Williams T (2005) Transmission dynamics of an iridescent virus in an experimental mosquito population: the role of host density. Ecol Entomol 30:376–382

Martínez G, Christian P, Marina CF, Williams T (2003) Sensitivity of *Invertebrate iridescent virus 6* to organic solvents, detergents, enzymes and temperature treatment. Virus Res 91:249–254

Mathieson WB, Lee PE (1981) Cytology and autoradiography of *Tipula* iridescent virus infection of insect suspension cell cultures, J Ultrastruct Res 74:59–68

McMillan N, Kalmakoff J (1994) RNA transcript mapping of the *Wiseana* iridescent virus genome. Virus Res 32:343–352

McMillan NA, Davison S, Kalmakoff J (1990) Comparison of the genomes of two sympatric iridescent viruses (types 9 and 16). Arch Virol 114:277–284

Mendelson JR, Lips KR, Gagliardo RW, Rabb GB, Collins JP et al (2006) Confronting amphibian declines and extinctions. Science 313:48

Miller DL, Rajeev S, Gray MJ, Baldwin CA (2007) Frog virus 3 infection, cultured American bullfrogs. Emerg Inf Dis 13:343

Miller LK (1996) Insect viruses. In: Fields BN, Knipe DM, Howley PM (eds) Fundamental virology, Lippincott-Raven, Philadelphia, pp 401–424

Moody NJG, Owens L (1994) Experimental demonstration of pathogenicity of a frog virus, bohle iridovirus, for a fish species, barramundi *Lates calcarifer*. Dis Aquat Org 18:95–102

Moore JB, Smith GL (1992) Steroid hormone synthesis by vaccinia virus enzymes: a new type of virus virulence factor. EMBO J 11:1973–1980

Morales HD, Robert J (2007) Characterization of primary and memory CD8 T cell responses against ranavirus (FV3) in *Xenopus laevis*. J Virol 81:2240–2248

Mullens BA, Velten RK, Federici BA (1999) Iridescent virus infection in *Culicoides variipennis sonorensis* and interactions with the mermithid parasite *Heleidomermis magnapapula*. J Invertebr Pathol 73:231–233

Murti KG, Goorha R (1989) Synthesis of FV3 proteins occurs on intermediate filament-bound polyribosomes. Biol Cell 65:205–214

Murti KG, Goorha R (1990) Virus-cytoskeleton interaction during replication of frog virus 3. In: Darai G (ed) Molecular biology of iridoviruses, Kluwer, Boston, pp 137–162

Murti KG, Goorha R, Klymkowsky MW (1988) A functional role for intermediate filaments in the formation of FV3 assembly sites. Virology 162:264–269

Nakajima K, Kunita J (2005) Red sea bream iridoviral disease. Uirusu 55:115–125

Nakajima K, MaenoY (1998) Pathogenicity of red sea bream iridovirus and other fish iridoviruses to red sea bream. Fish Pathol 33:143–144

Nakajima K, Inouye K, Sorimachi M (1998) Viral diseases in cultured marine fish in Japan. Fish Pathol 33:181–188

Nakajima K, Maeno Y, Honda A, Yokoyama K, Tooriyama T, Manabe S (1999) Effectiveness of a vaccine against red sea bream iridovirus disease in a field trial test. Dis Aquat Org 36:73–75

Nakajima K, Ito T, Kurita J, Kawakami H, Itano T, Fukuda Y, Aoi Y, Tooriyama, Manabe S (2002) Effectiveness of a vaccine against red sea bream iridoviral disease in various cultured marine fish under laboratory conditions. Fish Pathol 37:90–91

Nalcacioglu R, Marks H, Vlak JM, Demirbag Z, van Oers MM (2003) Promoter analysis of the *Chilo* iridescent virus DNA polymerase and major capsid protein genes. Virology 317:321–329

Netherton C, Moffet K, Brooks E, Wileman T (2007) A guide to viral inclusions, membrane rearrangements, factories and viroplasms produced during virus replication. Adv. Virus Res. 70: 101–182

Office International des Epizooties (2006) Aquatic Animal Health Code, 9th edn. OIE, Paris

Ohba M (1975) Studies on the pathogenesis of *Chilo* iridescent virus. 3. Multiplication of CIV in the silkworm *Bombyx mori* L. and field insects. Sci Bull Fac Agr Kyushu Univ 30:71–81

Ohba M, Aizawa K (1979) Multiplication of *Chilo* iridescent virus in noninsect arthropods. J Invertebr Pathol 33:278–283

Ohba M, Aizawa K (1982) Failure of *Chilo* iridescent virus to replicate in the frog *Rana limnocharis* (in Japanese). Proc Assoc Plant Protec Kyushu 28:164–166

Oliveros M, Garcia-Escudero R, Alejo A, Vinuela E, Salas ML, Salas J (1999) African swine fever virus dUTPase is a highly specific enzyme required for efficient replication in swine macrophages. J Virol 73:8934–8943

Otto H, Reche PA, Bazan F, Dittmar K, Haa F, Koch-Nolte F (2005) *In silico* characterization of the family of PARP-like poly(ADP-ribosyl) transferases (pARTs). BMC Genomics 6:139

Overholtzer M, Mailleaux AA, Mouneimne G, Normand G, Schnitt SJ, King RW, Cibas ES, Brugge JS (2007) A nonapoptotic cell death process, entosis, that occurs by cell-in-cell invasion. Cell 131:966–979

Pallister J, Goldie S, Coupar B, Hyatt A (2005) Promoter activity in the 5′ flanking regions of the Bohle iridovirus ICP18, ICP46 and major capsid protein genes. Arch Virol 150:1911–1919

Pallister J, Goldie S, Coupar B, Shiell B, Michalski WP, Siddon N, and Hyatt A (2007) Bohle iridovirus as a vector for heterologous gene expression. J Virol Methods 146:419–423

Pallister J, Gould A, Harrison D, Hyatt A, Jancovich J, Heine H (2007) Development of real-time PCR assays for the detection and differentiation of Australian and European ranaviruses. J Fish Dis 30:427–438

Paul ER, Chitnis NS, Henderson CW, Kaul RJ, D'Costa SM, Bilimoria SL (2007) Induction of apoptosis by iridovirus virion protein extract. Arch Virol 152:1353–1364

Paperna I, Vilenkin M, Alves de Matos AP (2001) Iridovirus infections in farm-reared tropical ornamental fish. Dis Aquat Org 48:17–25

Pearman PB, Garner TWJ (2005) Susceptibility of Italian agile frog populations to an emerging strain of *Ranavirus* parallels population genetic diversity. Ecol Lett 8:401–408

Pearman PB, Garner TWJ, Straub M, Greber UF (2004) Response of the Italian agile frog (*Rana latastei*) to a ranavirus, frog virus 3: a model for vial emergence in naïve populations. J Wildlife Dis 40:660–669

Picco AM, Brunner JL, Collins JP (2007) Susceptibility of the endangered California tiger salamander, *Ambystoma californiense*, to ranavirus infection. J Wildlife Dis 43:286–290

Pierce BE, Phillips PH, Jackson G (1991) Redfin virus (EHN) fish disease confirmed in South Australia. SA Fish 15:5–7

Pietrokovski S (1998) Identification of a virus intein and a possible variation in the protein-splicing reaction. Curr Biol 8:R634–R635

Pozet F, Moussa A, Torhy C, de Kinkelin P (1992) Isolation and preliminary characterization of a pathogenic icosahedral deoxyribovirus from the catfish *Ictalurus melas*. Dis Aquat Org 14:35–42

Radloff C, Juhl S, Vaia RA, Brunton J, Ward V, Kalmakoff J, Dokland T, Ha YH, Tomas EL (2005a) Bio-scaffolds for ordered nanostructures and metallodielectric nanoparticles. In: Lai WYC, Pau S, López OD (eds) Nanofabrication: technologies, devices and applications. Proc SPIE 5592:143–152

Radloff C, Vaia RA, Brunton J, Bouwer GT, Ward VK (2005b) Metal nanoshell assembly on a virus bioscaffold. Nano Lett 5:1187–1191

Raghow R, Granoff A (1979) Macromolecular synthesis in cells infected with frog virus 3. X. Inhibition of cell protein synthesis by heat-inactivated frog virus 3. Virology 98:319–327

Raghow R, Granoff A (1983) Cell-free translation of FV3 mRNA: initiation factors from infected cells discriminate between early and late viral mRNAs. J Biol Chem 258:571–578

Reading PC, Moore JB, Smith GL (2003) Steroid hormone synthesis by vaccinia virus suppresses the inflammatory response to infection. J Exp Med 197:1269–1278

Reddacliff LA, Whittington RJ (1996) Pathology of epizootic haematopoietic necrosis virus (EHNV) infection in rainbow trout (*Oncorhynchus mykiss* Walbaum) and redfin perch (*Perca fluviatilis* L). J Compar Pathol 115:103–115

Reyes A, Christian P, Valle J, Williams T (2004) Persistence of *Invertebrate iridescent virus 6* in soil. BioContr 49:433–440

Ricou G (1975) Production de *Tipula paludosa* Meig en prairie en fonction de l'humidité du sol. Rev Ecol Biol Sol 12:69–89

Robert J, Morales H, Buck W, Cohen N, Marr S, Gantress J (2005) Adaptive immunity and histopathology in frog virus 3-infected *Xenopus*. Virology 332:667–675

Rodger H D, Kobs M, Macarthney A, Frerichs GN (1997) Systemic iridovirus infection in freshwater angelfish, *Pterophyllum scalare* (Lichtenstein). J Fish Dis 20:69–72

Rojas S, Richards K, Jancovich JK, Davidson EW (2005) Influence of temperature on ranavirus infection in larval salamanders *Ambystoma tigrinum*. Dis Aquat Org 28:95–100

Sample RC, Bryan L, Long S, Majji S, Hoskins G, Sinning A, Chinchar VG (2007) Inhibition of protein synthesis and virus replication by antisense morpholino oligonucleotides targeted to the major capsid protein, the 18 kDa immediate early protein, and a viral homologue of RNA polymerase II. Virology 358:311–320

Schackelton LA, Parrish CR, Holmes EC (2006) Evolutionary basis of codon usage and nucleotide composition bias in vertebrate viruses. J Mol Evol 62:551–563

Schock DM, Bollinger TK, Chinchar VG, Jancovich JK, Collins JP (2008) Experimental evidence that amphibian ranaviruses are multihost pathogens. Copeia 133–143

Sieburth PJ, Carner GR (1987) Infectivity of an iridescent virus for larvae of *Anticarsia gemmatalis* (Lepidoptera: Noctuidae). J Invertebr Pathol 49:49–53

Sikorowski PP, Tyson GE (1984) *Per os* transmission of iridescent virus of *Helothis zea* (Lepidoptera: Noctuidae). J Invertebr Pathol 44:97–102

Speare R, Smith JR (1992) An iridovirus-like agent isolated from the ornate burrowing frog *Limnodynsastes ornatus* in northern Australia. Dis Aquat Org 14:51–57

Song WJ, Qin QW, Qiu J, Huang CH, Wang F, Hew CL (2004) Functional genomics analysis of Singapore grouper iridovirus: complete sequence determination and proteomic analysis. J Virol 78:12576–12590

Stasiak K, Demattei MV, Federici BA, Bigot Y (2000) Phylogenetic position of the *Diadromus pulchellus* ascovirus DNA polymerase among viruses with large double-stranded DNA genomes. J Gen Virol 81:3059–3072

Stasiak K, Renault S, Demattei MV, Bigot Y, Federici BA (2003) Evidence for the evolution of ascoviruses from iridoviruses. J Gen Virol 84:2999–3009

Stoltz DB (1971) The structure of icosahedral cytoplasmic deoxyriboviruses. J Ultrastruc Res 37:219–239

Stoltz DB (1973) The structure of icosahedral cytoplasmic deoxyriboviruses II. An alternative model. J Ultrastruc Res 43:58–74

Stoltz DB, Hilsenhoff WL, Stich HF (1968) A virus disease of *Chironomus plumosus*. J Invertebr Pathol 12:118–126

Storfer A, Alfaro ME, Ridenhour BJ, Jancovich JK, Mech SG, Parris MJ, Collins JP (2007) Phylogenetic concordance analysis shows an emerging pathogen is novel and endemic. Ecol Lett 10:1075–1083

Sudthongkong C, Miyata M, Miyazaki T (2001) Iridovirus disease in two ornamental tropical freshwater fishes: African lampeye and dwarf gourami. Dis Aquat Org 48:163–173

Sudthongkong C, Miyata M, Miyazaki T (2002) Viral DNA sequences of genes encoding the ATPase and the major capsid protein of tropical iridovirus isolates which are pathogenic to fishes in Japan, South China Sea, and Southeast Asian countries. Arch Virol 147:2089–2109

Sun W, Huang Y, Zhao Z, Gui JF, Zhang Q (2006) Characterization of the *Rana grylio* virus 3β-hydroxysteroid dehydrogenase and its novel role in suppressing virus-induced cytopathic effect. Biochem Biophysic Res Comm 351:44–50

Tan S-L, Katze MG (1999) The emerging role of the interferon-induced PKR protein kinase as an apoptotic effector: A new face of death? J Interferon Cytokine Res 19:543–554

Tan WGH, Barkman TJ, Chinchar VG, Essani K (2004) Comparative genomic analysis of frog virus 3, type species of the genus *Ranavirus* (family *Iridoviridae*). Virology 323:70–84

Tanaka H, Sato K, Saito Y, Yamashita T, Agoh M, Okunishi J, Tachikawa E, Suzuki K (2003) Insect diapause-specific peptide from the leaf beetle has consensus with a putative iridovirus peptide. Peptides 24:1327–1333

Tang KFJ, Redman RM, Pantoja CR, LeGroumellec M, Duraisamy P, Lightner DV (2007) Identification of an iridovirus in *Acetes erythraeus* (Sergestidae) and the development of *in situ* hybridization and PCR methods for its detection. J Invert Pathol 96:255–260

Tapiovaara H, Olesen NJ, Linden J, Rimaila-Parnanen E, von Bonsdorff CH (1998) Isolation of an iridovirus from pike-perch *Stizostedion lucioperca*. Dis Aquat Org 32:185–193

Tidona CA, Darai G (1997) The complete DNA sequence of lymphocystis disease virus. Virology 230:207–216

Tonka T, Weiser J (2000) Iridovirus infection in mayfly larvae. J Invertebr Pathol 76:229–231

Tortorella D, Gewurz BE, Furman MH, Schust DJ, Ploegh HL (2000) Viral subversion of the immune system. Annu Rev Immunol 18: 861–926

Tsai CT, Lin CH, Chang CY (2007) Analysis of codon usage bias and base compositional constraints in iridovirus genomes. Virus Res 126:196–206

Tweedel K, Granoff A (1968) Viruses and renal carcinoma of *Rana pipiens*. V. Effect of frog virus 3 on developing frog embryos and larvae. J Natl Cancer Inst 40:407–409

Undeen AH, Fukuda T (1994) Effects of host resistance and injury on the susceptibility of *Aedes taeniorhynchus* to mosquito iridescent virus. J Am Mosq Contr Assoc 10:64–66

Vetten HJ, Haenni A-L (2006) Taxon-specific suffixes for vernacular names. Arch Virol 151:1249–1250

Wang YQ, Lu L, Weng SP, Huang JN, Chan SM, He JG (2007) Molecular epidemiology and phylogenetic analysis of a marine fish infectious spleen and kidney necrosis virus-like (ISKNV-like) virus. Arch Virol 152:763–773

Ward VK, Kalmakoff J (1991) Invertebrate Iridoviridae. In: Kurstak E (ed) Viruses of invertebrates, Marcel Dekker, New York, pp 197–226

Watson LR, Groff JM, Hedrick RP (1998) Replication and pathogenesis of white sturgeon irido-virus (WSIV) in experimentally infected white sturgeon *Acipenser transmontanus* juveniles and sturgeon cell lines. Dis Aquat Org 32:173–184

Webby R, Kalmakoff J (1998) Sequence comparison of the major capsid protein gene from 18 diverse iridoviruses. Arch Virol 143:1949–1966

Weissenberg R (1965) Fifty years of research on the lymphocystis virus disease of fishes (1914–1964). Ann N Y Acad Sci 126:362–374

Werts C, Girardin SE, Philpott DJ (2006) TIR, CARD, and PYRIN: three domains for an antimicrobial triad. Cell Death Differen 13:798–815

Whittington RJ, Reddacliff GL (1995) Influence of environmental temperature on experi-mental infection of redfin perch (*Perca fluviatilis*) and rainbow trout (*Oncorhynchus mykiss*) with epizootic haematopoietic necrosis virus, an Australian iridovirus. Aust Vet J 11:421–424

Whittington RJ, Philbey A, Reddacliff GL, Macgowan AR (1994) Epidemiology of epizootic haematopoietic necrosis virus (EHNV) infection in farmed rainbow trout, *Oncorhynchus mykiss* (Walbaum): findings based on virus isolation, antigen capture ELISA and serology. J Fish Dis 17:205–218

Whittington RJ, Kearns C, Hyatt AD, Hengstberger S, Rutzou T (1996) Spread of epizootic haematopoietic necrosis virus (EHNV) in redfin perch (*Perca fluviatilis*) in southern Australia. Aust Vet J 73:112–114

Wileman T (2006) Aggresomes and autophagy generate sites for virus replication. Science 312:875–878

Williams BRG (1999) PKR: a sentinel kinase for cellular stress. Oncogene 18:6112–6120

Williams T (1993) Covert iridovirus infection of blackfly larvae. Proc R Soc B 251:225–230

Williams T (1996) The iridoviruses. Adv Virus Res 46:347–412

Williams T (1998) Invertebrate iridescent viruses. In: Miller LK, Ball LA (eds) The insect viruses, Plenum, New York, pp 31–68

Williams T, Cory JS (1994) Proposals for a new classification of iridescent viruses. J Gen Virol 75:1291–1301

Williams T, Hernández O (2006) Costs of cannibalism in the presence of an iridovirus pathogen of *Spodoptera frugiperda.* Ecol Entomol 31:106–113

Williams T, Barbosa-Solomieu V, Chinchar VG (2005) A decade of advances in iridovirus research. Adv Virus Res 65:173–248

Willis DB, Granoff A (1978) Macromolecular synthesis in cells infected by frog virus 3. IX. Two temporal classes of early viral RNA. Virology 86:443–453

Willis DB, Granoff A (1980) Frog virus 3 DNA is heavily methylated at CpG sequences. Virology 107:250–257

Willis DB, Granoff A (1985) Trans-activation of an immediate-early frog virus 3 promoter by a virion protein. J Virol 56:495–501

Willis DB, Goorha R, Granoff A (1984) DNA methyltransferase induced by frog virus 3. J Virol 49:86–91

Willis DB, Goorha R, Chinchar VG (1985) Macromolecular synthesis in cells infected by frog virus 3. Curr Topics Microbiol Immunol 116:77–106

Willis DB, Thompson JP, Essani K, Goorha R (1989) Transcription of methylated viral DNA by eukaryotic RNA polymerase II. Cell Biophys 15:97–111

Willis DB, Essani K, Goorha R, Thompson JP, Granoff A (1990a) Transcription of a methylated DNA virus. Nucleic Acid Methylation. UCLA Symp Mol Cell Biol 128:139–151

Willis DB, Thompson JP, Beckman W (1990b) Transcription of frog virus 3. In: Darai G (ed) Molecular biology of iridoviruses, Kluwer, Boston, pp 173–186

Wolf K (1988) Fish viruses and fish viral diseases. Cornell University Press, Ithaca.

Wrigley NG (1969) An electron microscope study of the structure of *Sericesthis* iridescent virus. J Gen Virol 5:123–134

Wrigley NG (1970) An electron microscope study of the structure of *Tipula* iridescent virus. J Gen Virol 6:169–173

Xie J, Lu L, Deng M, Weng S, Zhu J, Wu Y, Gan L, Chan S-M, He J (2005) Inhibition of reporter gene and Iridovirus-tiger frog virus in fish cell by RNA interference. Virology 338:43–52

Yan X, Olson NH, Van Etten JL, Bergoin M, Rossmann MG, Baker TS (2000) Structure and assembly of large lipid-containing dsDNA viruses. Nature Struct Mol Biol 7:101–103

Younghusband HB, Lee PE (1969) Virus-cell studies of *Tipula* iridescent virus in *Galleria mellonella* (L.). I. Electron microscopy of infection and synthesis of *Tipula* iridescent virus in hemocytes. Virology 38:247–254

Zemskov EA, Kang W, Maeda S (2000) Evidence for nucleic acid binding ability and nucleosome association of *Bombyx mori* nucleopolyhedrovirus BRO proteins. J Virol 74:6784–6789

Zhang QY, Xiao F, Li ZQ, Gui JF, Mao J, Chinchar VG (2001) Characterization of an iridovirus from the cultured pig frog (*Rana grylio*) with lethal syndrome. Dis Aquat Org 48:27–36

Zhang QY, Xiao F, Xie J, Li ZQ, Gui JF (2004) Complete genome sequence of lymphocystis disease virus isolated from China. J Virol 78:6982–6994

Zhang QY, Zhao Z, Xiao F, Li ZQ, Gui JF (2006) Molecular characterization of three *Rana grylio* virus (RGV) isolates and *Paralichthys olivaceus* lymphocystis disease virus (LCDV-C) in iridoviruses. Aquaculture 251:1–10

Zhang Y, Maley F, Maley GF, Duncan G, Dunigan DD, Van Etten JL (2007) Chloroviruses encode a bifunctional dCMP-dCTP deaminase that produces two key intermediates in dTTP formation. J Virol 81:7662–7671

Zhao Z, Ke F, Gui JF, Zhang Q (2007) Characterization of an early gene encoding for dUTPase in *Rana grylio* virus. Virus Res 123:128–137

Zheng FR, Sun XQ, Liu HZ, Zhang JX (2006) Study on the distribution and expression of a DNA vaccine against lymphocystis disease virus in Japanese flounder (*Paralichthys olivaceus*). Aquaculture 261:1128–1134

Zupanovic Z, Lopez G, Hyatt A, Shiell BJ, Robinson AJ (1998a) An improved enzyme linked immunosorbent assay for detection of anti-ranavirus antibodies in the serum of the giant toad (*Bufo marinus*). Dev Comp Immunol 22:573–585

Zupanovic Z, Musso C, Lopez C, Louriero CL, Hyatt AD, Hengstberger S, Robinson AJ (1998b) Isolation and characterization of iridoviruses from the giant toad *Bufo marinus* in Venezuela. Dis Aquat Org 33:1–9

Zwillenberg LO, Wolf K (1968) Ultrastructure of lymphocystis virus. J Virol 2:393–399

# Ascoviruses: Superb Manipulators of Apoptosis for Viral Replication and Transmission

B.A. Federici (✉), D.K. Bideshi, Y. Tan, T. Spears, Y. Bigot

## Contents

**Abstract** Ascoviruses are members of a recently described new family (*Ascoviridae*) of large double-stranded DNA viruses that attack immature stages of insects belonging to the order Lepidoptera, in which they cause a chronic, fatal disease. Ascoviruses have several unusual characteristics not found among other viruses, the most novel of which are their transmission by endoparasitic wasps and a unique cytopathology that resembles apoptosis. Cell infection induces apoptosis and in some species is associated with synthesis of a virus-encoded executioner caspase and several lipid-metabolizing enzymes. Rather than leading directly to cell death, synthesis of viral proteins results in the rescue of developing apoptotic bodies that are converted into large vesicles in which virions accumulate and continue to assemble. In infected larvae, millions of these virion-containing vesicles begin

B.A. Federici
Department of Entomology and Interdepartmental Programs in Microbiology, Genetics, and Molecular Biology, University of California, Riverside, Riverside, CA 92507, USA
brian.federici@ucr.edu

James L. Van Etten (ed.) *Lesser Known Large dsDNA Viruses.*
Current Topics in Microbiology and Immunology 328.
© Springer-Verlag Berlin Heidelberg 2009

to disperse from infected tissues 48–72 h after infection into the blood, making it milky white, a major characteristic of the disease. Circulation of virions and vesicles in the blood facilitates mechanical transmission by parasitic wasps. Although ascoviruses appear to be very common, only five species are currently recognized, with the type species being the *Spodoptera frugiperda ascovirus 1a*. Ascovirus virions are large, enveloped, typically bacilliform or reniform in shape, and, depending on the species, have genomes that range from 119 to 186 kbp. Molecular phylogenetic evidence indicates that ascoviruses evolved from iridoviruses (family *Iridoviridae*) that attack lepidopteran larvae and are likely the evolutionary source of ichnoviruses (family *Polydnaviridae*), which assist endoparasitic hymenopterans in overcoming the defense responses of their insect hosts. Thus, as other molecular evidence suggests that iridoviruses evolved from phycodnaviruses (family *Phycodnaviridae*), an evolutionary pathway is apparent from phycodnaviruses via iridoviruses and ascoviruses to ichnoviruses.

# Introduction

The family *Ascoviridae* is one of the newest families of viruses, erected in 2000 to accommodate several species of a newly recognized type of DNA virus that attacks larvae and pupae of insects belonging to the order Lepidoptera (Federici et al. 2005). Viruses of this family produce large, enveloped virions, measuring 130 nm in diameter by 300–400 nm in length, and when viewed by electron microscopy have a reticulated surface appearance. Virions are typically bacilliform or reniform in shape and contain a circular double-stranded DNA genome that, depending on the species, ranges from approximately 130 to 186 kbp. Whereas ascovirus virions are structurally complex, like those of other large DNA viruses that attack insects, such as those of iridoviruses (family *Iridoviridae*; Chinchar et al. 2005) and entomopoxviruses (family *Poxiviridae*; Buller et al. 2005), they differ from these in two significant aspects. First, the primary mechanism by which ascoviruses are transmitted from host to host is through mechanical transfer by female endoparasitic wasps when these lay eggs in their hosts. Second, ascoviruses have a unique cell biology and cytopathology in which shortly after infecting a cell, they induce apoptosis and then rescue the developing apoptotic bodies, converting these into virion-containing vesicles in which virions continue to assemble after the vesicles cleave from host cells. This aspect of viral reproduction apparently evolved to disseminate virions to the larval blood where they could contaminate the ovipositors of female wasps so that the virions and virion-containing vesicles could be transmitted to new hosts. Ascoviruses appear to occur worldwide, wherever there are endoparasitic wasps and larvae of lepidopteran species parasitized by these. The lepidopteran species attacked, as far as is known, are primarily those belonging to the family *Noctuidae*. The reason that ascoviruses were not discovered until the 1970s is that the disease they cause is chronic, with virtually no gross pathology apparent under field conditions to the untrained eye. Even today, only a handful of investigators are capable of recognizing the signs of disease in larval populations in the field, and thus only five ascovirus species have been recognized to date.

## Discovery and Geographic Distribution

The first ascoviruses were discovered during the late 1970s in southern California where they were found causing disease in larvae of moths belonging to the lepidopteran family *Noctuidae* (Federici et al. 1991). Diseased larvae were recognized by the presence of blood that was very white and opaque, in marked contrast to the blood of healthy larvae, which is translucent and slightly green (see Sect. 4). The color and opacity of the blood in diseased larvae is due to the presence of high concentrations of vesicles that contain virions. The white blood and virion-containing vesicles are diagnostic for the disease. Thus, the name for this group – ascoviruses – derived from the Greek *asco* meaning sac – was chosen on the basis of the latter characteristic. Since the discovery of the first ascovirus, they have been isolated as the cause of disease in larvae of many species of noctuids. In addition, an ascovirus that attacks the pupal stage of a lepidopteran species belonging to the family *Yponomeutidae* was discovered in the 1990s in France (Bigot et al. 1997a, 1997b).

Because ascoviruses have been discovered relatively recently, little is known about their distribution and taxonomy. Based on collections of infected larvae, ascoviruses are known to occur the United States, Europe (France), Australia, and Indonesia, and there is little doubt that they occur worldwide (Table 1). This is because their most common hosts, larvae of lepidopteran species belonging to the family *Noctuidae*, the largest family of the order Lepidoptera, as well as their most common known vectors, endoparasitic wasps of the families *Braconidae* and *Ichnuemonidae*, are distributed throughout the world. Although only a few ascovirus species have been described to date, there are probably many, and certainly many variants, that occur throughout the world. Thus, given the common occurrence of their hosts and vectors, it is likely that ascoviruses are very common insect viruses, perhaps the most common type

**Table 1** Lepidopteran hosts for ascoviruses and geographic origin

| Lepidopteran host species | Year isolated | Location |
|---|---|---|
| *Scotogramma trifolli* | 1977 | California |
| *Heliocoverpa zea* | 1977 | Mississippi |
| *Trichoplusia ni* | 1979 | California |
| *Heliothis virescens* | 1981 | South Carolina |
| *Spodoptera frugiperda* | 1982 | Florida |
| *Autographa precationis* | 1986 | Georgia |
| *Acrolepiosis assectella* | 1997 | France |
| *Spodoptera exigua* | 2000 | Indonesia |
| *Helicoverpa armigera* | 2003 | Australia |
| *Helicoverpa punctigera* | 2003 | Australia |

that occurs routinely. That they have not been discovered in large numbers, for example, like baculoviruses (family *Baculoviridae*; Theilmann et al. 2005), is, as noted above, because they exhibit few easily detectable signs of disease, making it difficult for individuals not familiar with the disease to recognize infected larvae in field populations.

## Recognized Species and Variants

Numerous ascovirus isolates, probably well over 100, have been collected over the past 20 years from these lepidopteran species (Federici et al. 1991; Hamm et al. 1998; Cheng et al. 2000; Newton 2003; Cheng et al. 2005). Analysis of the genomes of these isolates using restriction enzyme analysis has shown that many variants are present in each of these isolates. Nevertheless, these isolates can be grouped among five ascovirus species, even though the DNA of some isolates show a low level of cross-hybridization. Thus, at present, the five ascovirus species officially recognized based on a combination of properties including the relatedness of key genes such as the DNA polymerase and major capsid protein, the degree to which their genomic DNAs cross-hybridize in reciprocal Southern blot hybridization analyses, their lepidopteran host range, and their tissue tropism.

The type species is the *Spodoptera frugiperda ascovirus* (SfAV 1a), with the other species being *Tricoplusia ni ascovirus* (TnAV 2a), *Heliothis virescens ascovirus* (HvAV 3a), *Diadromus pulchellus ascovirus* (DpAV 4a), and *Spodoptera exigua ascovirus* (SeAV 5a). The Arabic numeral reflects the order in which each species was formally recognized, whereas the lower case letter indicates species variants. Variants from each type species are recognized by different consecutive lower case letters, e.g., TnAV 2b and 2c, would represent two different variants of TnAV 2a recognized subsequently. Here, we usually use acronyms without the numerical and lower case suffix, for example SfAV to refer to these viral species. An example of restriction enzyme patterns for several of these species are shown in Fig. 1, and their host range and tissue tropism is discussed in Sect. 4.

Ascoviruses have been isolated from many more insect species than those listed in Table 1, but most of these have turned out to be variants of known ascovirus species, and therefore they have not been named after the host from which they were isolated. For example, ascoviruses related to TnAV, HvAV, and SeAV have been isolated from noctuid species such as *Autographa precationis*, *Helicoverpa zea*, *H. armigera*, and *H. punctigera*. What this implies is that ascoviruses belonging to the species TnAV, HvAV, and SeAV have a broad and overlapping host range among different noctuid species, although this has only been tested experimentally to a limited extent. Nevertheless, because restriction enzyme analyses of ascovirus genomes indicate that numerous variants exist in most ascovirus infected larvae, and only very few of these have been plaque-purified and studied individually, it is likely that additional new species occur among these and that even more occur in nature. While the above species are currently recognized as valid species, the definition of an ascovirus species and the procedures for identification and differentiation of species and variants is currently

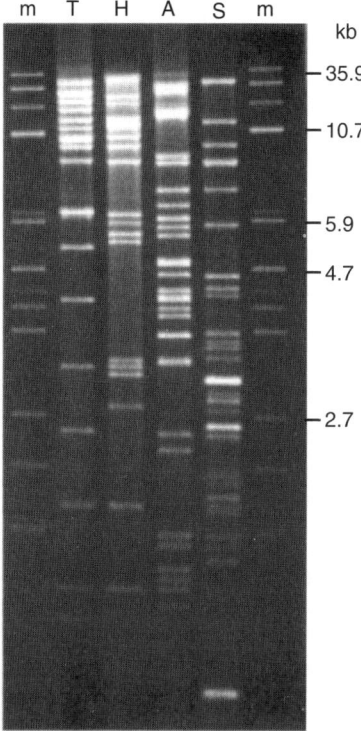

**Fig. 1** Restriction fragment (*Hin*dIII) patterns of ascovirus genomes from several ascovirus species. *T* is TnAV2, *H* is HvAv3, *A* is a variant of TnAV from *Autographa precationis*, and *S* is SfAV1a. m is a DNA ladder of size standards

under revision. In the near future, it is probable that from three to five conserved genes will be identified for taxonomic purposes, genes such as those encoding the major capsid protein and DNA polymerase, and that new species will be based on differences between a new species and existing ones of greater than 10% at the nucleotide level, as is done with papillomaviruses (de Villiers et al. 2004). For our purposes here, we refer to the ascoviruses studied using accepted and tentative species and their variants as reported in Federici et al. (2005).

## Ascovirus Disease

### *Signs of Disease and Gross Pathology*

The physical signs of ascovirus disease in infected lepidopteran larvae are very subtle. The most obvious sign of disease within 24 h of infection is a decrease in the normal rate of feeding. The feeding rate continues to slow as the disease progresses, and as a result, larvae gain little if any weight and fail to progress in development, especially in comparison to healthy larvae (Govindarajan and Federici 1990; Cheng et al. 2000; Hamm et al. 1998). Healthy larvae, particularly in the early stages of development, will easily quadruple their weight and size in a period of 3–4 days, whereas ascovirus-infected larvae cease to grow and may actually lose weight (Fig. 2). This feature of ascovirus disease is almost impossible to detect in infected larvae in the field, where mixtures of different instars typically occur in larval populations. However, it is very obvious under laboratory conditions when infected and healthy larvae are reared side by side over a period of a few days. A second feature easily noted in the laboratory is that ascovirus diseases are chronic, though usually fatal. When infected during early stages of development, ascovirus-infected larvae often survive for 2–3 weeks beyond the time at which most healthy larvae have completed their development and pupated (Govindarajan and Federici 1990). Signs of disease other than these are minor, but include the inability to completely cast the molted cuticle, a bloated thoracic region, and a white or creamy discoloration and hypertrophied appearance of the larval body at

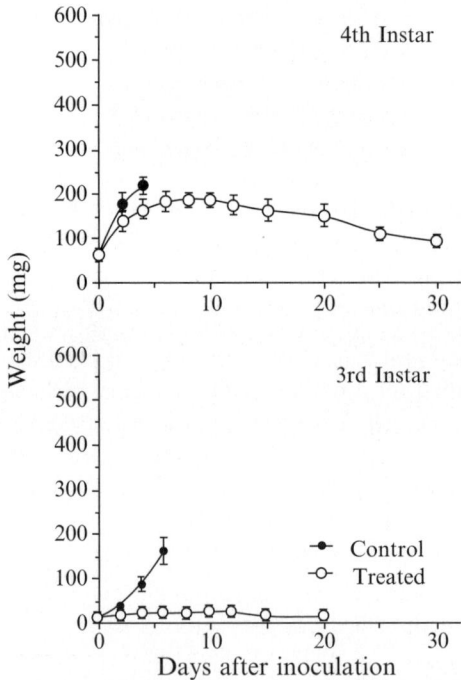

**Fig. 2** Effect of ascovirus infection on larval development. Third or fourth instars of the cabbage looper, *Trichoplusia ni* were inoculated with ten virion-containing vesicles 12 h prior to molting to the instar indicated and then reared individually. The endpoint in the control larvae indicates the time at which the larvae pupated. Note that in the treated larvae, growth, as indicated by weight gain, was severely retarded, primarily due to a cessation of feeding that began shortly after infection, especially in early instars

advanced stages of disease development, which is most apparent when viewing the ventral surface of larvae and the internal tissues through the intersegmental membranes (Federici and Govindarajan 1990; Cheng et al. 2000; Newton 2003).

Although these signs of disease may be difficult to detect by those who have no experience with ascovirus disease, it is probably the easiest of any of the viral diseases of insects to diagnose definitively. All one needs to do is examine a drop of blood from a caterpillar with the naked eye. If the blood is milky white in color, as no other known disease of caterpillars exhibits this characteristic, it is virtually certain the disease is caused by an ascovirus (Fig. 3). As described in the next section, this milky color is caused by the accumulation of virion-containing vesicles in the blood.

## Cytopathology and Formation of Virion-Containing Vesicles

In comparison to all other known viruses, the most unique property of ascoviruses is the unusual cytopathology that leads to the formation of the virion-containing vesicles. This process resembles apoptosis, and recent studies of the SfAV genome have shown that it encodes an executioner caspase, synthesized 9 h after infection, which by itself is capable of inducing apoptosis (Bideshi et al. 2005). However, the precise role of this caspase in ascovirus cytopathology has not been determined.

At the cellular level in vivo, the disease begins with extraordinary hypertrophy of the nucleus accompanied by invagination of sections of the nuclear envelope, followed by a corresponding enlargement of the cell (Fig. 4). Cells will typically grow from five to ten times the diameter of healthy cells. As the nucleus enlarges, the nuclear envelope ruptures and disintegrates into fragments (Federici 1983). At about this stage, the cell plasmalemma begins to invaginate along planes toward the now anucleate cell center. These planes are composed of bilinear arrays of mitochondria, as seen in cross-sections of infected cells by electron microscopy, between which small vesicles of lipid membrane are apparently synthesized de novo (Fig. 5). As this process continues, the small vesicles of lipid membrane coalesce, joining together and with the invaginating plasmalemma, thereby cleaving the cell into a cluster of from 20 to more than 30 vesicles. These range in size from 5 to 10 mm in diameter. This aspect of ascovirus cellular pathology resembles the formation of apoptotic bodies during apoptosis. However, rather than dissipate as the cell dies, the developing apoptotic bodies are rescued by the virus and progress to form vesicles in which virions continue to assemble. These virion-containing vesicles, also referred to as viral vesicles, typically remain in the tissue until the basement membrane ruptures, though on occasion cell hypertrophy can be so great that the enlarging cell erupts out through the basement membrane of the infected tissue, releasing large fragments of the infected cell directly into the blood. Analysis of the SfAV, TnAV, and HvAV genomes show that, unlike many other large DNA viruses, ascoviruses encode several lipid-metabolizing enzymes that are likely involved in the process of converting developing apoptotic bodies into virion-containing vesicles.

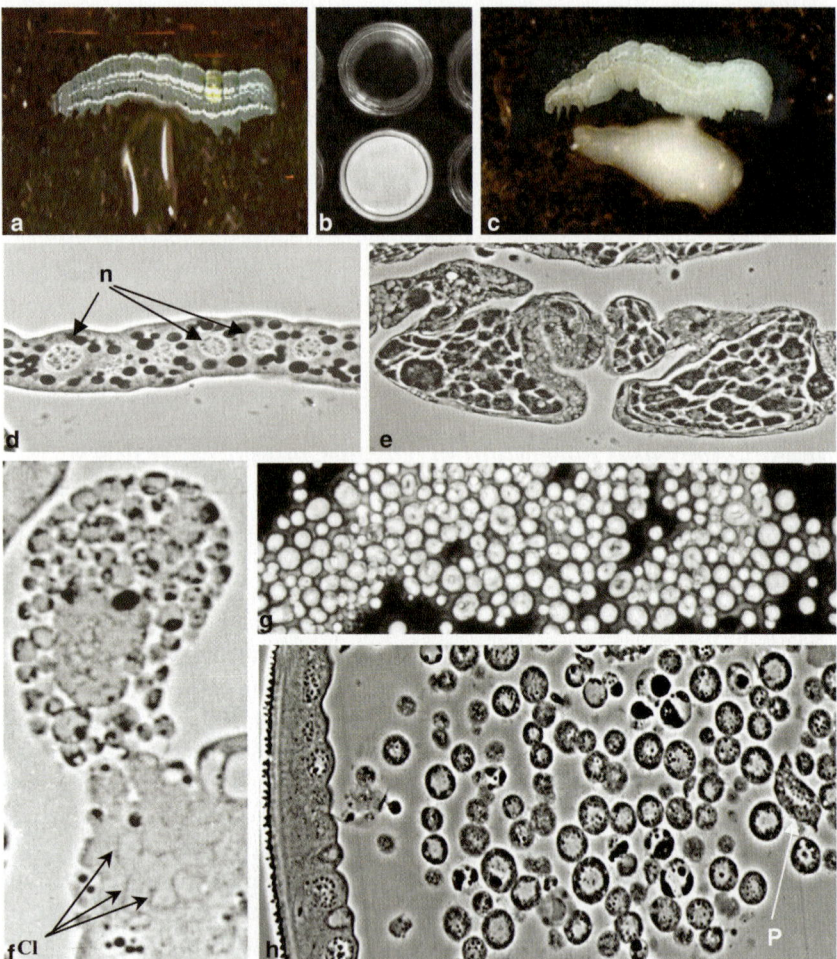

**Fig. 3a–h** Typical pathology associated with ascovirus infections. **a** Healthy larva of the cabbage looper, *Trichoplusia ni*. The clear pool of blood below the larva is normal hemolymph. **b** Comparison of the appearance of healthy hemolymph (*upper well*) with hemolymph from an ascovirus-infected larva (*lower well*) at 7 days postinfection. **c** Ascovirus-infected *T. ni* larvae showing typical appearance of infected hemolymph. The dense opacity is due to the accumulation of virion-containing vesicles in the hemolymph. **d** and **e**, respectively, sections through healthy and infected fat body tissue of a larva. Note the extensive hypertrophy of the ascovirus-infected cells in **e**. **f** Section through a lobe of fat body in which a greatly hypertrophied cell is cleaving into viral vesicles. **g** and **h**, respectively, wet mount of ascovirus-infected hemolymph shown in **b** and **c**, and the appearance of blood and spherical viral vesicles as observed is plastic sections. *n* nuclei, *Cl* cleavage planes throughout a cell, *P* plasmatocyte, a type of insect blood cell

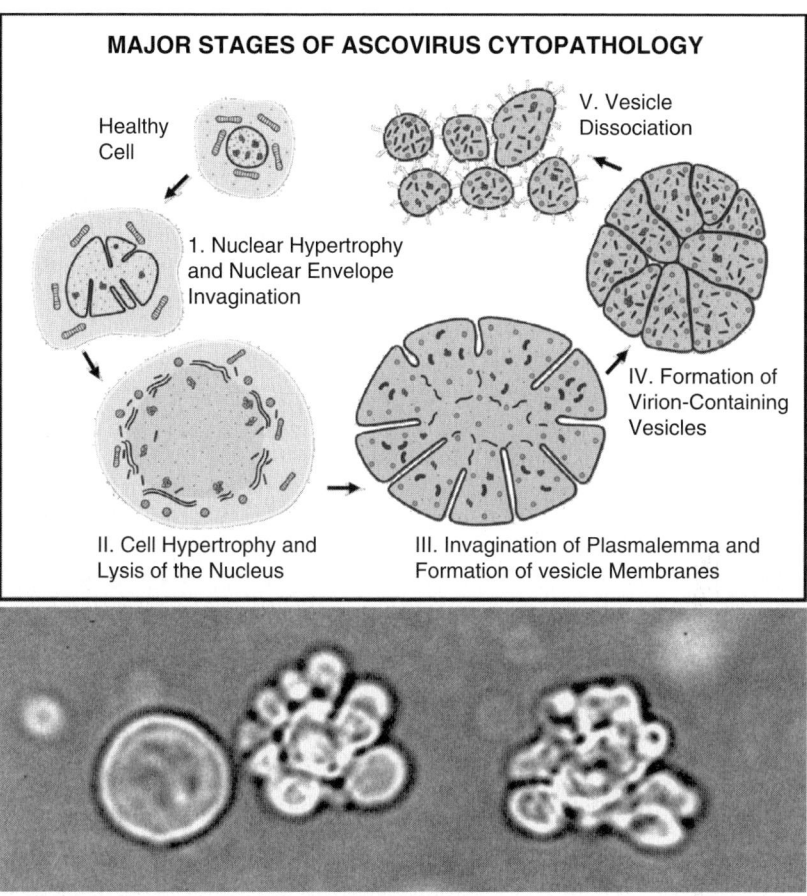

**Fig. 4** Ascovirus pathogenesis at the cellular level. *Top panel* Schematic illustration typical of ascovirus cytopathology. *Bottom panel* Vesicle formation in vitro. Phase-contrast micrograph showing an uninfected cell on the *left* and two infected Sf21 cells cleaving into vesicles approximately 12 h after infection with the *Spodoptera frugiperda ascovirus* (SfAV1a). Sf21 is a cell line derived from *Spodoptera frugiperda*. Ascovirus cytopathology resembles apoptosis, but the virus rescues developing apoptotic bodies and converts them into virion-containing vesicles for reproduction and transmission

This general pattern of vesicle formation occurs in vivo in all ascoviruses. However, variations do occur among different species, although these have received little study. For example, in some ascovirus species, such as SfAV and TnAV, the planes of mitochondria and developing small lipid membrane vesicles can extend from the periphery of the hypertrophied cell almost to the center. In other species, the planes first form near the cell periphery and cleave inward, progressively forming vesicles from the exterior to the interior regions of the cell. It may also be possible that the process varies depending on the developmental stage of the larvae at the time of infection, e.g., early instar versus late instar larvae. Regardless of the

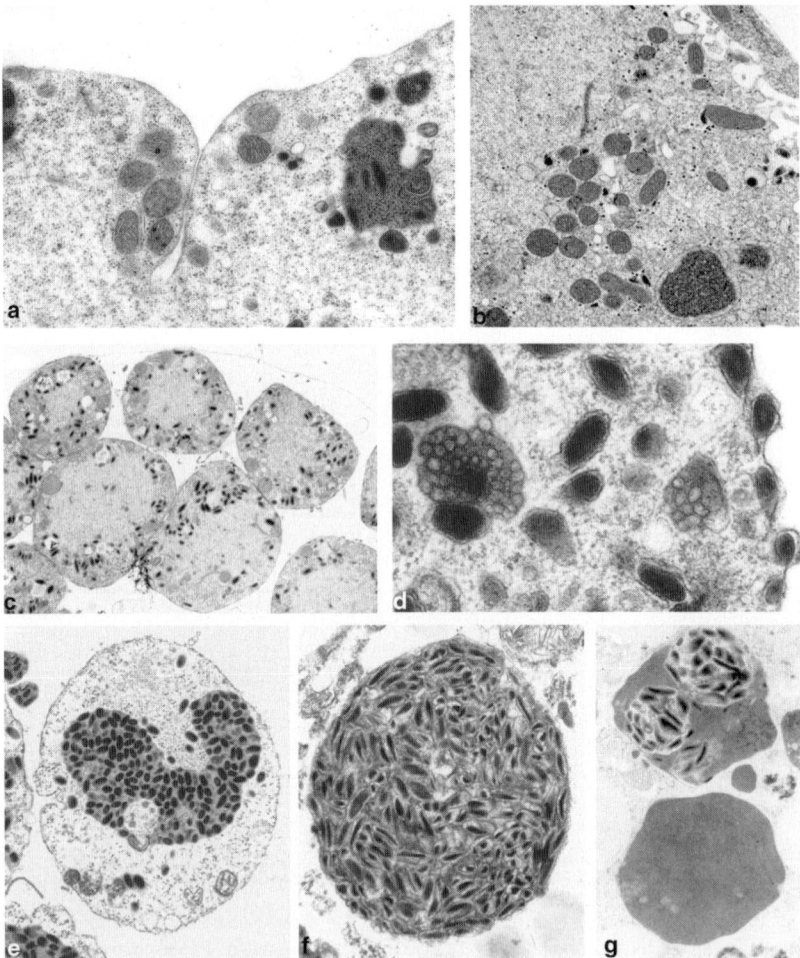

**Fig. 5a–g** Ultrastructural of ascovirus vesicle formation. **a** Initiation of plasmalemma invagination induced by ascovirus formation. **b** Small lipid vesicles developing along a plane in a hypertrophied cell. These small vesicles subsequently coalesce together and with the invaginating plasmalemma to form the outer membrane that delimits ascovirus virion-containing vesicles. In **a** and **b**, note the clusters of mitochondria near the developing lipid membranes. **c** Virion-containing vesicles that have cleaved from fat body cells in vivo but are still contained within fat body tissue. **d** Section through a portion of a SfAV virion-containing vesicles showing developing virions, occlusion of virions in the vesicular matrix characteristic of this ascovirus species, and virions at the vesicle periphery, apparently budding from the vesicle. **e–g** different types of mature virion-containing vesicles observed in ascovirus-infected insects. **e, f** SfAV and HvAV vesicles, respectively, in the blood of their hosts, *S. frugiperda* and *H. virescens*. **g** Virions of TnAV in a protein granule in the fat body of a late fifth-instar larva of its host, *Trichoplusia ni*

species or stage of development, all ascoviruses completely cleave infected cells into virion-containing vesicles. In vitro, i.e., in cell lines derived from *S. frugiperda*, *T. ni*, and *H. virescens*, essentially the same process occurs. However, compared to pathogenesis in vivo, relatively few vesicles are formed per cell, and the number of virions assembled per cell or vesicle is low. Thus, the cell lines can be used to study ascovirus cell biology, and these are useful because cell infection can be synchronized, but the in vitro systems currently available are not nearly as robust as in vivo infections.

Although the process by which viral vesicles are cleaved from cells can vary among different ascovirus species, the histopathology is similar among virtually all viruses. Vesicles accumulate in the tissues where they are formed, but as these tissues degenerate during disease progression, the basement membrane of infected tissues deteriorates and ruptures, allowing the vesicles to spill out into the blood. There they accumulate, reaching concentrations as high as $10^{7-8}$ vesicles/ml within 3–4 days of infection. There is some evidence that viral replication proceeds within the vesicles as they circulate in the blood, and therefore this tissue must also be considered one of the tissues attacked by ascoviruses. If fact, because such high concentrations of viral vesicles are found in the blood, this tissue could be considered a major site of infection, particularly if it is eventually shown that these viruses continue to replicate in the vesicles as they circulate in the blood.

Despite the chronic nature of the disease caused by ascoviruses, virion-containing vesicles are present in the blood within 2–3 days of infection. When the virus replicates in cells in vitro, the vesicles are formed within 12–16 h of infection (Bideshi et al. 2005). The rapid development and circulation of the viral vesicles in the blood probably evolved to enhance transmission of the virus by parasitic wasps.

## *Host Range and Tissue Tropism*

The cytopathology of ascoviruses is consistent among different viral species; however, considerable variation occurs with respect to the tissues attacked, i.e., in which replication occurs. TnAV, HvAV, and SeAV exhibit a relatively broad tissue tropism infecting the tracheal matrix, epidermis, fat body, and connective tissue. Differences exist between these species in that some HvAV variants infect the epidermis much more extensively than TnAV variants, whereas some of the latter can also replicate more extensively in fat body cells, but appear only to do this when larvae are infected early in their development. Alternatively, the type species, SfAV, and its variants have a very narrow tissue tropism, with the fat body being the primary site of infection (Federici and Govindarajan 1990; Hamm et al. 1998; Newton 2003; Cheng et al. 2005). DpAV occurs in the nuclei of all tissues of its wasp host, but appears to only produce progeny in ovarial tissues. In its lepidopteran pupal host, it attacks and replicates in most tissues (Bigot et al. 1997a, 1997b).

As noted above, the experimental host range of ascoviruses varies with the viral species (Table 2). TnAV, HvAV, and SeAV have a broad host range and are capable

**Table 2** Ascovirus species, their host range, and tissue tropism

| Ascovirus species | Host range[a] | Tissue tropism[b] |
| --- | --- | --- |
| *Spodoptera fugiperda ascovirus* (SfAV) | *Spodoptera* spp. | Fat body |
| *Trichoplusia ni ascovirus* (TnAV) | Noctuid spp. | Fat body, epidermis |
| *Heliothis virescens ascovirus* (HvAV) | Noctuid spp | Fat body, epidermis |
| *Diadromus pulchellus ascovirus* (DpAV) | *Acrolepiosis* spp. | Most pupal tissues |
| *Spodoptera exigua ascovirus* (SeAV) | Noctuid spp. | Fat body, epidermis |

[a] SfAV will infect other species of the genus *Spodoptera*, but not most other species of the family *Noctuidae*. TnAV, HvAV, and SeAV experimentally are capable of infecting a wide range of noctuid species, and even some non-noctuid lepidopterans

[b] Refers to the dominant tissue(s) infected, which produce the highest yields of virion-containing vesicles. The tracheal matrix is also heavily infected by TnAV, HvAV, and SeAV

of replication in a variety of noctuid species, as well as in selected species belonging to other families of the order Lepidoptera. Alternatively, the experimental host range of SfAV is limited to other species of the genus *Spodoptera*. DpAV can replicate in hymenopteran and lepidopteran hosts closely related to its natural host species, *A. assectella*. To propagate virus in the laboratory, all ascoviruses can be grown in their larval or pupal hosts. To infect caterpillars, they are injected with virus in the fourth or early fifth instar, and virion-containing vesicles are harvested from the blood 5–7 days later.

## Virion Structure, Assembly, and Composition

Depending on the species, the virions of ascoviruses are either bacilliform or reniform in shape, with complex symmetry, and very large, measuring about 130 nm in diameter by 300–400 nm in length. The virion consists of an inner particle surrounded by an outer envelope (Fig. 6). The inner particle is complex, containing a DNA/protein core as well as an apparent internal lipid bilayer surrounded by a distinctive layer of protein subunits. Thus, the virion appears to contain two lipid membranes, one associated with the inner particle, the other forming the lipid component of the envelope. In negatively stained preparations, virions have a distinctive reticulate appearance, which is thought to be due to superimposition of subunits on the surface of the internal particle with those in the envelope (Fig. 6).

Although there have been few biochemical studies of viral DNA replication or protein synthesis, studies carried out with ascoviruses in vivo and in vitro show that progeny virions first appear about 12 h after infection. Virion assembly is initiated after the nucleus ruptures and occurs prior to and during the cleavage of the cell into viral vesicles. The first recognizable structural component of the virion to form is the multilaminar layer of the inner particle (Fig. 6). Based on its ultrastructure, this layer consists of a unit membrane and an exterior layer of protein subunits. As the

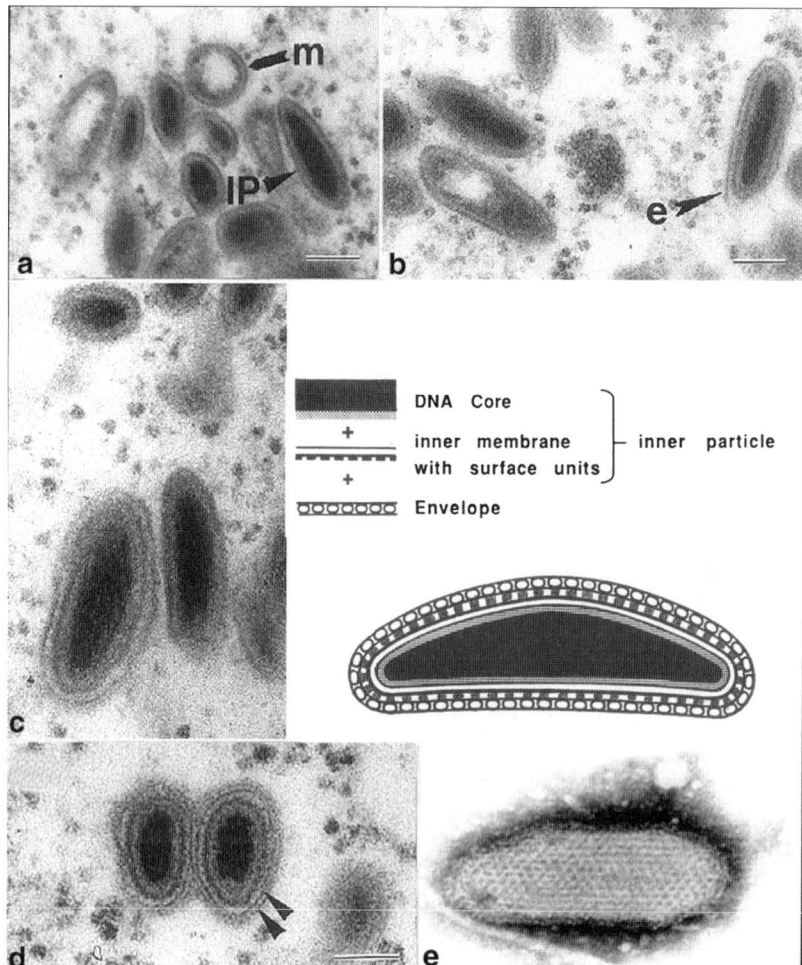

**Fig. 6a–e** Ascovirus virion assembly and structure. **a, b** Virion assembly begins with the development of the outer membrane (m) of the inner particle (*IP*), which as it forms occludes DNA and proteins from the vesicle cytoplasmic matrix. After inner particle formation is complete, the virion is enveloped in a membrane (*e*), apparently formed de novo within the cytoplamic matrix. **c** For comparison, a virion and an unenveloped inner particle. **d** Sections through two virions, with the *arrowheads* pointing to the inner and outer layers of the virion envelope. **e** Negatively stained ascovirus virion from the blood of an infected caterpillar. Note the reticulated, net-like appearance of the virion surface, which is characteristic of most ascovirus virions. **f** schematic interpretation of ascovirus virion structure based on the appearance of virions in ultrathin sections and negatively stained preparations. Bar in **a, b**, and **e** = 100 nm

multilaminar layer assembles, a dense nucleoprotein core aggregates on the interior surface. This process continues until the inner particle is complete. After formation, the inner particle is enveloped by membranes within the cell or vesicle. These membranes are apparently synthesized de novo. Thus, the assembly of the virions is reminiscent of that in other viruses with complex virions, such as the herpesviruses and poxviruses, where the virions differentiate after association of the precursors of virion structural components (Chinchar et al. 2005; Buller et al. 2005).

After formation, the virions of the TnAV and HvAV accumulate toward the periphery of the vesicle where they often form inclusion bodies, i.e., aggregations of virions (Federici et al. 1990; Asgari 2006). In SfAV, occlusion bodies form a foamy vesicular matrix that consists of a mixture of protein and minute spherical vesicles (Fig. 5D). When viewed with phase microscopy, these viral inclusion and occlusion bodies are largely responsible for the highly refractile appearance of the vesicles. Ascoviruses do not form the types of occlusion bodies characteristic of other types of DNA insect viruses, such as those formed by baculoviruses and entomopoxviruses (Buller et al. 2005; Theilmann et al. 2005).

As indicated by the size and complexity of the virions, the genomes of ascoviruses are large and consist of a single molecule of double-stranded circular DNA. The genomes of three species have been sequenced, the type species SfAV, TnAV, and HvAV. The SfAV genome is 157 kbp and codes for at least 120 proteins (Bideshi et al. 2006), whereas the TnAV 2a genome is slightly larger, 174 kbp, and codes for at least 134 proteins (Wang et al. 2006). The HvAV genome is 186 kbp and codes for potentially 186 proteins (Asgari et al. 2007).

Based on gel analyses (Fig. 7), ascovirus virions contain at least 12 structural polypeptides ranging in size from 12 to 200 kDa (Federici et al. 1990; Cui et al.

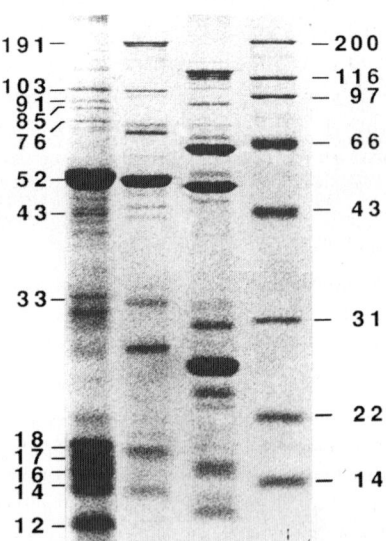

**Fig. 7** Protein composition of virions from three different ascoviruses as determined by sodium dodecyl sulfate polyacrylamide gel electrophoresis. The major protein at 50–52 kDa in all three virions is the major capsid protein. *HvAV, Heliothis virescens ascovirus* 3a; *TnAV, Trichoplusia ni ascovirus* 2a, and *SfAV, Spodoptera frugiperda ascovirus* 1a; *Std*, standard proteins used as markers for mass. Numbers indicate mass in kDa

2007). In addition to proteins and the DNA genome, the presence of an envelope as detected by electron microscopy, as well as experiments with detergents and organic solvents, indicate that virions contain a substantial lipid component. And, as in other enveloped viruses of eukaryotes, it is likely that the virion also contains carbohydrate in the form of glycoproteins, though none have been identified.

## Transmission and Ecology

One of the most interesting features of ascoviruses is that their transmission from host is largely dependent on their being vectored by female endoparasitic wasps belonging to the families *Braconidae* and *Ichneumonidae* (order Hymenoptera). Ascoviruses are extremely difficult to transmit per os, with typical infection rates averaging less than 15%, even when larvae are fed as many as $10^5$ virion-containing vesicles in a single dose. In contrast, infection rates for caterpillars injected with as few as ten virion-containing vesicles are typically greater than 90% (Hamm et al. 1985, 1986; Govindarajan and Federici 1990; Newton 2003). Moreover, experiments with parasitic wasps show that they are very effective in transmitting ascoviruses to their noctuid hosts. For example, when females are allowed to lay eggs in ascovirus-infected noctuid caterpillars, thereby contaminating their ovipositor, and then allowed to lay eggs in healthy larvae, the majority of the latter contract ascovirus disease (Hamm et al. 1985). Interestingly, though the parasite eggs hatch in their infected noctuid hosts, the parasite larvae die as the ascovirus disease develops in the caterpillar. Under field conditions, the prevalence of ascovirus disease in caterpillars is correlated with high rates of parasitization by endoparasitic wasps. When wasps from these populations are collected in the field and allowed to oviposit in healthy caterpillars reared in the laboratory, the latter often exhibit ascovirus disease within a few days (Hamm et al. 1985; Tillman et al. 2004). Thus, laboratory and field studies provide sound evidence that the primary mechanism for the transmission of ascoviruses attacking noctuid larvae is through being vectored mechanically by parasitic wasps. No evidence has been found in the lepidopteran hosts for transovum or transovarial transmission.

In the case of DpAV, the association of the virus with its wasp and caterpillar hosts is much more intimate (Bigot et al. 1997a, 1997b). DpAV DNA is carried in wasp nuclei as a circular molecule, and small numbers of virions are produced in the oviducts of females. However, the virus does not cause noticeable pathology in the wasp host. The females lay eggs in the pupal stage of the lepidopteran host, *Acrolepiopsis assectella*, introducing small numbers of ascovirus virions along with the wasp eggs. These virions invade lepidopteran host cells, replicate, and initiate destruction of major host tissues. The wasp larva then emerges from the egg and feeds on the host tissues and ascovirus virions. The DpAV genome is carried by both male and female wasps, where it is apparently transmitted from generation to generation transovarially. These observations make ascoviruses the only known group of viruses pathogenic to insects primarily dependent on vectors for their transmission.

Now that the characteristics of the disease are known, field studies in the southeastern United States and California are beginning to show that ascoviruses are probably the most common type of virus to occur during most of the year in populations of several important noctuid pests, including the cabbage looper, *Trichoplusia ni*, fall armyworm, *S. frugiperda*, and the corn earworm, *H. zea*. Prevalence rates range from 10% to 25%, depending on the species and time of the year, with the highest rates of infection, as noted above, being correlated with high levels of parasitization. In South Carolina, ascovirus infection rates as high as 60% have been reported in populations of noctuid larvae at the end of summer (Cheng et al. 2005).

# Phylogenetics

## *Phylogenetics of Species in the Genus Ascovirus*

As noted above in Sect. 3, initial studies during the early 1990s of the relationship of the three known ascovirus isolates, SfAV, TnAV, and HvAV, to one another based on a combination of studies including reciprocal DNA hybridization, host range, and differences in their tissue tropism showed that SfAV was a very distinct virus from TnAV and HAV (Federici et al. 1990). SfAV genomic DNA did not hybridize to TnAV or HvAV under stringent conditions, it had a host range limited to *Spodoptera* species, and it replicated almost exclusively in the fat body tissue. HvAV and TnAV DNA did hybridize, but it was clear that these viruses were not the same, as in reciprocal hybridization studies, the hybridization was much stronger under similar conditions between homologous DNA than heterologous mixtures, and some DNA fragments clearly did not hybridize. Moreover, though the host ranges of these two viruses were overlapping, and both had similar tissue tropisms, attacking the fat body as well as other tissues such as the epidermis and tracheal matrix, there were differences between these two viruses in the extent these tissues were attacked, which were confirmed in later studies (Hamm et al. 1998). Based on these studies, and the relatively few isolates available for study at the time, these viruses were considered three different species. Subsequently, several other ascovirus isolates and variants of these and the original three species became available for study (Bigot et al. 1997a, 1997b; Cheng et al. 2000), followed by even more isolates (Newton 2003; Cheng et al. 2005). In addition, genes such as those coding for d DNA polymerase (DNA pol) and major capsid protein (MCP) were cloned from these viruses and sequenced, and the relationships of the putative asco-virus species and isolates were determined based on these genes (Stasiak et al. 2000, 2003; Cheng et al. 2000; Cheng et al. 2005). The most recent assessment of these results suggests that there are probably four or five ascovirus species among the various isolates for which *dna pol* and *mcp* have been sequenced (Cheng et al. 2005; Federici et al. 2005), although only four of these are officially recognized at this time (Federici et al. 2005). The four species officially recognized are SfAV1a, TnAV2a, HvAV3a, and DpAV4a. As illustrated in Fig. 8, however, based on the deduced DNA pol and MCP sequences, the likely species are SfAV1a, TnAV2c, HvAV3a (as represented by the cluster of variants constituting the other TnAV and HvAV variants), and DpAV4a. It may be that among the numerous variants of HvAV3 there are other valid species, but this remains to be determined based on study of a greater range of *dna pol*, *mcp*, and other gene sequences from more isolates.

This delineation of ascovirus species reflects in general what we know about their different biologies and reciprocal DNA hybridization profiles. DpAV has the most genetic distance from the other ascoviruses, has the smallest genome (Table 3), and is the only one known to be vertically transmitted by its wasp vector, *D. pulchellus*. It is also the only ascovirus with the host target stage being the pupa. SfAV does not hybridize with the others to any significant extent, and its replication in normal hosts is restricted to the fat body. TnAV2c is similar to SfAV in its tissue tropism, but does not hybridize to any significant extent to the DNA of the other ascovirus species, and most of the other variants clustered under HvAV3 (including SeAV5a, TnAV2a, and most other variants similar to these), most of which cross-hybridize and have a broad tissue tropism, being capable of replication in the fat body, tracheal

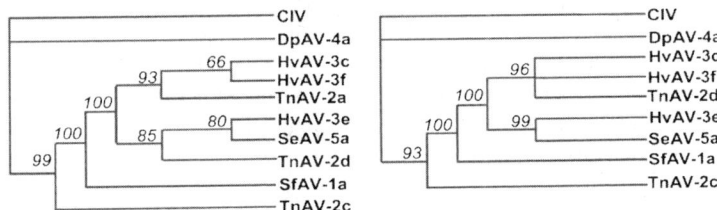

**Fig. 8** Maximum parsimony trees inferred from amino acid sequences of major capsid proteins (*A*) and DNA polymerase genes of various ascovirus isolates. Chilo iridescent virus, an iridovirus from the lepidopteran, *Chilo suppressalis*, was used as the outgroup for constructing these trees. From Cheng et al. (2005)

**Table 3** General features of ascovirus and the Chilo iridovirus (IIV6) genomes

| Virus | Size (bp) | ORFs | G+C% |
|-------|-----------|------|------|
| SfAV | 156,222 | 123 | 49.2 |
| TnAV | 174,059 | 165 | 35.2 |
| HvAV | 186,262 | 180 | 45.8 |
| DpAV | 119,343 | 119 | 49.7 |
| IIV6 | 212,482 | 468 | 28.6 |

Data from Chinchar et al. 2005 (IIV6); Asgari et al. 2007 (HvAV); Bideshi et al. 2006 (SfAV); Wang et al. 2006 (TnAV)

matrix, and epidermal tissues (Federici and Govindarajan 1990; Hamm et al. 1998; Cheng et al. 2005).

One of the problems that has plagued the taxonomy of many types of insect viruses, including the baculoviruses and cytoplasmic polyherdrosis viruses, is the practice of naming the virus after the species name of the insect host from which the isolate was obtained. This is a problem because most insect viruses are not species-specific, and thus a virus isolated, for example, from *H. virescens* is often capable or infecting not only other species of *Heliothis* and *Heliocoverpa*, but other species of noctuids as well. Unfortunately, ascovirus taxonomy also suffers from this problem, as is indicated by the viral names and relationships illustrated in Fig. 8. Current changes in ascovirus taxonomy are being aimed at finding an alternative to using the species name of the host of isolation to name ascoviruses.

## *Relationship of Ascoviruses to Other Large DNA Viruses*

The first insights into the phylogenetic relationship of ascoviruses to other types of large DNA viruses came from a study of the sequence of the SfAV d DNA polymerase gene in which it was shown that this virus represented a distinctly different type of virus, and one most closely related to iridoviruses (Pellock et al. 1996). The unique structure of ascovirus virions and their novel cytopathology suggested these viruses represented a new family of viruses, and this study provided good evidence for this. However, while it was realized at the time that similarities existed between the structure and assembly of ascovirus and iridovirus virions, the apparent close relationship of the ascoviruses to iridoviruses still came as somewhat of a surprise, primarily because the virions of the two viruses differ so much in symmetry. Ascovirus virions are reniform or bacilliform, depending on the species, with a distinctive surface pattern, whereas those of iridoviruses are always iscosahedral, with comparatively little surface structure when observed in negatively strained preparations (Chinchar et al. 2005; Federici et al. 2005). Nevertheless, a series of subsequent studies that included analyses of genes coding for thymidine kinase, and ATPase III in addition to the MCP and DNA polymerase, as shown in Fig. 9, have confirmed this relationship (Stasiak et al. 2000, 2003; Cheng et al. 2005).

Aside from the structural similarities in the way the virions assemble and the inferred relationship based on gene sequences, there are other properties that indicate iridoviruses and ascoviruses are closely related. For example, invertebrate iridoviruses and ascoviruses are both very difficult to transmit by feeding (Govindarajan and Federici 1990), and most, with the exception of DpAV, do not infect the midgut epithelium. It has also been shown that lepidopteran iridoviruses can infect and be transmitted by parasitic wasps that use lepidopteran larvae as their hosts (Lopez et al. 2002).

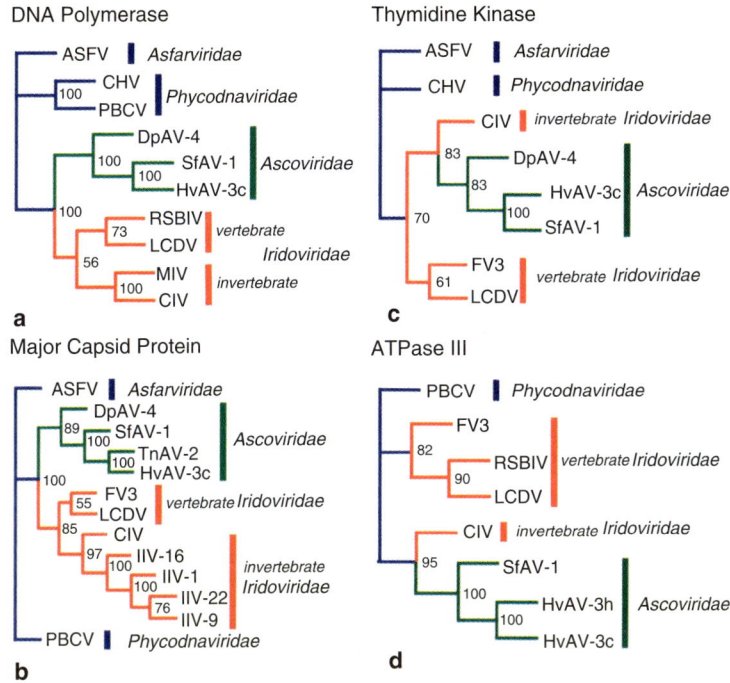

**Fig. 9a–d** Consensus trees resulting from phylogenetic analyses of four ascovirus and iridovirus proteins. **a** d DNA polymerase; **b** major capsid protein; **c** thymidine kinase; and **d** ATPase III. The sequences are from one asfarvirus, African swine fever virus (ASFV), two phycodnaviruses, Chorella virus (CHV) and *Paramecium bursaria* Chlorella virus (PBCV-1), nine iridoviruses, IV1, IV9, IV16, IV22, Chilo iridescent Virus (CIV), Frog Virus 3 (FV3), lymphocystis disease virus (LCDV), regular mosquito iridovirus (RMIV) and Red Sea bream iridovirus (RSBIV), and four ascoviruses, *Spodoptera frugiperda* Ascovirus 1a (SfAV-1a), *Trichoplusia ni* Ascovirus 2a (TnAV-2a), *Heliothis virescens* Ascovirus 3b and 3c (HvAV-3b and c), and *Diadromus pulchellus* Ascovirus 4a (DpAV-4a ). Consensus trees were rooted with the same proteins of African swine fever virus (ASFV, *Asfarviridae*) in **a**, **b** and **c**, or Chlorella virus (CHV, *Phycodnaviridae*) in **d**. For each of the analyzed proteins, the consensus trees obtained with parsimony and neighbor-joining programs were similar. The trees presented are those developed using the parsimony procedure. Numbers given at each node correspond to the bootstrap percent of values for 1,000 repetitions. (From Stasiak et al. 2003)

# Genomics

Over the past 2 years, the genomic sequences of three different ascoviruses have been reported, the type species, SfAV1a (Bideshi et al. 2006), TnAV2c (Wang et al. 2006), and HvAV3e (Asgari et al. 2007). The sequence of DpAV has been completed, but has not yet been completely annotated. Key characteristics of these four genomes and the Chilo iridovirus (IIV6), a lepidopteran iridovirus, are summarized in Table 3.

Ascovirus genomes vary considerably in size and to some extent in G+C content, but care must be taken in the implications of the size differences, because, for example, HvAV has 23 copies of bro-like genes (Asgari et al. 2007), whereas SfAV has only seven (Bideshi et al. 2006), and TnAV only three (Wang et al. 2006). Care must also be taken in viewing the number of ORFs, as in some of these viruses, stop codons are found in a least several of the genes, especially in the bro-like genes. As expected based on their similar biologies, the ascoviruses share many genes in common (Table 4), coding for proteins that function in virion structure (major capsid protein), nucleic acid metabolism ($\delta$ DNA polymerase, RNA polymerase, ATPases, RNAse III, and various nucleases), and inhibition of apoptosis (IAPs). In particular, with regard to the unique cytopathology of ascoviruses, the SfAV, TnAV, and HvAV also encode enzymes that potentially promote apoptosis and viral vesicle formation. These proteins include caspases and caspase-like proteins, which are not known to be encoded by any other virus, cathepsin B, and enzymes involved in lipid metabolism

**Table 4**  Genes shared by ascovirus genomes

| Protein encoded | Ascovirus ORF | | |
|---|---|---|---|
| | SfAV | HvAV | TnAV |
| DNA polymerase | 1 | 1 | 1 |
| DNA-directed RNA polymerase | 8 | 11 | 42 |
| DEAD-like helicase | 9 | 15 | 161 |
| Esterase-lipase | 13 | 19 | 132 |
| Zinc-dependent metalloprotease | 14 | 20 | 158 |
| RNAase III | 22 | 27 | 8 |
| IAP-like | 25 | 28 | 9 |
| Thymidine kinase | 40 | 55 | 154 |
| Major capsid protein | 41 | 56 | 145 |
| DNA-directed RNA polymerase II | 52 | 64 | 138 |
| DNA repair exonuclease | 59 | 71 | 121 |
| Serine/threonine protein kinase | 64 | 77 | 114 |
| FEN–1/FLAP-like endonuclease | 66 | 80 | 112 |
| DNA-directed RNA polymerase, b subunit | 67 | 82 | 110 |
| Caspase/caspase-like | 73 | 165 | 72 |
| S1/P1 nuclease | 75 | 134 | 135 |
| Baculovirus repeated ORF (Bro) | 79 | 157 | 165 |
| Lipopolysaccharide-modifying enzyme | 90 | 131 | 58 |
| Patatin-like phospholipase | 93 | 128 | 67 |
| Poxvirus D5-like ATPase | 99 | 119 | 78 |
| ATPase involved with DNA repair | 103 | 118 | 77 |
| Serine/threonine protein kinase | 104 | 117 | 88 |
| ATPase | 110 | 109 | 95 |
| PlsC, phosphate acyltransferase | 112 | 106 | 98 |
| Cathepsin B | 114 | 103 | 102 |

(sphingomyelin phosphodiesterase [SMase], patatin-like phospholipase $A_2$, PlsC phosphate acyltransferase, fatty acid elongase, and esterase/lipase).

Recently, it was demonstrated that the SfAV caspase is a functional homolog of executioner caspase 3/7, which apparently plays a direct role in initiating apoptosis (Bideshi et al. 2005) and may also be involved in maturation of virion proteins (Best and Bloom 2004). Whether the TnAV caspase 2-like protein plays similar roles in TnAV2c pathobiology is unknown. Although executioner caspases are the principle effectors of apoptosis, there is a growing body of evidence that lysosomal cathepsins, cysteine proteases of the C1 family of papain-like enzymes, can also contribute significantly to this response (Chwieralski et al. 2006), and this suggests a functional role for ascovirus cathepsin B. Upon activation, cathepsins released from lysosomes direct many cellular pathways, including protein degradation, metabolism, and apoptosis (Baek et al. 1997; Best and Bloom 2004; Ramanadhan et al. 2004). Recent evidence suggests that activated cathepsins destabilize mitochondrial membranes, causing the release of cytochrome $c$, which initiates the intrinsic apoptotic program (Chwieralski et al. 2006). In addition, cathepsins activate proapoptotic proteins of the Bcl2 superfamily, such as Bid and Bax. These proteins have been shown to activate initiator caspases, which in turn activate executioner caspases, and possibly the ascovirus caspases.

The presence of at least five ascovirus genes coding for enzymes involved in lipid metabolism is unusual, as this is not known to occur collectively in other eukaryotic and prokaryotic viruses. The ascovirus patatin-like phospholipase $A_2$ and coesterase/JHE alpha/beta hydrolases likely play important roles in the reorganization of existing membranes, de novo cellular membrane synthesis, and viral membrane biogenesis and egress of virus from the cell, functions clearly observed with the p37 lipase encoded by the vaccinia virus (Baek et al. 1997). Importantly, these enzymes could also function in initiating and mediating the apoptotic response through oxidative stress induced by endogenous ceramides synthesized by cellular and ascovirus-encoded SMases, or from long chain fatty acid precursors for sphingoid lipogenesis synthesized by ascovirus fatty acid elongases (Andrieu-Abadie and Lavade 2002; Baudry et al. 2001; Birbes et al. 2001; Kolesnick and Hannun 1999; Siskind 2005; Zhao et al. 2002). It is interesting to note that ceramide-induced oxidative stress can destabilize lysosomal membranes, which results in release of cathepsin B. Furthermore, it has been shown that apoptosis induced by endoplasmic reticulum stress is amplified by overexpression of phospholipase $A_2$ ($iPLA_2B$; Ramanadhan et al. 2004). Thus, the collective occurrence of ascovirus caspases, cathepsin B, SMase, and patatin-like $PLA_2$ genes strongly suggests that these function cooperatively in coordinated pathways that lead to the rapid onset of the apoptotic response observed during ascovirus-induced cytopathology (Bideshi et al. 2005; Federici 1983). Indeed, in the presence of inhibitors of cathepsin B and SMases, and antioxidants such as α-tocopherol acetate and reduced glutathione, we have observed marked reductions in SfAV1-induced apoptosis and vesicle formation in vitro (B.A. Federici et al., unpublished data).

# Origin and Evolution

The subject of viral evolution over millions of years has received relatively little study due to the lack of a fossil record. Moreover, viruses are considered polyphyletic and therefore many if not most of the more than 70 families of viruses are thought to have originated independently. In this regard, the ascoviruses may provide a unique opportunity to obtain insights into virus evolution over long periods. Phylogenetic comparisons of ascovirus genes sequenced to date including those coding DNA polymerase and major capsid protein, as well as several enzymes indicate these viruses evolved from a lepidopteran iridovirus (family *Iridoviridae*). Iridoviruses, in turn, appear to have originated from phycodnaviruses (family *Phycodnaviridae*, Wilson et al. 2005), which attack certain ciliates and algae (Fig. 9). On the other end of the evolutionary scale, ascovirus virions are structurally and morphologically similar to the particles formed by ichnoviruses of the family *Polydnaviridae*. Ichnovirus particles are produced in the reproductive tracts of endoparasitic wasps of the family *Ichneumonidae*, and the wasp vector and host of DpAV is a member of this family. Thus, there is a reasonable possibility that ascoviruses and ichnoviruses are related phylogenetically and share a common ancestor. This possibility is currently under investigation and should be resolved over the next few years through a comparative analysis of the molecular evolution of genes of ascoviruses and ichnoviruses, after more structural genes from the latter viruses have been cloned and sequenced. A major question to be addressed is whether the DpAV represents an early ascovirus branch that evolved from an iridovirus or is representative of an ascovirus branch that eventually led to the origin of ichnovirus particles.

With respect to the ichnoviruses and bracoviruses, recent genomic data on the DNA contained by the particles of these putative viruses suggest that these are not viruses but rather are an unusual highly evolved type of organelle, which evolved from DNA viruses by symbiogenesis in endoparasitic wasps to suppress the internal defense responses of their insect hosts (Federici 1991; Federici and Bigot 2003). It has been known for many years that ichnoviruses and bracoviruses do not have many of the characteristics typical of viruses (Webb et al. 2005). For example, the so-called virions of both these virus types are produced only in specialized cells in the follicular epithelium, most commonly the calyx region, of the female wasps that produce these particles. The assembly of the particles resembles the assembly of virions, and DNA is packaged in the particles, as is characteristic for viruses. The particles are secreted from wasp cells and enter the ovipositional fluid, where they accompany eggs into caterpillar hosts during oviposition. Once inside the caterpillar host, the particles enter host cells, most commonly hemocytes and fat body cells, where the DNA is released by the particles in host cell nuclei and the encoded genes are subsequently expressed. All of this resembles the biology of viruses. However, the similarities between these particles and true viral virions end at this point. No particle DNA is replicated in caterpillar host cells, nor are any particle structural proteins synthesized, with rare exception. As a result, there are no particle progeny produced in the caterpillar hosts. Based primarily on this unusual biology, it was suggested many years ago (Federici 1991) and hypothesized in considerably more detail recently (Federici and Bigot 2003) that the particles produced by these wasps are organelles that arose from insect viruses. The basic hypothesis is that these organelles evolved by symbiogenesis, a process initiated with symbiosis between two biologic entities that eventually fuse to become a more complex organism. Classic examples of this phenomenon are the origin of mitochondria and plastids from bacteria.

Until recently, there was virtually no molecular evidence for this hypothesis. However, recent genomic studies of bracoviruses (Espagne et al. 2004) and ichnoviruses (Webb et al. 2006; Tanaka et al. 2007) provide very strong if not overwhelming evidence for the organelle as opposed to the viral hypothesis. The key findings in all of these studies show that the DNA in the particles (i) is to a very large extent wasp genomic DNA encoding genes involved in suppression of caterpillar host defenses, (ii) encodes no enzymes for viral replication, (iii) encodes few if any particle structural proteins, and (iv) unlike the members of every other known viral family, which typically have coding efficiencies of 95% or greater, the coding efficiencies of bracoviruses and ichnoviruses are rarely greater than 30%. Thus, the evidence suggests that what happened during the symbiogenic process is that the genomes of pathogenic insect viruses were integrated into the genomes of certain wasp species, after which numerous wasp genes involved in suppression of host defenses replaced the viral genes needed for viral replication. The original viral genes for structural proteins and particle assembly, essential for particle production and vectoring wasp genes into their hosts, now reside in the wasp genomes. In essence, this is an excellent example of the extent to which symbiogenesis can lead to more complex organisms, as unlike mitochondria and plastids, which have maintained and replicate some of the original bacterial DNA, bracovirus and ichnovirus particles apparently contain few if any of the original viral genes.

The viruses from which the bracovirus and ichnoviruses evolved are not known. However, as bracovirus particles resemble baculoviruses and nudiviruses that can replicate in wasp oviducts (Hamm et al. 1988, 1990), and ichnovirus particles resemble DpAV in both structure and transmission biology, these are the best candidates for viruses from which these particles originated. Future wasp genomic studies, especially identification and phylogenic analysis of the genes coding for particle structural proteins and their assembly should enable identification of the viruses that served as the evolutionary source of these particles.

## Discussion and Future Studies

At present, too little is known about ascoviruses to assess whether they are or will turn out to be of economic importance. Their poor infectivity per os makes it highly unlikely they will even be developed as viral insecticides, especially given the successful advent of insect-resistant transgenic crops. However, as more entomologists become familiar with the disease caused by ascoviruses, it may be shown that in habitats rarely treated with chemical insecticides, such as transgenic crops, these viruses are responsible for significant levels of natural pest suppression, particularly where parasitic wasps are abundant. Such findings would encourage even greater emphasis on the development of biological control and other more environmentally sound methods of pest control. With respect to the cell biology of viral vesicle formation, ascoviruses provide an interesting model for how apoptosis can be manipulated at the molecular level. Additionally, study of the unusual process by which ascoviruses rescue the developing apoptotic bodies to form viral vesicles could lead to insights into how cells manipulate the cytoskeleton and mitochondria. Important points that remain to be resolved are the identification of the viral genes involved with the rescue of the developing apoptotic bodies and their conversion into viral vesicles. Iridoviruses also induce and take advantage of apoptosis for virus transmission (see Chapter 4), but the process is much more developed and complex in most ascoviruses. Lastly, it is possible that viral vesicles will provide a unique anucleate cellular system for studying the replication of a complex type of enveloped DNA viruses in vitro.

# References

Andrieu-Abadie N, Lavade T (2002) Sphingomyelin hydrolysis during apoptosis. Biochim Biophys Acta 1585:126–134

Asgari S (2006) Replication of *Heliothis virescens* ascovirus in insect cell lines. Arch Virol 151:1689–1699

Asgari S (2007) A caspase-like gene from the *Heliothis virescens* ascovirus (HvAV3e) is not involved in apoptosis but is essential for virus replication. Virus Res. 128:99–105

Asgari S, Davis J, Wood D, Wilson P, McGrath A (2007) Sequence and organization of the *Heliothis virescens* ascovirus genome. J Gen Virol 88:1120–1132

Baek, SH, Kwak JY, Lee SH, Lee T, Ryu SH, Uhlinger DJ, Lambeth JD (1997) Lipase activities of p37, the major envelope protein of vaccinia virus. J Biol Chem 272:32042–32049

Baudry K, Swain E, Rahier A, Germann M, Batta A, Rondet S, Mandala S, Henry K, Tint GS, Edlind T, Kurtz M, Nickels TJ Jr (2001) The effects of the erg26-1 mutation on the regulation of lipid metabolism in *Saccharomyces cerevisiae*. J Biol Chem 16:12702–12711

Best SM, Bloom ME (2004) Caspase activation during virus infection: more than just a kiss of death? Virology 320:191–194

Bideshi DK, Tan Y, Bigot Y, Federici BA (2005) A viral caspase contributes to modified apoptosis for virus transmission. Genes Dev 19:1416–1421

Bideshi DK, Demattei MV, Rouleux-Bonnin F, Stasiak K, Tan Y, Bigot S, Bigot Y, Federici BA (2006) Genomic sequence of *Spodoptera frugiperda ascovirus* 1a, an enveloped, double-stranded DNA insect virus that manipulates apoptosis for viral reproduction. J Virol 80:11791–11805

Bigot Y, Rabouille A, Doury G, Sizaret P-Y, Delbost F, Hamelim M-H, Periquet G (1997a) Particle and genome characterization of a new member of Ascoviridae, *Diadromus pulchellus* ascovirus. J Gen Virol 78:1139–1147

Bigot Y, Rabouille A, Doury G, Sizaret P-Y, Delbost F, Hamelim M-H, Periquet G (1997b) Biological and molecular features of the relationships between *Diadromus pulchellus* ascovirus, a parasitoid hymenopteran wasp (*Diadromus pulchellus*) and its lepidopteran host, *Acrolepiopsis assectella*. J Gen Virol 78:1149–1163

Birbes HS, Bawab S, Hannun Y, Obied LM (2001) Selective hydrolysis of a mitochondrial pool of spingomyelin induces apoptosis. FASEB J. 14:2669–2679

Buller RM, Arif BM, Black DM, Dumbell KR, Espisito JJ, Lefkowitz EJ, McFadden G, Moss B, Mercer AA, Moyer RW, Skinner MA, Tripathy DN (2005) *Family Poxviridae*. In: Fauquet CM, Mayo MA, Maniloff J, Desselberger U, Ball LA (eds) Virus taxonomy, eighth report of the International Committee on Taxonomy of Viruses. Elsevier, Academic, San Diego, pp 117–143

Cheng XW, Carner GR, Arif BM (2000) A new ascovirus from *Spodoptera exigua* and its relatedness to the isolate from *Spodoptera frugiperda*. J Gen Virol 81:3083–3092

Cheng XY, Wang L, Carner GR, Arif BM (2005) Characterization of three ascovirus isolates from cotton insects. J Invertebr Pathol 89:193–202

Chwieralski CE, Welte T, Buhling F (2006) Cathepsin-regulated apoptosis. Apoptosis 11:143–149

Chinchar VG, Essbauer S, He JG, Hyatt A, Miyazaki T, Seligy V, Williams T (2005) *Family Iridoviridae*. In: Fauquet CM, Mayo MA, Maniloff J, Desselberger U, Ball LA (eds) Virus taxonomy, eighth report of the International Committee on Taxonomy of Viruses. Elsevier, Academic, San Diego, pp 145–162

Cui L, Cheng X, Li L, Li J (2007) Identification of *Trichoplusia ni* ascovirus 2c virion structural proteins. J Gen Virol 88:2194–2197

de Villiers EM, Fauquet C, Broker TR, Bernard HU, Hausen H (2004) Classification of papillomaviruses. Virology 324:17–27

Espagne E, Dupuy C, Huguet E, Cattolico L, Provost B, Martines N, Poirie M, Periquet G, Drezen JM (2004) Genome sequence of a polydnavirus: insights into symbiotic virus evolution. Science 306:286–289

Federici BA (1983) Enveloped double-stranded DNA insect virus with novel structure and cytopathology. Proc Nat Acad Sci U S A 80:7664–7668

Federici BA (1991) Viewing polydnaviruses as gene vectors of endoparasitic hymenoptera. Redia 74:387–392

Federici BA, Bigot Y (2003) Origin and evolution of polydnaviruses by symbiogenesis of insect DNA viruses in endoparasitic wasps. J Insect Physiol 49:419–432

Federici BA, Govindarajan R (1990) Comparative histopathology of three ascovirus isolates in larval noctuids. J Invertebr Pathol 56:300–311

Federici BA, Vlak JM, Hamm JJ (1990) Comparison of virion structure, protein composition, and genomic DNA of three Ascovirus isolates. J Gen Virol 71:1661–1668

Federici BA, Hamm JJ, Styer EL (1991) Ascoviruses. In: Adams J, Bonami G (eds) An atlas of invertebrate viruses. CRC Press, Boca Raton, FL, pp 339–349

Federici BA, Bigot Y, Granados RR, Hamm JJ, Miller LK, Newton I, Stasiak K, Vlak JM (2005) *Family Ascoviridae*. In: Fauquet CM, Mayo MA, Maniloff J, Desselberger U, Ball LA (eds) Virus taxonomy, eighth report of the International Committee on Taxonomy of Viruses. Elsevier, Academic, San Diego, pp 269–274

Govindarajan R, Federici BA (1990) Ascovirus infectivity and the effects of infection on the growth and development of Noctuid larvae. J Invertebr Pathol 56:291–299

Hamm JJ, Nordlung DA, Marti OG (1985) Effects of a nonoccluded virus of *Spodoptera frugiperda* (Lepidoptera: Noctuidae) on the development of a parasitoid, *Cotesia marginiventris* (Hymenoptera: Braconidae). Environ Entomol 14:258–261

Hamm JJ, Pair SD, Marti OG (1986) Incidence and host range of a new ascovirus isolated from the fall armyworm, *Spodoptera frugiperda* (Lepidoptera: Noctuidae). Fla Entomol 69:524–541

Hamm JJ, Styer EL, Lewis, WJ (1988) A baculovirus pathogenic to the parasitoid *Microplitus croceipes* (Hymenoptera: Braconidae). J Invertebr Pathol 52:189–191

Hamm JJ, Styer EL, Lewis JJ (1990) Comparative virogenesis of filamentous virus and polydnavirus in the female reproductive tract of *Cotesia marginiventris* (Hymenoptera: Braconidae). J Invertebr Pathol 55:357–360

Hamm JJ, Styer EL, Federici BA (1998) Comparison of field-collected ascovirus isolates by DNA hybridization, host range, and histopathology. J Invertebr Pathol 72:138–146

Kolesnick R, Hannun YA (1999) Ceramide and apoptosis. Trends Biochem Sci 24:224–227

Lopez M, Rojas JC, Vandame R, Williams T (2002) Parasitoid-mediated transmission of an irido-virus. J Invertebr Pathol 80:160–170

Newton IR (2003) The biology and characterisation of the ascoviruses (Ascoviridae: Ascovirus) of *Helicoverpa armigera* Hubner and *Helicoverpa punctigera* Wallengren (Lepidoptera: Noctuidae) in Australia. PhD Dissertation, University of Queensland, Queensland, Australia

Pellock BJ, Lu A, Meagher RB, Weise MJ, Miller LK (1996) Sequence, function, and phyloge-netic analysis of an ascovirus DNA polymerase gene. Virology 216:146–157

Ramanadhan S, Hus FF, Zhang S, Jin C, Bohrer A, Song H, Bao S, Ma Z, Turk J (2004) Apoptosis of insulin-secreting cells induced by endoplasmic reticulum stress is amplified by overexpres-sion of Group VIA calcium-independent phospholipase A2 (iPLA2B) and suppressed by inhi-bition of iPLA2B. Biochem 43:918–930

Stasiak K, Demattei MV, Federici BA, Bigot Y (2000) Phylogenetic position of the DpAV-4a ascovirus DNA polymerase among viruses with a large double-stranded DNA genome. J Gen Virol 81:3059–3072

Stasiak K, Renault S, Demattei MV, Bigot Y, Federici BA (2003) Evidence for the evolution of ascoviruses from iridoviruses. J Gen Virol 84:2999–3009

Siskind LJ (2005) Mitochondrial ceramide and the induction of apoptosis. J Bioenerg Biomembr 37:143–153

Tanaka K, Lapointe R, Barney WE, Makkay AM, Stoltz D, Cusson M, Webb BA (2007) Shared and species-specific features among ichnovirus genomes. Virology 363:26–35

Theilmann DA, Blissard GW, Bonning B, Jehle J, O'Reilly, Rohrmann, GF, Thiem S, Vlak J (2005) Family Baculoviridae. In: Fauquet CM, Mayo MA, Maniloff J, Desselberger U, Ball LA (eds) Virus taxonomy, eighth report of the International Committee on Taxonomy of Viruses. Elsevier, Academic, San Diego, pp 177–185

Tillman PG, Styer EL, Hamm JJ (2004) Transmission of an ascovirus from *Heliothis virescens* (Lepidoptera, Noctuidae) and effects of the pathogen on survival of a parasitoid, *Cardiochiles nigriceps* (Hymenopterea, Braconidae). Environ Entomol 33:633–643

Wang L, Xue J, Seaborn CP, Arif BM, Cheng XW (2006) Sequence and organization of the *Trichoplusia ni* ascovirus 2c (*Ascoviridae*) genome. Virology 354:167–177

Webb BA, Beckage NE, Hayakaya Y, Lanzrein B, Stoltz DB, Strand MR, Summers MD (2005) *Family Polydnaviridae*. In: Fauquet CM, Mayo MA, Maniloff J, Desselberger U, Ball LA (eds) Virus taxonomy, eighth report of the International Committee on Taxonomy of Viruses. Elsevier, Academic, San Diego, pp 255–267

Webb B, Strand MR, Dickey SE, Beck MH, Hilgarth RS, Barney WE, Kadash K, Kroemer JA, Lindstrom KG, Tattanadechakul W, Shelby KS, Thoetkiattikul H, Turnbull, MW, Witherell RA (2006) Polydnavirus genomes reflect their dual roles as mutulists and pathogens. Virology 347:160–174

Wilson WH, Van Etten JL, Schroeder DS, Nagasaki K, Brussaard C, Bratbak G, Suttle C (2005) Family *Phycodnaviridae*. In: Fauquet CM, Mayo MA, Maniloff J, Desselberger U, Ball LA (eds) Virus taxonomy, eighth report of the International Committee on Taxonomy of Viruses. Elsevier, Academic, San Diego, pp 163–175

Zhao S, Du XY, Chai, MQ, Chen JS, Zhou YC, Song JG (2002) Secretory phospholipase A$_2$ induces apoptosis via a mechanism involving ceramide generation. Biochim Biophys Acta 1581:75–88

# Whispovirus

J.-H. Leu, F. Yang, X. Zhang, X. Xu, G.-H. Kou, C.-F. Lo (✉)

**Contents**

**Abstract** During the last two decades, a combination of poor management practices and intensive culturing of penaeid shrimp has led to the outbreak of several viral diseases. White spot disease (WSD) is one of the most devastating and it can cause massive death in cultured shrimp. Following its first appearance in 1992–1993 in Asia, this disease spread globally and caused serious economic losses. The causative

C.-F. Lo
Institute of Zoology, National Taiwan University, Taipei, 106 Taiwan ROC
gracelow@ntu.edu.tw

James L. Van Etten (ed.) *Lesser Known Large dsDNA Viruses.*
Current Topics in Microbiology and Immunology 328.
© Springer-Verlag Berlin Heidelberg 2009

agent of WSD is white spot syndrome virus (WSSV), which is a large, nonoccluded, enveloped, rod- or elliptical-shaped, dsDNA virus of approximately 300 kbp. WSSV has a very broad host range among crustaceans. It infects many tissues and multiplies in the nucleus of the target cell. WSSV is a lytic virus, and in the late stage of infection, the infected cells disintegrate, causing the destruction of affected tissues. The WSSV genome contains at least 181 ORFs. Most of these encode proteins that show no homology to known proteins, although a few ORFs encode proteins with identifiable features, and these are mainly involved in nucleotide metabolism and DNA replication. Nine homologous regions with highly repetitive sequences occur in the genome. More than 40 structural protein genes have been identified, and other WSSV genes with known functions include immediate early genes, latency-related genes, ubiquitination-related genes, and anti-apoptosis genes. Based on temporal expression profiles, WSSV genes can be classified as early or late genes, and they are regulated as coordinated cascades under the control of different promoters. Both genetic analyses and morphological features reveal the uniqueness of WSSV, and therefore it was recently classified as the sole species of a new monotypic family called *Nimaviridae* (genus *Whispovirus*).

## Introduction

The first reported appearance of white spot syndrome virus (WSSV) occurred in 1992–1993 in the southern provinces of mainland China and also in the northern counties of Taiwan, and then it quickly spread to other areas in Southeast Asia and subsequently to shrimp farming areas all over the world, including the Americas, Europe, and the Middle East. The virus causes serious economic damage to the shrimp culture industry. WSSV is highly lethal to most of the commercially cultivated penaeid shrimp species. An acute outbreak of WSD can cause a cumulative mortality of up to 100% within 3–10 days in cultured shrimp. WSSV is a large (80–120×250–380 nm), nonoccluded, rod- to elliptical-shaped DNA virus. The virions consist of a rod-shaped nucleocapsid enveloped by a trilaminar membrane and they often have a unique, tail-like extension at one end. The virus genome is a circular dsDNA of approximately 300 kbp. Although this virus has a remarkably broad host range among crustaceans, including many species of shrimp, crayfish, crab, and lobster, WSSV is highly pathogenic and virulent only on penaeid shrimp. WSSV mainly infects cells in tissues of ectodermal and mesodermal origins, and the virus replicates and assembles the virions in the hypertrophied nuclei of infected cells without the production of occlusion bodies. WSSV undergoes lytic infection, and at the late stage of infection, the infected nuclei/cells disintegrate, which causes the affected tissues/organs to become seriously damaged and necrotic. Complete WSSV genome sequence analyses have revealed that most of the predicted gene products show no homology to known proteins, and even though some gene products have identifiable features, they are more similar to eukaryotic proteins than to viral proteins. The identifiable proteins are mainly involved in nucleotide metabolism and DNA replication.

Based on its unique morphological and genetic features, the International Committee on Taxonomy of Viruses (ICTV) assigned WSSV as the only member of the genus *Whispovirus* within a new virus family called *Nimaviridae*. This name refers to the thread-like extension at one end of the virus particle (*nima* is Latin for "thread"). Although various geographical isolates with genotypic variability (variants) have been identified, they are all currently classified as a single species within the genus *Whispovirus*.

The huge impact of this virus on the shrimp culture industry – and even on the entire economies of some countries – has driven research into how WSD might be prevented, controlled and even cured. To study a causative viral agent, a cell line is an invaluable tool, but unfortunately, no cell lines susceptible to WSSV infection are available. Although some primary cultures established from various tissues and organs of penaeid shrimp have been reported to be susceptible to WSSV infection, the procedures for establishing primary cultures are cumbersome. The quality and availability of these cultures also depend on the quality of the individual shrimp, and differences between individual shrimp mean that experimental results may vary greatly. Therefore, the usefulness of primary cultures to study WSSV is limited. Furthermore, no successful transgene-related study on WSSV has been conducted in a primary culture. Recently, however, a primary hematopoietic cell culture (Hpt cells) derived from hematopoietic tissue from freshwater crayfish has been reported to be susceptible to WSSV infection (Jiravanichpaisal et al. 2006), and a dsRNA transfection system has been established that can be applied to this culture to reduce the expression of two genes that are constitutively expressed in Hpt cells (Liu and Soderhall 2007). Progress is therefore being made to develop the tools to study WSSV in vitro, and the next step should be the development of a DNA transfection system that can deliver and express cellular or viral genes in Hpt or other crustacean cells. Such a transfection system could in principle use the WSSV immediate-early gene (*ie1*) promoter (Liu et al. 2005) to drive gene expression. For in vivo study, an increasing body of evidence indicates that RNAi technology is feasible for reducing expression of both cellular and viral genes in shrimp (Robalino et al. 2004, 2005), although to date low-level nonspecific antiviral activity has interfered with the experiments. A few successful in vivo transgene studies have also been reported, but compared to other experimental organisms, it is difficult to apply this technology to shrimp.

## Transmission and Host Range

WSSV can be transmitted horizontally either orally by feeding on diseased individuals or contaminated food, or through gills and/or other body surfaces by direct exposure to virus particles in the water. Experimentally, the animals are infected through feeding, immersion, or direct injection. The virus is also transmitted from brooder to offspring. However, brooder-to-offspring transmission is more likely to be caused by viral contamination of the egg mass and not the direct infection of oocytes or sperms with the virus (Lo et al. 1997).

Although penaeid shrimp are highly susceptible to WSSV, several studies indicate that larval and early postlarval shrimp are resistant to WSSV, whether they are challenged by immersion or orally (Venegas et al. 1999; Yoganandhan et al. 2003). Although no disease occurred in these larvae and early postlarvae, the shrimp tested positive for WSSV by a nested PCR assay and they showed abnormal mortality when they developed to the late postlarval or juvenile stages. A similar pattern of viral pathogenicity occurs in other shrimp viral diseases as well (Yoganandhan et al. 2003), and the relative resistance of shrimp larvae might be due to their rapid growth, metabolic processes, and incomplete development of the target tissues and organs during the shrimp's early developmental stages (Venegas et al. 1999). Differential susceptibility of WSSV at different development stages has also been described for the giant freshwater prawn *Macrobrachium rosenbergii*, which showed relatively high mortality in postlarvae and juveniles but lower mortalities in subadults and adults (Pramod Kiran et al. 2002).

WSSV has a remarkably broad host range among crustaceans. Almost every species of penaeid shrimp is susceptible to WSSV. Moreover, the virus can infect other marine, brackish water, and freshwater crustaceans, including crayfishes, crabs, spiny lobsters, and even hermit crabs (Lo et al. 1996; Flegel 1997, 2006). However, in contrast to penaeid shrimp, the infection is often not lethal for these species, and consequently they may serve as reservoirs of the virus. In WSSV-infected, moribund penaeid shrimp, many nuclei become hypertrophied and most of the affected tissues become necrotic. By contrast, although some WSSV-infected crabs and nonpenaeid shrimp showed high numbers of hypertrophied nuclei, the tissues are less damaged and no mortality occurs. The reason for this discrepancy between the responses of penaeid shrimp and other nonsusceptible crustaceans to WSSV infection is unknown, but Flegel (1997) has proposed that WSSV-triggered apoptosis in penaeid shrimp might be the cause. Although WSSV infection occurs in wild crustaceans, including penaeid shrimp, there is no evidence that the virus causes significant mortalities in these animals. Relevant factors presumably include environmental stress, WSSV infection levels, and host densities, all of which are considerably lower in the wild, so that a large outbreak of WSD in wild animals is much less likely to happen. Finally, copepods collected from WSSV-infected farms as well as the larvae of at least one insect were diagnosed as WSSV-positive by PCR, suggesting that, even if these animals are not infected, they may serve as reservoir hosts of WSSV.

## Clinical Features and Pathology

WSSV was named for the white spots on the carapace, appendages and inside the epidermis of diseased shrimp. These spots range from barely visible to 3 mm in diameter, and they sometimes coalesce into larger plates. The spots may result from the abnormal deposition of calcium salts in the cuticular epidermis (Wang et al. 1997) or else from disruption in the transfer of exudates from the epithelial cells to the cuticle (Wang et al. 1999). However, it should be noted that environmental stress factors, such as high alkalinity, or bacterial diseases can also cause white spots on the carapace of shrimp, and that moribund shrimp with WSD often have few, if any, white spots. Therefore, the appearance of white spots is not a definitive diagnostic sign of WSSV infection. Furthermore, other crustaceans, such as crayfish, show no white spots when infected with WSSV (Jiravanichpaisal et al. 2001).

In the early stage of WSD, the infected animals become lethargic, gather around the edges of ponds at the surface during the day, reduce their food consumption, and change color to pink or reddish brown due to the expansion of the cuticular chromatophores. In moribund shrimp, there is systemic destruction of target tissues, and most of the infected cells have homogeneous hypertrophied nuclei. As the infection progresses, virus particles are continuously released from the infected tissues into the hemolymph. An abundance of virions in the hemolymph is known as viremia, and for experimental purposes, viremic hemolymph is a convenient source of virus inoculum (Wu et al. 2002).

## Histopathology and Tissue Tropism

Normal histology with hematoxylin and eosin staining show that moribund shrimp exhibit basophilic inclusions in infected nuclei, widespread focal necrosis in tissues of ectodermal and mesodermal origins, and hemocytic infiltration of the gills, hepatopancreatic hemal sinuses, and hemocoel (Wongteerasupaya et al. 1995). Other studies indicate that WSSV mainly infects cells in tissues of ectodermal (cuticular epidermis, fore- and hindgut, gills, and nervous tissue) and mesodermal (lymphoid organ, antennal gland, connective tissue, and hematopoietic tissue) origins, whereas tissues of endodermal origin (hepatopancreatic tubule epithelium and midgut epithelium) are refractory to WSSV infection (Chang et al. 1996; Durand et al. 1996; Wang et al. 1999).

WSSV infection induces similar cellular changes in all target tissues (Wongteerasupaya et al. 1995; Wang et al. 1999). In the first stage of infection, the infected cells display nuclear hypertrophy, nucleoli dissolution, chromatin margination, and a more homogeneous eosinophilic central region. The infected cells then proceed to develop an intranuclear eosinophilic Cowdry A-type inclusion, which is amorphous and surrounded by a clear halo beneath the nuclear membrane. This eosinophilic inclusion subsequently becomes a light basophilic, denser inclusion separated by a transparent zone from the marginated chromatin. During this time, the cytoplasm becomes less dense and more lucent. At the late stage of infection, the nuclear membrane is disrupted, so that the intranuclear transparent zone between the intranuclear inclusion and the marginated chromatin disappears, fusing with the lucent cytoplasm. At the end of cellular degeneration, the nucleus or whole cell disintegrates, leading to vacant areas in the sections. In moribund shrimp, most tissues and organs are heavily infected with the virus and show massive multifocal necrosis.

Chang et al. (1996) conducted a detailed temporal study to identify the major target tissues as well as the initial infection sites of WSSV in experimentally infected shrimp. In this study, the shrimp were infected with WSSV through feeding, and at various times postinfection (p.i.) (0, 16, 22, 40, 52 and 64 h p.i.), tissue samples were taken and assayed by in situ hydridization. WSSV-positive signals were first detected in the stomach, gill, cuticular epidermis, and connective tissue of the hepatopancreas at 16 h p.i. At this time, some shrimp were WSSV-positive in the stomach but not in the gills, whereas other shrimp were WSSV-positive in the gills but not in the stomach. This suggests that WSSV infection can occur either via the oral pathway or via water to the gill or cuticular epidermis. At 22 h p.i., the lymphoid organ, antennal gland, muscle tissue, hematopoietic tissue, heart, midgut, and hindgut also became positive. The nervous tissue and compound eyes showed no positive signals until 40 h p.i. As infection proceeded (52–64 h p.i.), the stomach, gill, hematopoietic tissue, lymphoid organ, antennal gland, and cuticular epidermis were all heavily infected with WSSV and became seriously damaged and necrotic.

The failure of these vital organs at the late infection stage presumably contributed to the death of the infected shrimp. Organs that were more lightly infected by WSSV included the hepatopancreas, nerve ganglia, compound eye, muscle tissue (mostly connective tissue cells), and the connective tissue of the midgut and hindgut. The integrity of these organs was maintained up to the late stage of infection. However, a comparative analysis of the ESTs from normal and WSSV-infected postlarvae cDNA libraries showed that the transcriptional profiles of the hepatopancreas, compound eye, and muscle changed significantly (Leu et al. 2007). These findings suggest that even though these lightly infected organs remain structurally and morphologically intact, WSSV infection nevertheless has a dramatic effect on the interior of their constituent cells, at least at the transcriptional levels.

Tissue tropism in the reproductive organs has also been studied using in situ hybridization and transmission electron microscopy (TEM) (Lo et al. 1997). In testes, WSSV-positive cells were located in the connective tissue layer surrounding the seminiferous tubules, but no germ cells were infected. In the spermatophore, only muscle and connective tissue cells were WSSV positive. In the ovary, follicle cells, oogonia, oocytes, and connective tissues were WSSV-positive. However, the fact that no mature eggs were infected suggested that infected egg cells were killed by the virus before maturation.

# WSSV Morphology

## External Morphology

When examined under TEM after negative staining, WSSV virions purified from penaeid shrimp tissues or hemolymph show no difference from virions from the hemolymph of infected crayfish. WSSV virions (Fig. 1a and c) are enveloped,

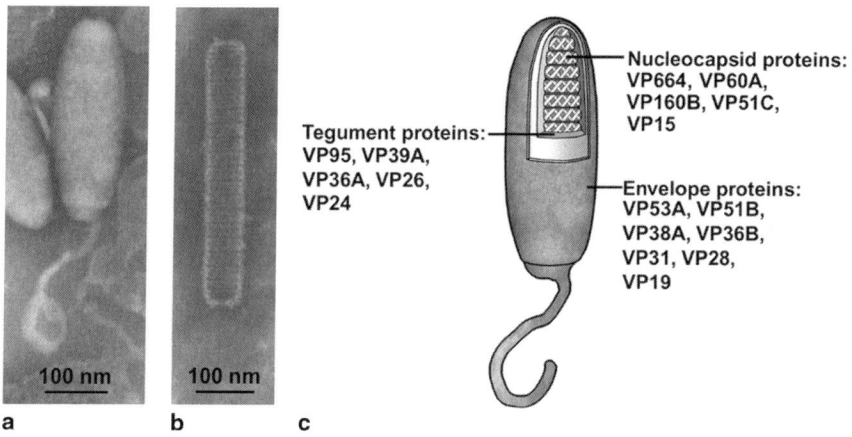

**Fig. 1a–c** Morphology of the WSSV virion. Negative contrast electron micrographs of (**a**) an intact WSSV virion with tail-like extension and (**b**) nucleocapsid. **c** A schematic diagram based on **a** that shows the layered structures of a WSSV virion, i.e., envelope, tegument, and nucleocapsid. The major proteins that constitute these layers are indicated. (**a** and **b** from Leu et al. 2005, with permission)

cylindrical to elliptical in shape, with one extremity slightly fatter than the other. They measure 80–120 nm in width and 250–380 nm in length (OIE 2003a). Some purified virions contain a filamentous appendage at the narrow end (Wang et al. 1995; Wongteerasupaya et al. 1995). This long, tail-like structure might represent a long envelope extension and measures 270–310 nm in length (Durand et al. 1996). The nonenveloped nucleocapsid (Fig. 1b) is rod-shaped and its size is 330–350×58–67 nm, which is longer and thinner than the intact virus particle (Wang et al. 1995). The nucleocapsid has a superficially segmented appearance, with the segments running perpendicular to the longitudinal axis of the nucleocapsid. An intact nucleocapsid usually consists of 14–19 segments (J.-H. Leu et al., unpublished observation). The segments are roughly 23 nm thick and separated from each other by an electron-dense band of approximately 6 nm. These segments are actually ring-like structures, as revealed by the exposed termini of the separated segments in degraded or broken nucleocapsids (Durand et al. 1997; Huang et al. 2001). Therefore, the nucleocapsid seems to be made of a series of stacked ring structures (Wang et al. 1995). Each segment (or ring) is composed of two parallel rows of 12–14 globular subunits (this subunit is now thought to be composed of protein VP664; See Sect. 7), each of which is approximately 8–10 nm in diameter. TEM sometime shows some fatter, more ovoid-shaped nucleocapsids with a detached fragmented envelope and a cross-hatched appearance (Durand et al. 1996, 1997). The different surface patterns on the differently shaped nucleocapsids (rod vs ovoid) suggest that the arrangement of the globular subunits determines the overall shape of the nucleocapsid.

## Internal WSSV Ultrastructure

The enveloped virions are cylindrical to elliptical in the longitudinal section and round to somewhat pentagonal or hexagonal in the transverse section (Fig. 2B; Wongteerasupaya et al. 1995; Durand et al. 1997). The complete assembled virion is 275–335×116–138 nm, while the capsid measures 246–296×75–93 nm and the envelope is 7–9 nm thick (Wang et al. 1999). The core of the nucleocapsid is highly electron-dense and measures 176–232×55–77 nm (Wongteerasupaya et al. 1995). The envelope is trilaminar, consisting of two electron-opaque layers separated by one electron-lucent layer (Wongteerasupaya et al. 1995; Durand et al. 1997; Wang et al. 1999).

## Ultrastructural Cytopathology and Viral Morphogenesis

Present understanding of WSSV morphogenesis is based on early TEM studies on naturally and experimentally infected penaeid shrimp (Durand et al. 1997; Wang et al. 1999). WSSV replicates and assembles within the nucleus (Fig. 2a), and in an acute infection, its life cycle is completed within 24 h. Based on ultrastructural changes in infected cells, Wang et al. (1999) described four distinct morphological profiles that correspond to the different stages of viral infection.

In profile 1, the nuclei of infected cells are slightly hypertrophied, and most of the electron-dense chromatin is discontinuously located along the nuclear membrane. Given the fusion of the nucleolus with the chromatin, the central area of the

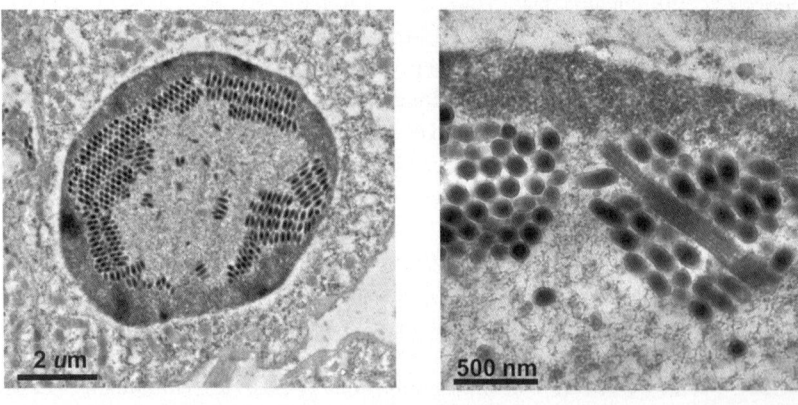

a                                            b

Fig. 2a, b Transmission electron micrographs of WSSV-infected intestine tissues from crayfish (*Procambarus clarkii*) at 48 h post-WSSV injection. a Virion particles in paracrystalline arrangement are located at the periphery of the nucleus. b Virion particles clustered around a long, rod-shaped structures (LRS) in the nucleus

nucleus becomes thin and homogenous. Also in the nucleus, viral envelope material appears as fibrillar membranous fragments. In the cytoplasm, the endoplasmic reticulum (ER) is enlarged and there are many free ribosomes.

In profile 2, the infected nuclei become grossly hypertrophied and rounded, and the chromatin forms an electron-dense ring zone around the nucleus membrane. The more electron-lucent central area is the virogenic region that contains both viral envelope material in membrane or vesicle forms and viral particles. However, at this early stage, many of these particles are immature virions with envelopes that are open at one end. In the cytoplasm, the mitochondria deteriorate, and vacuolated membrane networks form and cover most of the cytoplasm, resulting in an abnormal appearance. Long, rod-shaped structures (LRS) (Fig. 2b; Durand et al. 1997; Wang et al. 1999) sometimes appear in infected nuclei. A LRS is a bundle of long, cylindrical tubules, each of which is several times longer than a WSSV virion (up to 2 μm). The tubules themselves consist of two electron-dense longitudinal bands (20 nm in width) separated by an electron-clear band (6 nm in width; Durand et al. 1997). In the longitudinal section, each tubule displays a segmentation pattern with regular spacing, suggesting that it is composed of numerous repeating units in a stacked series. Single tubules can be observed, but in most cases, two or three tubules tend to lie side by side. Because these tubules have a diameter relatively close to that of an empty nucleocapsid, as well as a similar segmentation pattern, they are presumed to be nucleocapsid precursors (Durand et al. 1997).

In profile 3, the nucleus contains many, evenly distributed viral particles. Generally, the immature, developing virions are located centrally, whereas mature virions are found in the peripheral region. The chromatin ring zone disappears, the nuclear membrane breaks down, and the nucleus coalesces with the cytoplasm. In the cytoplasm, most cellular organelles appear abnormal, and they disintegrate into granules or membranous structures. The plasma membrane is partially disrupted.

In profile 4, infected cells are grossly damaged. The sloughing of viral particles and cellular components produces many voids in infected, lesioned tissues.

EM observations suggest the following sequences for WSSV morphogenesis. Morphogenesis begins with the de novo synthesis of the viral envelope material in the nucleus, which is manifested by the appearance of fibrillar membranous structures. These fibrillar membranous structures then fuse to form large, linear or circular membranes that subsequently wrap around the empty nucleocapsids. The development of the nucleocapsid begins with the formation of the LRS. Once formed, the assembled LRS tubules fragment to form the shorter, naked empty capsids. Each capsid is then partially wrapped by viral envelope membrane so that one extremity of the envelope remains open. The nucleoprotein, which appears as a thin electron-dense filament, enters the capsid through this opening, and at this stage, the viral DNA-packed nucleocapsid is a cylinder-shaped structure. The nucleoprotein filament gradually fills the capsid, expanding its cross-sectional diameter until the capsid is completely filled with the electron-dense nucleoprotein. Probably, to accommodate the increased nucleoprotein material, the center part of the nucleocapsid becomes slightly distended so that the entire nucleocapsid becomes shorter and fatter. Finally, the open end of the nucleocapsid is closed, and the corresponding end of the envelope narrows and extends to form WSSV's characteristic tail. The mature virion has an olive-like shape.

EM studies of WSSV morphogenesis suggest that WSSV might also use another pathway for virion assembly (Wang et al. 1999, 2000). Instead of packing nucleoprotein into a partially enveloped empty capsid as described above, the alternative pathway proposes that the electron-dense nucleocapsid is assembled first and then enveloped by viral membrane. This suggests that at least some steps of viral morphogenesis need not necessarily occur within the viral envelope.

## Structural Proteins

To study viral structural proteins, large quantities of purified virus particles are needed. In initial studies, WSSV virions were purified from the hemolymph or epithelium tissues of infected peaneid shrimp, but the low quantity and quality of these virus preparations meant that only three major structural proteins – VP28, VP26, and VP24 – were identified using peptide sequencing and antibody recognition (van Hulten et al. 2000a, 2000b). Later, an alternative, more suitable host, the freshwater crayfish, was established for WSSV virus amplification and purification. Compared to penaeid shrimp, freshwater crayfish are cheaper, easier to culture, and more readily available. Most importantly, for unknown reasons, freshwater crayfish can survive WSSV infection better than penaeid shrimp, and the infected crayfish accumulates many WSSV virions in its hemolymph, from which the virions can be purified with high yield and high quality (Huang et al. 2001). Subsequently, after the WSSV genome was sequenced and proteomic approaches were applied (i.e., one- or two-dimensional gel electrophoresis coupled with mass spectrometry), the identification of structural proteins became easier. Thanks to all of these advances, at least 40 WSSV structural proteins have now been identified, ranging in size from 6,077 amino acids (a.a.) to 68 a.a. ( Table 1; Huang et al. 2002; Zhang et al. 2004; Tsai et al. 2004; Xie et al. 2006). When purified WSSV virions are subjected to 8%–18% gradient SDS-PAGE analysis, more than 30 protein bands can be distinguished, and the identities of the major protein bands – VP664, VP28, VP26, VP24, VP19, and VP15 – are now known. However, the biological functions of most of WSSV's structural proteins are unknown.

When viewed by EM, the purified virions appear as intact enveloped virus particles, or as naked nucleocapsids, or as nucleocapsids that are partially wrapped in a fragmented envelope. In the absence of any visible evidence of a third structural component, this led to the belief that the WSSV virion consisted of an envelope and a nucleocapsid. The first study to indicate the presence of another structure in the virion was published by Tsai et al. (2006), who used different concentrations of NaCl, combined with Triton X-100, to differentially solubilize structural proteins from WSSV virions, and their results suggested that, in addition to the envelope and capsid proteins, there are a group of proteins that should be classified as tegument proteins, including VP24, VP26, and at least three others (Fig. 1c). In contrast to

**Table 1** WSSV structural protein genes

| ORF name | Size (a.a.) | Protein name | Function/characteristics | References |
|---|---|---|---|---|
| wsv009 | 95 | VP95 | Structure protein | Huang et al. 2002 |
| wsv026 | 507 | VP507 | Structure protein | Zhang et al. 2004 |
| wsv115 | 968 | VP53B | Structure protein | Tsai et al. 2004 |
| wsv129 | 357 | VP357 | Structure protein | Huang et al. 2002 |
| wsv137 | 337 | VP337 | Structure protein | Zhang et al. 2004 |
| wsv198 | 278 | VP32 | Structure protein | Tsai et al. 2004; Xie et al. 2006 |
| wsv199 | 856 | VP320 | Structure protein | Zhang et al. 2004 |
| wsv249 | 216 | VP216 | Structure protein | Zhang et al. 2004 |
| wsv260 | 387 | VP387 | Structure protein | Zhang et al. 2004 |
| wsv269 | 489 | VP53C | Structure protein | Tsai et al. 2004 |
| wsv284 | 100 | VP13A | Structure protein | Tsai et al. 2004 |
| wsv293 | 60 | VP14 | Structure protein | Xie et al. 2006 |
| wsv303 | 184 | VP184 | Structure protein | Huang et al. 2002 |
| wsv332 | 786 | VP75 | Structure protein | Tsai et al. 2004 |
| wsv338 | 433 | VP11 | Structure protein | Tsai et al. 2004 |
| wsv390 | 321 | ORF390 | Structure protein | Tsai et al. 2004 |
| wsv465 | 1243 | VP136B | Structure protein | Tsai et al. 2004 |
| wsv502 | 362 | VP362 | Structure protein | Zhang et al. 2004 |
| wsv526 | 448 | VP448 | Structure protein | Huang et al. 2002 |
| wsv001 | 1684 | VP1684,Collagen-like | Structure protein, envelope | Huang et al. 2002; Li et al. 2004 |
| wsv011 | 1301 | VP53A, VP150 | Structure protein, envelope | Tsai et al. 2006; Xie et al. 2006 |
| wsv035 | 972 | VP110 | Structure protein, envelope | Tsai et al. 2004; Li et al. 2006a; Xie et al. 2006 |
| wsv209 | 1606 | VP187 | Structure protein, envelope | Li et al. 2006b; Xie et al. 2006 |
| wsv216 | 1194 | VP124 | Structure protein, envelope | Xie et al. 2006 |
| wsv237 | 292 | VP41A | Structure protein, envelope | Huang et al. 2002; Tsai et al. 2004; Xie et al. 2006 |
| wsv238 | 486 | VP51A,VP52A | Structure protein, envelope | Tsai et al. 2004; Xie et al. 2006 |
| wsv242 | 300 | VP300, VP41B | Structure protein, envelope | Huang et al. 2002; Tsai et al. 2004; Xie et al. 2006 |
| wsv254 | 281 | VP281,VP36B,VP33 | Structure protein, envelope | Huang et al. 2002; Tsai et al. 2004; Xie et al. 2006 |

(continued)

**Table 1** (continued)

| ORF name | Size (a.a.) | Protein name | Function/characteristics | References |
|---|---|---|---|---|
| wsv256 | 384 | VP384, VP51B, VP52B | Structure protein, envelope | Huang et al. 2002; Tsai et al. 2004; Xie et al. 2006 |
| wsv259 | 309 | VP38A,VP38 | Structure protein, envelope | Tsai et al. 2004; Xie et al. 2006 |
| wsv321 | 117 | VP13B,VP16 | Structure protein, envelope | Tsai et al. 2004; Xie et al. 2006 |
| wsv325 | 465 | VP60A, VP56 | Structure protein, envelope | Tsai et al. 2004; Xie et al. 2006 |
| wsv327 | 856 | VP90 | Structure protein, envelope | Xie et al. 2006 |
| wsv339 | 283 | VP39B,VP39 | Structure protein, envelope | Tsai et al. 2004; Xie et al. 2006 |
| wsv340 | 261 | VP31 | Structure protein, envelope | Tsai et al. 2004; Xie et al. 2006 |
| wsv386 | 68 | VP68, VP12B | Structure protein, envelope | Huang et al. 2002; Tsai et al. 2004; Zhang et al. 2004 |
| wsv414 | 121 | VP19 | Structure protein, envelope | Huang et al. 2002; van Hulten et al. 2002 |
| wsv421 | 204 | VP28 | Structure protein, envelope | van Hulten et al. 2000b |
| wsv002 | 208 | VP24 | Structure protein, tegument | van Hulten et al. 2000a; Tsai et al. 2006 |
| wsv077 | 297 | VP36A | Structure protein, tegument | Tsai et al. 2004; 2006 |
| wsv306 | 419 | VP39A | Structure protein, tegument | Tsai et al. 2004; 2006 |
| wsv311 | 204 | VP26 | Structure protein, tegument | van Hulten et al. 2000; Tsai et al. 2006; Xie et al. 2006 |
| wsv442 | 800 | VP95 | Structure protein, tegument | Huang et al. 2002; Tsai et al. 2004, 2006; Xie et al. 2006 |
| wsv037 | 1280 | VP160B | Structure protein, capsid | Tsai et al. 2006 |
| wsv214 | 80 | VP15 | Structure protein, capsid, DNA-binding | Zhang et al. 2001; Witteveldt et al. 2005 |
| wsv220 | 674 | VP76,VP73 | Structure protein, capsid | Huang et al. 2002; Tsai et al. 2006; Xie et al. 2006 |
| wsv271 | 1218 | VP136,VP136A | Structure protein, capsid | Tsai et al. 2006; Xie et al. 2006 |
| wsv289 | 1565 | VP160A, VP190 | Structure protein, capsid | Tsai et al. 2006; Xie et al. 2006 |
| wsv308 | 466 | VP466, VP51C, VP51 | Structure protein, capsid | Huang et al. 2002; Tsai et al. 2006; Xie et al. 2006 |
| wsv360 | 6077 | VP664 | Structure protein, capsid | Leu et al. 2005; Tsai et al. 2004, 2006 |

the envelope proteins, which were easily dissolved in Triton X-100 buffer without NaCl, the tegument proteins were only solublized in salt-containing Triton X-100 buffer, suggesting that they are loosely associated with both the envelope and nucleocapsid. The notion that VP26 is a tegument protein is supported by other reports (Xie and Yang 2005; Xie et al. 2006).

Most of the WSSV structural proteins are envelope proteins, and these proteins should play pivotal roles in virus binding, entry, and assembly (Chazal et al. 2003). Since VP28 is the most abundant WSSV envelope protein, it is generally thought to play an important role in the life cycle of WSSV. In a study designed to investigate its role in viral infection, a polyclonal antibody against VP28 was reported to neutralize WSSV's infectivity, suggesting that VP28 was involved in the systemic infection of WSSV in shrimp (van Hulten et al. 2001b). However, a later study (Robalino et al. 2006) demonstrated that this WSSV-neutralizing activity was actually due to a nonspecific inhibitor present in the rabbit serum used in the original study. The interaction of VP28 with shrimp hemocyte membrane proteins has, however, been demonstrated. At least three membrane proteins can interact with recombinant VP28 protein, and one of them is shrimp Rab7 protein (Sritunyalucksana et al. 2006). Rab7 is a small GTP-binding protein, and in mammalian cells, it mainly functions in exocytosis and endocytosis. The importance of this protein in WSSV infection was investigated using an in vivo neutralization assay. In contrast to the WSSV-injected group, most of the shrimp that were injected with WSSV plus PmRab7 or with WSSV plus anti-Rab7 antibody survived to the end of the study, and histopathologically they showed no sign of WSSV infection. Thus, Sritunyalucksana et al. (2006) demonstrated that PmRab7 is involved in WSSV infection in shrimp, although unfortunately they did not present any in vivo evidence of VP28-PmRab7 binding, e.g., by using immunofluorescence to show the co-localization of both proteins in WSSV-infected shrimp cells.

The interaction of VP28 with other WSSV structural proteins has been demonstrated. Using far-Western analysis and co-immunoprecipitation, the tegument proteins VP26 and VP24 bound to VP28 (Xie et al. 2006). Interestingly, these three structural proteins have a high degree of similarity in both their nucleotide and amino acid sequences. This similarity strongly suggests that they arose from the same ancestor, and after gene duplications, they diverged and evolved to have different structural functions (van Hulten et al. 2000a). All three of these proteins have several conserved regions in common, including a strong hydrophobic region with an α-helix that possibly functions as a transmembrane region, and another region that may be involved in mediating the interaction between these proteins (van Hulten et al. 2000a). Overall, these studies suggest that VP28 is a multifunctional envelope protein that not only binds to host membrane proteins (thereby mediating the infection of WSSV), but also interacts with VP24 and VP26 to anchor the envelope onto the underlying tegument layer. In addition, VP26 interacts with the nucleocapsid (see below; Xie and Yang 2005).

The nucleocapsid protein VP15 is a highly basic protein with DNA binding activity, and it is presumed to be involved in the condensation and packing of the WSSV genome into the nucleocapsid (Zhang et al. 2001; van Hulten et al. 2002). VP664 is another major nucleocapsid protein. It is remarkable for its size, consisting of a long polypeptide of 6,077 a.a. encoded by an intron-less giant open reading frame (ORF) of 18,234 nucleotides. This huge protein has a calculated molecular weight of 664 kDa (from which it derived its name, VP664) and it is the largest viral structural protein ever found. This huge structural protein appears to form the stacked ring structures that are visible in the nucleocapsid under TEM (Leu et al. 2005). So far, VP664 is the only nucleocapsid protein that can be unambiguously identified through immunogold electron microscopy (IEM) analysis. When WSSV virions were treated with detergent under conditions that removed the envelope but not the tegument, both VP26 and VP664 were detected by IEM. Conversely, when the envelope was intact, no gold particles for either VP26 or VP664 were detected (Tsai et al. 2006). These results suggest that either some part of VP664 protrudes through the tegument or that the tegument is a relatively permeable structure that allows antibodies to freely diffuse through the tegument layer and interact with the VP664. In addition to VP15 and VP664, the WSSV nucleocapsid contains at least five other proteins.

The tegument protein VP26 also appears to interact with actin (Xie and Yang 2005). It is generally recognized that actins play important roles in the life cycle of many viruses, especially in the transportation of viruses in the infected cells from the initial attachment sites to the replication and gene expression sites, as well as in the final assembly and egress of progeny virus (Radtke et al. 2006). Since Xie and Yang's study (2005) also showed that purified nucleocapsids bound to recombinant VP26, the authors proposed that after WSSV fusion and uncoating, the nucleocapsid might use VP26 to bind to actin, thus allowing the nucleocapsid to travel along the actin filament to the nucleus (Xie and Yang 2005).

Some posttranslational modifications have also been identified in WSSV structural proteins. Both VP28 and another abundant envelope protein, VP19, show threonine residue phosphorylation (Xie et al. 2006). On SDS-PAGE, VP28 migrates as three protein bands with slightly different molecular masses, and Xie et al. (2006) interpreted these bands as corresponding to different phosphorylation levels. Proteolytic processing of viral structural proteins is important for the assembly and morphogenesis of many viruses, and at least three WSSV structural proteins, VP150 (Xie et al. 2006), VP53A, and VP36A (see Fig. 3 in Tsai et al. 2006), appear to be processed. For example, Xie et al. (2006) reported that, in addition to full-length VP150, four smaller related proteins were detected with anti-VP150 antibody, although they mentioned that these products could be due to protease degradation. None of the WSSV structural proteins appear to be glycosylated (van Hulten et al. 2002; Xie et al. 2006), which is very unusual for an animal virus, because glycosylated envelope proteins are often involved in receptor attachment and membrane fusion.

The Arg-Gly-Asp (RGD) motif is known to interact with integrins, and many viruses use this structural motif to bind to integrins on the surface of the host cell in order to effect viral entrance. At least six WSSV structural proteins contain this motif, and one of these, VP110 (an envelope protein), interacts with crayfish hemocytes. Since this interaction is blocked by synthetic RGDT peptides, Li et al. (2006) suggested that the RGD motif in VP110 plays a role in WSSV infection. However, there are no reports of synthetic RGDT peptides affecting the infectivity of WSSV virions.

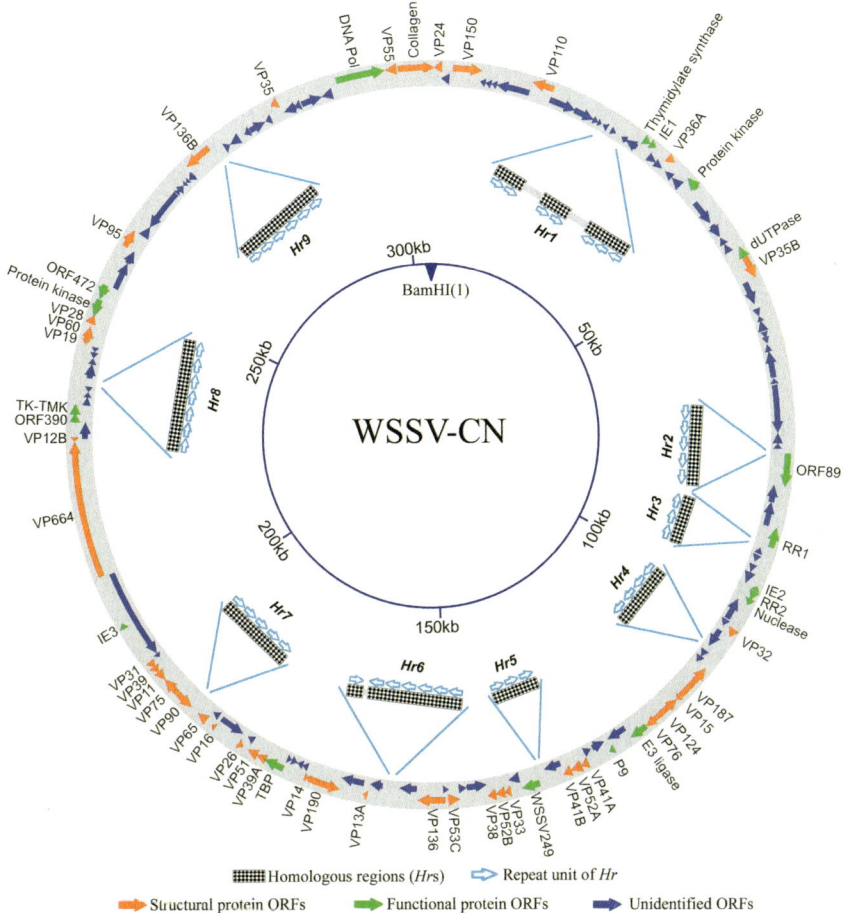

**Fig. 3** Schematic diagram showing the genomic organization of the circular double-stranded WSSV-CN genome. The positions and direction of transcription of corresponding genes are indicated with *solid arrows*. The *G* at the start (GGATCC) of the largest BamHI fragment is designated position 1

The crystallographic structures of VP26 and VP28 were determined recently (Tang et al. 2007). Both proteins have β-barrel architecture with a protruding N-terminal region, and further studies showed that both proteins assume a trimeric form in the envelope. Structural comparisons revealed that both VP26 and VP28 are evolutionarily related to structural proteins from other viruses, including the bacteriophage PRD1 major coat protein P3 and adenovirus major coat protein hexon as well as other proteins cited by Tang et al. (2007). Based on the structural features, combined with evidence from the literature, Tang et al. (2007) hypothesized that the putative N-terminal transmembrane region of VP26 and VP28 anchors the viral envelope membrane in such a way that the core β-barrel protrudes outside the envelope, where it would be free to interact with host receptors or membranes to initiate viral infection. It is also possible that this β-barrel may mediate interactions with other structural proteins, such as VP26 and VP24.

In summary, most of the WSSV structural proteins have been identified. Future studies are needed to determine how these proteins interact with each other, how they assemble to form the virion structure, and how they interact with host proteins during infection.

## WSSV Genome Structure

The WSSV genome is a large circular dsDNA of approximately 300 kbp (Fig. 3). Three WSSV isolates from China (WSSV-CN, accession no. AF332093), Thailand (WSSV-TH, accession no. AF369029), and Taiwan (WSSV-TW, accession no. AF440570), have been completely sequenced, and their genome sizes are 305, 297, and 307 kbp, respectively. The total G+C content of the WSSV genome is roughly 41%, and this dinucleotide (G+C) is uniformly distributed throughout the genome (van Hulten et al. 2001b; Yang et al. 2001). Presently, each of the three sequenced WSSV genomes has its own gene or ORF nomenclature system, but the ICTV whispovirus study group committee recently chose the China isolate, WSSV-CN, as the type strain. In this review, we will use the WSSV-CN gene nomenclature system.

Most of the WSSV genome sequences are unique; only 3% of the genome consists of highly repetitive sequences. These repetitive sequences are organized into nine homologous regions, which are distributed throughout the genome and are mainly located in intergenic regions. The homologous regions contain 47 repeated mini-fragments that include direct repeats, atypical inverted repeat sequences, and imperfect palindromes. Many baculovirus genomes contain homologous regions, which function as enhancers of early gene transcription and as initiation sites for DNA replication. They have also been implicated as sites of DNA recombination. Although the organization of the WSSV homologous regions is similar to that of the baculoviruses, the function(s) of WSSV homologous regions has not been determined. However, at least one WSSV protein, WSV021, an early gene product, binds to the homologous regions in vitro (Zhu et al. 2007).

An open reading frame (ORF) is a region of an organism's genome that could potentially encode a protein. Sequence analysis identified a total of 531 putative ORFs that consist of at least 60 codons in the WSSV-CN isolate. Roughly one-third of these (181 ORFs) are nonoverlapping. Approximately 80% of the nonoverlapping ORFs have a potential polyadenylation site (AATAAA) downstream of the ORF. The sizes of the proteins encoded by these predicted ORFs range from 60 to 6,077 a.a. (The 6,077-a.a. ORF encodes the major nucleocapsid protein, VP664; see Sect. 7). Only 45 of the nonoverlapping ORFs encode proteins that resemble known proteins (>20% a.a. identity) or motifs (Yang et al. 2001). Conversely, some WSSV ORFs encode proteins that share significant similarities (40% or even higher) to each other, and these ORFs can be classified into the same gene family. A total of 27 ORFs can be classified into ten putative WSSV gene families, and it has been proposed that these families arose from gene duplications in the WSSV genome (van Hulten et al. 2001).

Comparison of the complete genome sequences of the three isolates identified five types of genetic variations:

1. A large 13 kbp deletion in the WSSV-TH genome relative to the WSSV-CN and WSSV-TW genomes;
2. A genetically variable region of approximately 750 bp in WSSV-TH;
3. A transposase sequence with 1,337 bp that is only present in WSSV-TW;
4. Variation in the number of repeat units within homologous regions and direct repeats;
5. Single nucleotide mutations, including insertions/deletions and single nucleotide polymorphisms (Marks et al. 2004).

If the variations in the number of repeats and single nucleotide mutations are ignored, then these three isolates have 99.32% nucleotide identify. In WSSV-CN, the region corresponding to the 13-kbp deletion site in WSSV-TH contains 13 ORFs; the genetic variations in this region are related to differential virulence in different WSSV isolates (Lan et al. 2002). Thus, although this region may not be important for WSSV survival, it might contain some virulence factors. This region also encodes a nucleocapsid protein (VP35; Chen et al. 2002a) that is presumably nonessential, based on its absence in the WSSV-TH isolate.

## WSSV Genes Identified by Homology Searches

Homology searches reveal that most of the proteins (73%) encoded by the WSSV ORFs have no significant similarity to proteins from viruses or other organisms and that they lack identifiable functional domains (Table 2, Fig. 3). The few WSSV ORFs that are identifiable encode gene products involved in nucleotide metabolism, specifically both subunits of ribonucleotide reductase (RR, WSV172, and WSV188), a chimeric thymidylate/thymidine kinase (TK-TMK, WSV395), thymidylate synthase (TSY, WSV067), and dUTP pyrophosphatase (dUTPase, WSV112). RR and TK-TMK are important for the synthesis of deoxynucleotide precursors for DNA replication. The chimeric WSSV TK-TMK is unusual because in other large DNA viruses TK and TMK are normally encoded by separate ORFs. Only a few viruses contain TSY, and this enzyme is important in the de novo pathway of pyrimidine biosynthesis. However, many large DNA viruses encode dUTPase; this enzyme is responsible for regulating cellular levels of dUTP. Possession of these genes might allow WSSV to replicate efficiently in nondividing cells. The enzymatic activities of RR, TK-TMK, TSY, and dUTPase have been verified by assaying purified recombinant proteins (Lin et al. 2002; Tzeng et al. 2002; Li et al. 2004b; Liu and Yang 2005). The RR activity in shrimp gills increases after WSSV infection (Lin et al. 2002).

Bioinformatic searches identified only one WSSV ORF related to DNA replication. This ORF encodes WSSV DNA polymerase (DNA pol, WSV514), which contains all seven conserved DNA pol sequence motifs. However, the spacer regions surrounding the conserved motifs are expanded; consequently, WSSV DNA pol is about twice as large as other viral DNA pols (Chen et al. 2002b). Two serine/threonine protein kinases (PK; WSV083 and WSV423) have also been identified.

**Table 2** WSSV ORFs with confirmed functions/characteristics, not including structural proteins

| ORF name | Size (a.a.) | Protein name | Function/characteristics | References |
|---|---|---|---|---|
| wsv021 | 200 | WSV021 | Hr-binding protein | Zhu et al. 2007 |
| wsv390 | 321 | ORF390 | Anti-apoptosis | Wang et al. 2004 |
| wsv514 | 2195 | DNA polymerase (DNA pol) | DNA replication | Chen et al. 2002b |
| wsv069 | 224 | IE1 | Immediate-early gene | Liu et al. 2005; 2007 |
| wsv187 | 108 | IE2 | Immediate-early gene | Liu et al. 2005 |
| wsv359 | 60 | IE3 | Immediate-early gene | Liu et al. 2005 |
| wsv151 | 1436 | WSV151, ORF89 | Latency-related gene | Khadijah et al. 2003 |
| wsv366 | 84 | WSSV366 | Latency-related gene | Khadijah et al. 2003 |
| wsv427 | 623 | WSV427 | Latency-related gene | Khadijah et al. 2003; Lu and Kwang 2004 |
| wsv230 | 82 | P9 (ICP11) | Nonstructural, highly abundant protein | Liu et al. 2006; Wang et al. 2007 |
| wsv112 | 461 | dUTPase | Nucleotide metabolism | Liu et al. 2005 |
| wsv172 | 848 | Ribonucleotide reductase large subunit (RR1) | Nucleotide metabolism | Lin et al. 2002 |
| wsv188 | 413 | Ribonucleotide reductase large subunit(RR2) | Nucleotide metabolism | Lin et al. 2002 |
| wsv191 | 311 | Nonspecific nuclease | Nucleotide metabolism | Li et al. 2005 |
| wsv395 | 398 | Chimeric thymidine and thymidylate kinase (TK-TMK) | Nucleotide metabolism | Tsai et al. 2000b; Tzeng et al. 2002 |
| wsv067 | 289 | Thymidylate synthase | Nucleotide metabolism | Li et al. 2004 |
| wsv083 | 581 | Protein kinase 2 | Protein modification | Van Hulten and Vlak 2001 |
| wsv423 | 730 | Protein kinase 1 | Protein modification | Liu et al. 2001 |
| wsv222 | 844 | WSV222 | Ubiquitination-related protein | He et al. 2006 |
| wsv249 | 783 | WSSV249 | Ubiquitination-related protein | Wang et al. 2005 |

These two PK genes have 45% amino acid sequence similarity, and phylogenetic analysis suggests that the two genes were probably derived from a common ancestor through a gene duplication event (Liu et al. 2001; van Hulten and Vlak 2001; van Hulten et al. 2001a). However, their kinase activities and their potential substrates have not been determined. A nonspecific nuclease (WSV191) was also identified, based on several conserved structural motifs, and its nuclease activity was demonstrated with recombinant protein (Li et al. 2005). Homology searches showed that WSSV encodes a collagen-like protein (WSV001). This is the first collagen gene to be identified in a virus genome, and the encoded protein resembles human collagen type VII (42% a.a. identity). RT-PCR showed that this gene was expressed at an early stage of infection, and IEM indicates that the protein was located on the outside of the viral envelope (Li et al. 2004a).

Several WSSV ORFs contain identifiable protein motifs, from which the functions of these proteins can be predicted. Two proteins have an EF-hand calcium-binding motif (WSV079 and WSV427; WSV427 is a latency-related gene), suggesting that WSSV might modulate calcium levels in infected cells. The RING-H2 finger motif is involved in ubiquitin-conjugating enzyme (E2)-dependent ubiquitination, and many proteins containing a RING finger play a key role in the ubiquitination pathway. Four WSSV proteins (WSV199, WSV222, WSV249, and WSV403) contain this motif, and ubiquitination activity has been demonstrated for two of them (WSV222, WSV249; Wang et al. 2005; He et al. 2006). The host proteins that interact with these two viral proteins have been identified. WSV249 interacts with a shrimp ubiquitin-conjugating enzyme, PvUbc, and mediates ubiquitination through its RING-H2 motif in the presence of E1 and PvUbc. WSV222 binds to and ubiquitinates a shrimp tumor suppressor-like protein (TSL), and the ubiquitinated TSL then undergoes proteasome-dependent degradation. Since transient expression of TSL in BHK cells leads to apoptosis, and apoptosis can be blocked by WSV222, He et al. (2006) proposed that WSV222 might use ubiquitin-mediated degradation of TSL to function as an anti-apoptosis protein and thus promote WSSV propagation. Several other WSSV proteins contain zinc finger or leucine zipper motifs, both of which are involved in DNA–protein interactions and in the regulation of transcriptional activation. One of these ORFs, WSV069, has been identified as an immediate early gene (see Sect. 10.1).

# WSSV Genes Identified Through Functional Studies

## *Immediate-Early Genes*

By definition, transcription of a viral immediate early gene should be insensitive to cycloheximide, and based on this criterion, three WSSV immediate early genes, *ie1* (*wsv069*), *ie2* (*wsv178*) and *ie3* (*wsv359*), were identified (Table 2). After large-scale screening using microarrays, RT-PCR showed that the corresponding transcripts of these three genes were consistently present in cycloheximide-treated, WSSV-infected shrimp

(Liu et al. 2005). IE1 contains a Cys2/His2-type zinc finger motif, which is involved in DNA binding and therefore suggests that IE1 might function as a transcription factor. The ORFs corresponding to *ie2* and *ie3* do not contain any recognizable functional motifs. The roles of these two genes are unknown; possibly they encode regulatory RNAs.

The three *ie* genes have different promoter sequences. The *ie1* gene promoter is similar to other WSSV early gene promoters and it has a canonical TATA box and a downstream Inr element (see Sect. 11.3). In contrast, the *ie2* and *ie3* genes lack both a TATA box and an Inr element. *ie1* has very strong promoter activity in cultured SF9 insect cells (SF9) and does not require other viral factors (Liu et al. 2005). The *ie1* promoter contains a STAT-binding motif that is important for this high level of promoter activity; both insect and shrimp STATs bind this motif in vitro (Liu et al. 2007). Liu et al. (2007) extrapolated evidence from the insect cell to hypothesize that STATs probably directly activate WSSV *ie1* gene expression and also contribute to its high promoter activity in shrimp cells.

## Latency-Related Genes

Three latency-related genes, *wsv151*, *wsv427*, and *wsv366*, were isolated from specific pathogen-free (SPF) shrimp in a study to test whether these SPF shrimp were WSSV asymptomatic carriers (Table 2) (Khadijah et al. 2003). The SPF shrimp were diagnosed as WSSV-free by a commercial WSSV detection kit, but when their RNAs were analyzed using WSSV DNA microarrays, several WSSV transcripts were detected. After further confirmation by RT-PCR, the transcribed genes were identified as *wsv151*, *wsv427*, and *wsv366*. These three genes were highly expressed in SPF shrimp relative to *vp15*, *vp26*, and *vp24*, while the reverse was true in WSSV-infected shrimp. Yeast two-hybrid screening showed that WSV427 interacted with a novel shrimp serine/threonine protein phosphatase, and co-immunoprecipitation experiments established that these two proteins interacted both in vivo and in vitro (Lu and Kwang 2004). When WSV151 was expressed in SF9 cells, it had an apparent molecular mass of roughly 165 kDa and was localized in the nucleus. A nuclear localization signal was identified at amino acids 678–683 ([678]KMKRKR[683]). In Sf9 cells, WSV151 showed some degree of gene regulatory activity by repressing its own promoter activity as well as the promoter activities of a protein kinase and the WSSV thymidine-thymidylate kinase genes (Hossain et al. 2004). However, it is unknown if WSV151 represses these promoters directly or if it acts indirectly by first interacting with other insect proteins.

## Anti-apoptosis Gene

WSSV infection induces apoptosis in shrimp, and the infected tissues display characteristic signs of apoptosis, i.e., nuclear disassembly, fragmentation of DNA into a ladder, and increased caspase-3 activity (Wongprasert et al. 2003). WSSV-induced apoptosis occurs in bystander, noninfected cells, whereas infected cells are nonapoptotic (Wongprasert et al. 2003). Although the factors that induce apoptosis are currently unknown, an anti-apoptosis protein (ORF390; WSV390) was recently identified in WSSV. Wang et al. (2004) used a genetic complementation strategy to show that ORF390 restored the replication ability of a *p35*-deficient *Autographa californica nucleopolyhedrovirus* (AcMNPV) in SF-9 cells. From this evidence, Wang et al. (2004) proposed that ORF390 might function as an apoptotic suppressor like P35 in AcMNPV. ORF390 has three putative cleavage sites near its C-terminus, two for caspase-9 and one for caspase-3. Our studies (J.-H. Leu et al., unpublished data) have shown that ORF390 is a direct *Penaeus monodon* caspase inhibitor and that the caspase-3 cleavage site is important for this function. Therefore, it is likely that ORF390 functions by using a mechanism similar to that of AcMNPV P35. Specifically, the target caspase cleaves the caspase-3 site in ORF390, the cleaved ORF390 forms a complex with the target caspase, and consequently the activity of the bound caspase is blocked. Since it is reasonable to suppose that shrimp use apoptosis as a protective response to prevent the spread of WSSV, it is equally likely that WSSV produces the anti-apoptosis protein, ORF390, to block apoptosis and thus facilitate virus replication.

## The Highly Expressed icp11 *(or* vp9*) Gene*

The existence of the *icp11* gene (*wsv230*) first became apparent from an analysis of a WSSV-infected shrimp postlarvae cDNA library. This gene had the highest number of corresponding cDNA clones of all the WSSV genes in the library (Wang et al. 2007a). A proteomic study to observe protein profile changes during WSSV infection subsequently identified ICP11 in the stomach of WSSV-infected shrimp, and in fact ICP11 was the only WSSV protein to be detected in this study (Wang et al. 2007b). Because of its high protein levels, ICP11is potentially a promising target for an immunology-based WSSV detection assay. Confirmation of its high abundance at transcription and translational levels was provided by microarray and Western blot analysis, respectively (Wang et al. 2007a). Immunofluorescence staining detected ICP11 in both the cytoplasm and the nucleus of WSSV-infected shrimp hemocytes (Wang et al. 2007a). X-ray and nuclear magnetic resonance (NMR) showed that ICP11 is folded in a way that is similar to the DNA-binding domain of the papillomavirus E2 protein (Liu et al. 2006). From this structural evidence, it was hypothesized that like the E2 protein in papillomavirus, ICP11 might be involved in transcriptional regulation (Liu et al. 2006).

## WSSV Gene Expression and Regulation

### Gene Expression Studies Using RT-PCR

Transcriptional activities of WSSV genes have been studied in penaeid shrimp and crayfish. To achieve synchronized viral infection, the animals are infected by intramuscular injection. Temporal transcription studies on various WSSV genes that were presumed to be early (*rr1, rr2, pk1, tk- tmk, dnapol, ie1, ie2 ,and ie3)* and on several major structural protein genes that were expected to be late (*vp664, vp28, vp26, vp24, vp19,* and *vp15*), confirmed that these two classes of genes had different temporal expression profiles (Tsai et al. 2000a, 2000b; Liu et al. 2001, 2005; Chen et al. 2002b; Marks et al. 2003; Leu et al. 2005). RT-PCR detected WSSV early genes in *P. monodon* as early as 2–4 h p.i., with transcription levels that slowly and steadily increased until the end of the experiment (48 or 60 h p.i.) or else reached a plateau phase at 12–18 h p.i. For the major viral structural protein genes, although very low transcription levels were detected as early as 2–4 h p.i., the transcription levels surged at 12–18 h .p.i and steadily increased thereafter.

In crayfish, expression of the structural protein genes was not observed until at least 16–24 h p.i. (Marks et al. 2003). *vp15* was the first gene to be detected at 16 h p.i., and the early expression of this protein supports the assumption that VP15's DNA-binding activity is involved in WSSV genome binding, which occurs after the WSSV genome is replicated and before the genome is packaged into the viral particles. Expression of *vp28, vp26, vp24,* and *vp19* was first detected at 1 day p.i. Most of these genes had strong and relatively constant expression levels, but the expression level of *vp24* was low and declined gradually. This low expression level is consistent with the low quantity of VP24 protein in the virions. *rr1* gene transcription has also been studied in crayfish (Marks et al. 2003). *rr1* transcripts were first detected at 6 h p.i., significantly increased at 2 days p.i. and then slowly increased to the end of the experimental period (7 days p.i.). Taken together, these studies in penaeid shrimp and crayfish reveal that the transcriptional activities of WSSV genes are generally delayed and progress more slowly in crayfish, which might explain why crayfish is more tolerant to WSSV infection. Another point to note is that no RNA splicing events have been identified in WSSV genes.

## Gene Expression Analyses Using DNA Microarrays

WSSV gene expression has also been globally analyzed using DNA microarrays. In the first study to use this technology, Wang et al. (2003) analyzed the expression profile of the 532 putative genes of the WSSV-TW isolate in gills of experimentally infected *P. monodon*. Their analysis showed that at least 89.5% (476/532) of the WSSV ORFs were expressed in the WSSV-infected gill tissues. Among these genes, 23.5%, including *rr1* and *rr2*, began to express at 2 h p.i., and these genes were considered to be early genes. For the other WSSV genes, 4.2% expressed at 6 h p.i., 17.7% at 12 h p.i. and 47.9% at 24 h p.i. Most of the WSSV structural protein genes were first expressed either at 12 h p.i. (nine WSSV structural protein genes) or at 24 h p.i. (14 WSSV structural protein genes); these and other genes that also began to be expressed at these time points were considered to be late genes. Interestingly, several structural protein genes expressed at 2 h p.i., suggesting that in addition to their function as structural proteins, these proteins might also have other important roles in the early infection stage (Wang et al. 2003; Tsai et al. 2004).

In another microarray study, Marks et al. (2005) analyzed the transcription profiles of 184 putative genes encoded by the WSSV-TH isolate in WSSV-infected
*P. monodon* gill tissues. They found that 79% of the putative ORFs were detected
and that these genes could be clustered into two major classes: the first class
reached maximal expression at 20 h p.i. and the second class at 2 days p.i. Most of
the known WSSV early genes, including *rr1*, *rr2*, *tk-tmk*, and both *pk* genes, were
found in the first class, and therefore all the genes in this class were classified by
the authors as early genes. Conversely, the second class included all major and most
minor structural protein genes, all of which are predicted to be expressed late during infection; therefore, genes that were expressed at the same time as the structural
protein genes were considered to be late genes. However, a gene classification
scheme based only on the times of maximal expression is too simplistic and prone
to error, and in the Marks et al. (2005) study, two functionally early genes, *tsy* and
*dna pol*, were inappropriately classified as late genes. A third microarray study
analyzed the transcription activities of WSSV-CN genes in WSSV-infected crayfish
hepatopancreas (Lan et al. 2006). In this study of 151 putative genes, 81.1% were
detected, and expression of 47 of these began at 6 h p.i. These 47 genes were classified as early genes.

Most of the data from these early microarray studies still need to be confirmed
by additional experiments. Meanwhile, functional and temporal studies (RT-PCR
and microarray) suggest that WSSV resembles other large dsDNA viruses in that
gene expression is regulated in a coordinated series of cascades, and its genes fall
into at least three classes (i.e., immediately-early, early, and late genes).

## *WSSV Gene Promoter Features*

The 5′RACE procedure was used to determine the transcription initiation site (TIS),
from which the promoter regions of WSSV genes can be deduced and analyzed.
Initial studies showed that WSSV early genes, including *rr1*, *rr2* (Tsai et al. 2000a),
and *dnapol* (Chen et al. 2002b), shared consensus features in their promoter
regions. A TATA box was present 25–27 nucleotides upstream of the TISs, and the
consensus initiator (Inr) motif, a/tCAc/g/tT, closely matched the a/c/t/CAg/tT Inr
motif of arthropods (Cherbas and Cherbas 1993). Based on these early studies, the
structure of the promoter regions of these three WSSV early genes seems to mimic
the RNA polymerase II promoter of arthropods. On the other hand, the initial studies on the promoters of several major WSSV structural protein genes only showed
the presence of an A/T-rich sequence 25 nt upstream of the TIS and failed to identify a consensus sequence. Nevertheless, a short sequence, AATAAC, was identified near the TISs of both *vp664* and *vp28*, and it may be significant that these two
genes encode the most abundant nucleocapsid and envelope protein, respectively
(Marks et al. 2003; Leu et al. 2005).

With the identification and characterization of more WSSV genes, more data
became available for WSSV gene promoter analysis. In a microarray study, Marks

et al. (2005) grouped the WSSV genes as early and late genes according to their temporal expression profiles. Later, Marks et al. (2006) used an *in silico* method to analyze the upstream regions of early and late WSSV genes to search for conserved promoter motifs, and their results were then validated by alignments of the empirically determined 5′ ends of various WSSV mRNAs. As with the earlier studies (Tsai et al. 2000a; Chen et al. 2002b), this study also found that the upstream region of the WSSV early genes contained a TATA box and a consensus Inr element. The Inr element was located near the TISs of the early genes and its sequence was similar to that of the *Drosophila* Inr. All the evidence, therefore, indicates that the WSSV early gene promoter is similar to the *Drosophila* RNA polymerase II core promoter sequence, and this in turn suggests that the WSSV early genes are under the control of the cellular transcription machinery. The deduced structure of the WSSV early gene promoter is that a consensus TATA box (TATAa/tA) is located 20–30 nt upstream of the Inr (a/cTCANT), and the Inr is between 20 and 85 nucleotides upstream of the translational start codon of the early genes. The results of the Marks et al. (2006) study also confirmed the Inr motif (a/tCAc/g/tT) as deduced from *rr1*, *rr2*, and *dnapol*. However, not all early or immediate early genes conform to this structure. For example, as noted above, *ie2* and *ie3* have neither a TATA box nor an Inr. Interestingly, *ie3* has two different TISs, which it uses at the early and late stages of infection (24 h vs 60 h p.i.), respectively (Liu et al. 2005).

Promoter regions in WSSV late genes have also been investigated (Marks et al. 2006). The TISs of eight major structural protein genes have been experimentally determined, and alignment of upstream regions revealed a degenerate motif (ATNAC) within or near the TISs, and an A/T rich region located 20–25 nt upstream of the TISs. The ATNAC motif is consistent with an earlier motif (AATAAC) that had been predicted from *vp664* and *vp28* (Liu et al. 2005). The distance from the TIS to the translational start codon varied from gene to gene and ranged from 30 to 220 nt. *In silico* analysis found that the ATNAC motif was present in the first 100 nt upstream of the start codon in 40% of WSSV's late genes (as classified by microarray data). This result suggests that the ATNAC motif might play an important role in late gene expression. Among the eight major structural protein genes, only *vp15* has a functional TATA box, and it is located within the A/T rich region (Marks et al. 2005). The presence of this TATA box might explain why *vp15* is transcribed earlier than WSSV's other major structural protein genes. *In silico* analysis further showed that about half of WSSV's putative late genes (including its structural protein genes and those genes classified as late based on microarray data) contained a consensus TATA box. Transcription of WSSV's late genes would therefore seem to involve at least two different promoters, namely one with, and one without, a TATA box. Since the WSSV genome has no homologs of RNA polymerase or its subunits, at least one novel transcription factor would need to be induced at the early infection stage. These as yet unidentified factors (candidates might include IE1, IE2 or IE3) would presumably recognize a late gene specific motif (such as the ATNAC motif), and then recruit the host cellular transcription machinery, particularly the RNA polymerase II system, to transcribe the late genes. The WSSV genome has two annotated transcription factors, a TATA-box binding protein (TBP; WSV303) and a CREB-binding protein (CBP; WSV100), but the similarity of these two ORFs to target proteins is weak, and their transcriptional functions (if any) remain to be identified.


## WSSV Might Use Polycistronic mRNA and IRES for Gene Expression

The 3'RACE procedure was used to determine the polyadenylation sites of WSSV genes. All the identified sites from early and late genes are within the expected range for polyadenylation in eukaryotic mRNAs, i.e., 15–25 nt downstream of the sequence AAUAAA. This result suggests that WSSV uses the regular cellular enzymes for polyadenylation of its mRNA. However, our own experiments indicate that, for some genes, the 3'RACE procedure was unable to identify the 3' end of a gene transcript due to PCR failure. One possible explanation is that these genes produce transcripts without a poly(A) tail. Alternatively, for at least some genes, it may be that the 3' end of the transcript is not easily amplified under normal conditions. For instance, as shown in Fig. 3, genes with the same transcription orientation are sometimes clustered together (*vp31/vp39/vp11* or *vp41A/vp52A/vp41B*), and the intergenic region between these genes can be as short as 1 nt. In a preliminary study on the transcription of one of these gene clusters (*vp31/vp39/vp11*), we used 5'RACE and Northern blot analysis to show that a single mRNA molecule encompassing all three genes was transcribed. It is reasonable to expect that other gene clusters might also use this same transcription strategy, and if a gene is located at the 5' end of such a cluster, then 3'RACE might fail to identify the 3' end of the transcript, because the corresponding 3' end of the cDNA would be very long and thus difficult to amplify.

This same preliminary study also found that internal ribosome entry site (IRES) elements might be used to translate the protein of the second gene in the *vp31/vp39/vp11* cluster. The possibility that WSSV might use an IRES to regulate translation was first proposed by Han and Zhang (2006), who identified a 180 bp IRES in the upstream region of *vp28*. When this IRES was inserted into a baculovirus genome under the control of a *polyhedrin* promoter, it drove the efficient translation of a downstream reporter protein in insect cells. However, although this IRES works efficiently in insect cells, its importance in regulating the translation of *vp28* in WSSV-infected shrimp has not been established, and it is not clear if this is a functional IRES in vivo. Han and Zhang (2006) found this 180-bp fragment in a *vp28* cDNA clone (Zhang et al. 2002), in which it separated an in-frame minicistron from the downstream *vp28* coding region. However, 5'RACE analysis by Marks et al. (2003) showed that the 5'UTR (untranslated region) of *vp28* was only approximately 35 bp. This is obviously much shorter than the upstream region of the Zhang et al. (2002) cDNA clone, and it is too short to function properly as an IRES. Nevertheless, while this conflicting evidence for a *vp28* IRES remains unresolved, it seems likely that polycistronic mRNAs coupled with IRESs are used to regulate the expression of at least some of the clustered WSSV structural protein genes. If WSSV does use IRESs, then it would be the first marine DNA virus reported to do so. This kind of gene regulation mechanism is analogous to the operon in prokaryotes; that is, functionally similar or related genes are transcribed as a polycistronic, single mRNA molecule.

## WSSV Modulates the Expression of Host Cell Genes

Knowledge of the interactions between virus and host is important for understanding the pathogenesis of a viral disease. One way to study virus–host interactions is to examine the expression profile changes of host genes after virus infection. A variety of approaches have been applied to investigate the effect of WSSV infection on gene expression profiles of shrimp host cells, including a mRNA differential display technique (Astrofsky et al. 2002), suppression subtractive hybridization (SSH; Pan et al. 2005), SSH and differential hybridization (He et al. 2005), cDNA microarrays (Dhar et al. 2003; Wang et al. 2006), and ESTs (Rojtinnakorn et al. 2002). All of these studies focused on the gene expression changes in immune-related organs, specifically the hepatopancreas and hemocytes. EST analysis of shrimp hemocytes (Rojtinnakron et al. 2002) showed that WSSV infection stimulates the expression of defense-related genes; the putative defense genes increased from 2.7% of the total ESTs in a normal shrimp library to 15.7% of the total ESTs in an infected library. To look for genes that might be involved in viral resistance in WSSV-resistant shrimp, Pan et al. (2005) used SSH to identify differentially expressed genes in the hepatopancreas. The highly abundant genes in their subtractive library included a $\beta$-1–3-D-glucan-binding protein, hemocyanin, lectin, ferritin, oxygenase, and chitinase. Several genes encoding apoptotic-related proteins, antioxidant enzymes, and small GTPases were also expressed at higher levels in virus-resistant shrimp. Genes that are differentially expressed in the hemocytes of virus-resistant shrimp have also been identified (He et al. 2005). These include genes encoding an interferon-like protein, a $(2'-5')$ oligo(A) synthetase-like protein, several redox-related factors, a C-type lectin, a laminin-like protein and a translationally controlled tumor protein (TCTP).

A proteomic approach was used to observe protein expression profile changes in a primary WSSV target organ after WSSV infection (Wang et al. 2007b). In this study, proteins were extracted from the stomachs of WSSV-infected or uninfected shrimp, and then separated using two-dimensional gel electrophoresis (2-DE). The profiles were then compared, and protein spots of interest were identified using mass/mass spectrometry. The results showed that WSSV infection increased protein levels for several key glycolytic enzymes, several proteins in the electron transport chain, and a kinase involved in nucleic acid synthesis, suggesting that WSSV upregulates the syntheses of ATP and nucleic acids to promote its rapid multiplication. WSSV infection also appears to upregulate a protein involved in calcium homoeostasis (sarco/ER-type calcium pump-ER $Ca^{2+}$-ATPase), a cellular signaling protein (14–3-3b) and a voltage-dependent anion channel (VDAC). Proteins with decreased levels included several digestive enzymes, two calcium-binding proteins (SCP and calponin), and a small ubiquitin-like modifier. Accumulated protein levels were unchanged for proteins related to cellular components, the ATP buffering system/resistance to environmental stresses, protein folding activity, antioxidant activity, and amino acid catabolism. This study provided the first definitive global view of host protein expression levels and showed how WSSV subverts the cellular processes.

The RNA and protein level changes in a virus-infected cell are the outcome of interactions between the host cells and the replicating viruses. As mentioned above, many differentially expressed genes have been identified in WSSV-infected shrimp, but in most cases their biological functions and importance in infected shrimp remain unknown. Nevertheless, it seems likely that WSSV modulates the expression of some host genes to promote its multiplication, while the host cells regulate the expression of genes with antiviral effects to combat the WSSV infection. As yet, no overexpression techniques have been developed and used to evaluate the importance of these differentially expressed genes in WSSV-infected shrimp; conversely, RNAi seems to be a promising tool to silence the expression of genes of interest for the related studies, since it has been successfully used both in vitro (Liu and Soderhall 2007) and in vivo (Robalino et al. 2004, 2005).

**Acknowledgements** We are indebted to Paul Barlow for his helpful criticism.

# References

Astrofsky KM, Roux MM, Klimpel KR, Fox JG, Dhar AK (2002) Isolation of differentially expressed genes from white spot virus (WSV) infected Pacific blue shrimp (*Penaeus stylirostris*). Arch Virol 147:1799–1812

Chang PS, Lo CF, Wang YC, Kou GH (1996) Identification of white spot syndrome associated baculovirus (WSBV) target organs in the shrimp *Penaeus monodon* by *in situ* hybridization. Dis Aquat Org 27:131–139

Chazal N, Gerlier D (2003) Virus entry, assembly, budding, and membrane rafts. Microbiol Mol Biol Rev 67:226–237

Chen LL, Leu JH, Huang CJ, Chou CM, Chen SM, Wang CH, Lo CF, Kou GH (2002a) Identification of a nucleocapsid protein (VP35) gene of shrimp white spot syndrome virus and characterization of the motif important for targeting VP35 to the nuclei of transfected insect cells. Virology 293:44–53

Chen LL, Wang HC, Huang CJ, Peng SE, Chen YG, Lin SJ, Chen WY, Dai CF, Yu HT, Wang CH, Lo CF, Kou GH (2002b) Transcriptional analysis of the DNA polymerase gene of shrimp white spot syndrome virus. Virology 301:136–147

Cherbas L, Cherbas P (1993) The arthropod initiator: the capsite consensus plays an important role in transcription. Insect Biochem Mol Biol 23:81–90

Dhar AK, Dettori A, Roux MM, Klimpel KR, Read B (2003) Identification of differentially expressed genes in shrimp (*Penaeus stylirostris*) infected with white spot syndrome virus by cDNA microarrays. Arch Virol 148:2381–2396

Durand S, Lightner DV, Nunan LM, Redman RM, Mari J, Bonami JR (1996) Application of gene probes as diagnostic tools for white spot baculovirus (WSBV) of penaeid shrimp. Dis Aquat Org 27:59–66

Durand S, Lightner DV, Redman RM, Bonami JR (1997) Ultrastructure and morphogenesis of white spot syndrome baculovirus (WSSV). Dis Aquat Org 29:205–211

Flegel TW (1997) Special topic review: major viral diseases of the black tiger prawn (*Penaeus monodon*) in Thailand. World J Microbiol Biotechnol 13:433–442

Flegel TW (2006) Detection of major penaeid shrimp viruses in Asia, a historical perspective with emphasis on Thailand. Aquaculture 258:1–33

Han F, Zhang X (2006) Internal initiation of mRNA translation in insect cell mediated by an internal ribosome entry site (IRES) from shrimp white spot syndrome virus (WSSV). Biochem Biophys Res Commun 344:893–899

He F, Fenner BJ, Godwin AK, Kwang J (2006) White spot syndrome virus open reading frame 222 encodes a viral E3 ligase and mediates degradation of a host tumor suppressor via ubiquitination. J Virol 80:3884–3892

He N, Qin Q, Xu X (2005) Differential profile of genes expressed in hemocytes of white spot syndrome virus-resistant shrimp (*Penaeus japonicus*) by combining suppression subtractive hybridization and differential hybridization. Antivir Res 66:39–45

Hossain MS, Khadijah S, Kwang J (2004) Characterization of ORF89-a latency-related gene of white spot syndrome virus. Virology 325:106–115

Huang C, Zhang L, Zhang J, Xiao L, Wu Q, Chen D, Li JK (2001) Purification and characterization of white spot syndrome virus (WSSV) produced in an alternate host: crayfish, *Cambarus clarkii*. Virus Res 76:115–125

Huang C, Zhang X, Lin Q, Xu X, Hu Z, Hew CL (2002) Proteomic analysis of shrimp white spot syndrome viral proteins and characterization of a novel envelope protein VP466. Mol Cell Proteomics 1:223–231

Jiravanichpaisal P, Bangyeekhun E, Soderhall K, Soderhall I (2001) Experimental infection of white spot syndrome virus in freshwater crayfish *Pacifastacus leniusculus*. Dis Aquat Orga 47:151–157

Jiravanichpaisal P, Soderhall K, Soderhall I (2006) Characterization of white spot syndrome virus replication in in vitro-cultured haematopoietic stem cells of freshwater crayfish, *Pacifastacus leniusculus*. J Gen Virol 87:847–854

Khadijah S, Neo SY, Hossain MS, Miller LD, Mathavan S, Kwang J (2003) Identification of white spot syndrome virus latency-related genes in specific-pathogen-free shrimps by use of a microarray. J Virol 77:10162–10167

Lan Y, Lu W, Xu X (2002) Genomic instability of prawn white spot bacilliform virus (WSBV) and its association to virus virulence. Virus Res 90:269–274

Lan Y, Xu X, Yang F, Zhang X (2006) Transcriptional profile of shrimp white spot syndrome virus (WSSV) genes with DNA microarray. Arch Virol 151:1723–1733

Leu JH, Tsai JM, Wang HC, Wang AH, Wang CH, Kou GH, Lo CF (2005) The unique stacked rings in the nucleocapsid of the white spot syndrome virus virion are formed by the major structural protein VP664, the largest viral structural protein ever found. J Virol 79:140–149

Leu JH, Chang CC, Wu JL, Hsu CW, Hirono I, Aoki T, Juan HF, Lo CF, Kou GH, Huang HC (2007) Comparative analysis of differentially expressed genes in normal and white spot syndrome virus infected *Penaeus monodon*. BMC Genomics 8:120

Li H, Zhu Y, Yang F (2006) Identification of a novel envelope protein (VP187) gene from shrimp white spot syndrome virus. Virus Res 115:76–84

Li L, Lin S, Yang F (2005) Functional identification of the non-specific nuclease from white spot syndrome virus. Virology 337:399–406

Li L, Lin S, Yang F (2006) Characterization of an envelope protein (VP110) of white spot syndrome virus. J Gen Virol 87:1909–1915

Li Q, Chen Y, Yang F (2004a) Identification of a collagen-like protein gene from white spot syndrome virus. Arch Virol 149:215–223

Li Q, Pan D, Zhang JH, Yang F (2004b) Identification of the thymidylate synthase within the genome of white spot syndrome virus. J Gen Virol 85:2035–2044

Lin ST, Chang YS, Wang HC, Tzeng HF, Chang ZF, Lin JY, Wang CH, Lo CF, Kou GH (2002) Ribonucleotide reductase of shrimp white spot syndrome virus (WSSV): expression and enzymatic activity in a baculovirus/insect cell system and WSSV-infected shrimp. Virology 304:282–290

Liu H, Soderhall I (2007) Histone H2A as a transfection agent in crayfish hematopoietic tissue cells. Dev Comp Immunol 31:340–346

Liu WJ, Yu HT, Peng SE, Chang YS, Pien HW, Lin CJ, Huang CJ, Tsai MF, Huang CJ, Wang CH, Lin JY, Lo CF, Kou GH (2001) Cloning, characterization, and phylogenetic analysis of a shrimp white spot syndrome virus gene that encodes a protein kinase. Virology 289:362–377

Liu WJ, Chang YS, Wang CH, Kou GH, Lo CF (2005) Microarray and RT-PCR screening for white spot syndrome virus immediate-early genes in cycloheximide-treated shrimp. Virology 334:327–341

Liu WJ, Chang YS, Wang AH, Kou GH, Lo CF (2007) White spot syndrome virus annexes a shrimp STAT to enhance expression of the immediate-early gene ie1. J Virol 81:1461–1471

Liu X, Yang F (2005) Identification and function of a shrimp white spot syndrome virus (WSSV) gene that encodes a dUTPase. Virus Res 110:21–30

Liu Y, Wu JL, Song J, Sivaraman J, Hew CL (2006) Identification of a novel nonstructural protein, VP9, from white spot syndrome virus: its structure reveals a ferredoxin fold with specific metal binding sites. J Virol 80:10419–10427

Lo CF, Ho CH, Peng SE, Chen CH, Hsu HC, Chiu YL, Chang CF, Liu KF, Su MS, Wang CH, Kou GH (1996) White spot syndrome baculovirus (WSBV) detected in cultured and captured shrimps, crabs and other arthropods. Dis Aquat Org 27:215–225

Lo CF, Ho CH, Chen CH, Liu KF, Chiu YL, Yeh PY, Peng SE, Hsu HE, Liu HC, Chang CF, Su MS, Wang CH, Kou GH (1997) Detection and tissue tropism of white spot syndrome baculovirus (WSBV) in captured brooders of *Penaeus monodon* with a special emphasis on reproductive organs. Dis Aquat Org 30:53–72

Lu L, Kwang J (2004) Identification of a novel shrimp protein phosphatase and its association with latency-related ORF427 of white spot syndrome virus. FEBS Lett 577:141–146

Marks H, Mennens M, Vlak JM, van Hulten MC (2003) Transcriptional analysis of the white spot syndrome virus major virion protein genes. J Gen Virol 84:1517–1523

Marks H, Goldbach RW, Vlak JM, van Hulten MC (2004) Genetic variation among isolates of white spot syndrome virus. Arch Virol 149:673–697

Marks H, Vorst O, van Houwelingen AM, van Hulten MC, Vlak JM (2005) Gene-expression profiling of white spot syndrome virus in vivo. J Gen Virol 86:2081–2100

Marks H, Ren XY, Sandbrink H, van Hulten MC, Vlak JM (2006) In silico identification of putative promoter motifs of white spot syndrome virus. BMC Bioinformatics 7:309

OIE (World Organisation for Animal Health, formerly Office International des Epizooties) (2003a) Manual of diagnostic tests for aquatic animals, 4th edn. OIE, Paris.

Pan D, He N, Yang Z, Liu H, Xu X (2005) Differential gene expression profile in hepatopancreas of WSSV-resistant shrimp (*Penaeus japonicus*) by suppression subtractive hybridization. Dev Comp Immunol 29:103–112

Pramod Kiran RB, Rajendran KV, Jung SJ, Oh MJ (2002) Experimental susceptibility of different life-stages of the giant freshwater prawn, *Macrobrachium rosenbergii* (de Man), to white spot syndrome virus (WSSV). J Fish Dis 25:201–207

Radtke K, Dohner K, Sodeik B (2006) Viral interactions with the cytoskeleton: a hitchhiker's guide to the cell. Cell Microbiol 8:387–400

Robalino J, Browdy CL, Prior S, Metz A, Parnell P, Gross P, Warr G (2004) Induction of antiviral immunity by double-stranded RNA in a marine invertebrate. J Virol 78:10442–10448

Robalino J, Bartlett T, Shepard E, Prior S, Jaramillo G, Scura E, Chapman RW, Gross PS, Browdy CL, Warr GW (2005) Double-stranded RNA induces sequence-specific antiviral silencing in addition to nonspecific immunity in a marine shrimp: convergence of RNA interference and innate immunity in the invertebrate antiviral response? J Virol 79:13561–13571

Robalino J, Payne C, Parnell P, Shepard E, Grimes AC, Metz A, Prior S, Witteveldt J, Vlak JM, Gross PS, Warr G, Browdy CL (2006) Inactivation of white spot syndrome virus (WSSV) by normal rabbit serum: implications for the role of the envelope protein VP28 in WSSV infection of shrimp. Virus Res 118:55–61

Rojtinnakorn J, Hirono I, Itami T, Takahashi Y, Aoki T (2002) Gene expression in heamocytes of kuruma prawn, *Penaeus japonicus*, in response to infection with WSSV by EST approach. Fish Shell Immunol 13:69–83

Sritunyalucksana K, Wannapapho W, Lo CF, Flegel TW (2006) PmRab7 is a VP28-binding protein involved in white spot syndrome virus infection in shrimp. J Virol 80:10734–10742

Tang X, Wu J, Sivaraman J, Hew CL (2007) Crystal structures of major envelope proteins VP26 and VP28 from white spot syndrome virus shed light on their evolutionary relationship. J Virol 81:6709–6717

Tsai JM, Wang HC, Leu JH, Hsiao HH, Wang AH, Kou GH, Lo CF (2004) Genomic and proteomic analysis of thirty-nine structural proteins of shrimp white spot syndrome virus. J Virol 78:11360–11370

Tsai JM, Wang HC, Leu JH, Wang AH, Zhuang Y, Walker PJ, Kou GH, Lo CF (2006) Identification of the nucleocapsid, tegument, and envelope proteins of the shrimp white spot syndrome virus virion. J Virol 80:3021–3029

Tsai MF, Lo CF, van Hulten MC, Tzeng HF, Chou CM, Huang CJ, Wang CH, Lin JY, Vlak JM, Kou GH (2000a) Transcriptional analysis of the ribonucleotide reductase genes of shrimp white spot syndrome virus. Virology 277:92–99

Tsai MF, Yu HT, Tzeng HF, Leu JH, Chou CM, Huang CJ, Wang CH, Lin JY, Kou GH, Lo CF (2000b) Identification and characterization of a shrimp white spot syndrome virus (WSSV) gene that encodes a novel chimeric polypeptide of cellular-type thymidine kinase and thymidylate kinase. Virology 277:100–110

Tzeng HF, Chang ZF, Peng SE, Wang CH, Lin JY, Kou GH, Lo CF (2002) Chimeric polypeptide of thymidine kinase and thymidylate kinase of shrimp white spot syndrome virus: thymidine kinase activity of the recombinant protein expressed in a baculovirus/insect cell system. Virology 299:248–255

van Hulten MC, Vlak JM (2001) Identification and phylogeny of a protein kinase gene of white spot syndrome virus. Virus Genes 22:201–207

van Hulten MC, Goldbach RW, Vlak JM (2000a) Three functionally diverged major structural proteins of white spot syndrome virus evolved by gene duplication. J Gen Virol 81:2525–2529

van Hulten MC, Westenberg M, Goodall SD, Vlak JM (2000b) Identification of two major virion protein genes of white spot syndrome virus of shrimp. Virology 266:227–236

van Hulten MC, Witteveldt J, Peters S, Kloosterboer N, Tarchini R, Fiers M, Sandbrink H, Lankhorst RK, Vlak JM (2001a) The white spot syndrome virus DNA genome sequence. Virology 286:7–22

van Hulten MC, Witteveldt J, Snippe M, Vlak JM (2001b) White spot syndrome virus envelope protein VP28 is involved in the systemic infection of shrimp. Virology 285:228–233

van Hulten MC, Reijns M, Vermeesch AM, Zandbergen F, Vlak JM (2002) Identification of VP19 and VP15 of white spot syndrome virus (WSSV) and glycosylation status of the WSSV major structural proteins. J Gen Virol 83:257–265

Venegas CA, Nonaka L, Mushiake K, Shimizu K, Nishizawa T, Muroga K (1999) Pathogenicity of penaeid rod-shaped DNA virus PRDV to Kuruma prawn in different development stages. Fish Pathol 34:19–23

Wang B, Li F, Dong B, Zhang X, Zhang C, Xiang J (2006) Discovery of the genes in response to white spot syndrome virus (WSSV) infection in *Fenneropenaeus chinensis* through cDNA microarray. Mar Biotechnol 8:491–500

Wang CH, Lo CF, Leu JH, Chou CM, Yeh PY, Chou HY, Tung MC, Chang CF, Su MS, Kou GH (1995) Purification and genomic analysis of baculovirus associated with white spot syndrome (WSBV) of *Penaeus monodon*. Dis Aquat Org 23:239–242

Wang CH, Yang HN, Tang CY, Lu CH, Kou GH, Lo CF (2000) Ultrastructure of white spot syndrome virus development in primary lymphoid organ cell cultures. Dis Aquat Org 41:91–104

Wang CS, Tang KFJ, Kou GH, Chen SN (1997) Light and electron microscopic evidence of white spot disease in the giant tiger shrimp, *Penaeus monodon* (Fabricius), and the kuruma shrimp, *Penaeus japonicus* (Bate), cultured in Taiwan. J Fish Dis 20:323–331

Wang HC, Lin AT, Yii DM, Chang YS, Kou GH, Lo CF (2003) DNA microarrays of the white spot syndrome virus genome: genes expressed in the gills of infected shrimp. Proceedings of Marine Biotechnology Conference 2003 (P1-045)

Wang HC, Wang HC, Kou GH, Lo CF, Huang WP (2007a) Identification of *icp11*, the most highly expressed gene of shrimp white spot syndrome virus (WSSV). Dis Aquat Org 74:179–189

Wang HC, Wang HC, Leu JH, Kou GH, Wang AH, Lo CF (2007b) Protein expression profiling of the shrimp cellular response to white spot syndrome virus infection. Dev Comp Immunol 31:672–686

Wang YG, Hassan MD, Sharriff M, Zamri SM, Chen X (1999) Histopathology and cytopathology of white spot syndrome virus (WSSV) in cultured *Penaeus monodon* from peninsular Malaysia with emphasis on pathogenesis and the mechanism of white spot formation. Dis Aquat Org 39:1–11

Wang Z, Hu L, Yi G, Xu H, Qi Y, Yao L (2004) ORF390 of white spot syndrome virus genome is identified as a novel anti-apoptosis gene. Biochem Biophys Res Commun 325:899–907

Wang Z, Chua HK, Gusti AA, He F, Fenner B, Manopo I, Wang H, Kwang J (2005) RING-H2 protein WSSV249 from white spot syndrome virus sequesters a shrimp ubiquitin-conjugating enzyme, PvUbc, for viral pathogenesis. J Virol 79:8764–8772

Witteveldt J, Vermeesch AM, Langenhof M, de Lang A, Vlak JM, van Hulten MC (2005) Nucleocapsid protein VP15 is the basic DNA binding protein of white spot syndrome virus of shrimp. Arch Virol 150:1121–1133

Wongprasert K, Khanobdee K, Glunukarn SS, Meeratana P, Withyachumnarnkul B (2003) Time-course and levels of apoptosis in various tissues of black tiger shrimp *Penaeus monodon* infected with white-spot syndrome virus. Dis Aquat Org 55:3–10

Wongteerasupaya C, Vickers JE, Sriuairatana S, Nash GL, Akarajamorn A, Boonsaeng V, Panyim S, Tassanakajon A, Withyachumnarnkul B, Flegel TW (1995) A non-occluded, systemic baculovirus that occurs in cells of ectodermal and mesodermal origin and causes high mortality in the black tiger prawn *Penaeus monodon*. Dis Aquat Org 21:69–77

Wu JL, Suzuki K, Arimoto M, Nishizawa T, Muroga K (2002) Preparation of an inoculum of white spot syndrome virus for challenge tests in *Penaeus japonicus*. Fish Pathol 37:65–69

Xie X, Yang F (2005) Interaction of white spot syndrome virus VP26 protein with actin. Virology 336:93–99

Xie X, Xu L, Yang F (2006) Proteomic analysis of the major envelope and nucleocapsid proteins of white spot syndrome virus. J Virol 80:10615–10623

Yang F, He J, Lin X, Li Q, Pan D, Zhang X, Xu X (2001) Complete genome sequence of the shrimp white spot bacilliform virus. J Virol 75:11811–11820

Yoganandhan K, Narayanan RB, Sahul Hameed AS (2003) Larvae and early post-larvae of *Penaeus monodon* (Fabricius) experimentally infected with white spot syndrome virus (WSSV) show no significant mortality. J Fish Dis 26:385–391

Zhang X, Xu X, Hew CL (2001) The structure and function of a gene encoding a basic peptide from prawn white spot syndrome virus. Virus Res 79:137–144

Zhang X, Huang C, Xu X, Hew CL (2002) Identification and localization of a prawn white spot syndrome virus gene that encodes an envelope protein. J Gen Virol 83:1069–1074

Zhang X, Huang C, Tang X, Zhuang Y, Hew CL (2004) Identification of structural proteins from shrimp white spot syndrome virus (WSSV) by 2DE-MS. Proteins 55:229–235

Zhu Y, Ding Q, Yang F (2007) Characterization of a homologous-region-binding protein from white spot syndrome virus by phage display. Virus Res 125:145–152

# Jumbo Bacteriophages

R.W. Hendrix

## Contents

**Abstract** There is currently a handful of genome sequences available for tailed bacteriophages with genomes of more than 200 kbp of DNA, designated here as giant or jumbo phages. The majority of the proteins predicted from the genome sequences of these phages have no matches in the current sequence databases, and the genomes themselves are diverse enough to preclude the sorts of detailed comparative analysis that has benefited study of the smaller phages, for which hundreds of genome sequences are available. However, it is informative to extrapolate the better known genome organizations and mechanisms of evolution seen in the smaller phages to the jumbo phages. In this way, we see that the jumbo phages encode the same functions as the smaller phages, supplemented with large numbers of mostly small genes of mostly undiscovered functions. A case can be made that the jumbo phages evolved from smaller tailed phages, possibly in a process mediated by the constraints imposed on genome size by capsid size.

R.W. Hendrix
Pittsburgh Bacteriophage Institute and Department of Biological Sciences, University of Pittsburgh, Pittsburgh, PA 15260, USA
rhx@pitt.edu

James L. Van Etten (ed.) *Lesser Known Large dsDNA Viruses.*
Current Topics in Microbiology and Immunology 328.
© Springer-Verlag Berlin Heidelberg 2009

# Introduction

In-depth studies of a small number of model bacteriophages over the past 60 years or so have been central to the development of the discipline of molecular biology. Phages as model systems or experimental tools have had a leading role in revealing fundamental aspects of the ways genes are expressed and regulated and how biological structures are built and function. The advent of high-throughput DNA sequencing has allowed comparative genomic approaches with a substantial broadening of our view of the world of phages and, especially, of their evolution. This increase in our technical capacity to study phages has fortuitously coincided with the realization that phages are much more than convenient model systems for laboratory studies. We now understand that phages constitute a majority of biological organisms on the planet, that they have had and continue to have a major influence on the evolution of their bacterial and archaeal hosts, and that they are responsible for a sizable fraction of the carbon and energy cycling in the oceans and no doubt in the rest of the biosphere as well (Suttle 2005).

In this discussion of giant phages, I will consider only the dsDNA tailed phages, since the smallest of them have genomes of a size (~19 kbp) that is bigger than the biggest of the phages known from other groups. The first tailed phage to be completely sequenced, *Escherichia coli* phage lambda, has a genome of 48.5 kbp, and several dozen genome sequences in this size range are now known. Several genomes are now available in the 100- to 200-kbp size range, including the well-studied *Bacillus subtilis* phage SPO1 (132 kbp), and *E. coli* phages T5 (135 kbp) and T4 (166 kbp), as well as several phages of *Mycobacterium smegmatis* and T4-like phages of several different Gram-negative hosts. There is a handful of phages with sequenced genomes in the 200- to 400-kbp range, including KVP40 (*Vibrio*, 245 kbp), KZ (*Pseudomonas*, 280 kbp), EL (*Pseudomonas*, 211 kbp), N3 (*Sinorhyzobium*, 207 kbp), PAU (*Sphingomonas*, 220 kbp), 121Q (*Proteus*, 341 kbp), and PBS1 (*Bacillus*, 252 kbp); with the exception of PBS1, none of these has been studied extensively by biochemical or genetic methods, and an analysis of the last four sequences listed has not yet been published as of this writing. The biggest bacteriophage known, phage G of *B. megaterium*, has a genome of 498 kbp. A weakness in the current state of the data for the phages in the 200+ kbp size range is that most of them have no close relatives, which severely limits the possibilities for the sort of multisequence comparative analysis that has proved so powerful in the smaller phages. (The exceptions are N3, KVP40, and 121Q, which have head and tail genes that put them in the T4-like group, but are otherwise quite different from each other. Also, the two *Pseudomonas* phages, KZ and EL, share moderate amino acid sequence similarity for about one-third of their genes, and a comparative analysis based on this pair has been published (Hertveldt et al. 2005).) There is, however, hope that this situation could change soon. Serwer et al. (2007) showed that phage G fails to make visible plaques unless an unusually low concentration of agarose is used in its plating. This is thought to be because the G virions are too big to diffuse well in stiffer gels. The implication of this finding is that for the 90+ years since the discovery of phages, biologists have been isolating new phages from the environment under conditions that select against very large phages. Armed with this knowledge, it should be possible to isolate new phages of exceptionally large sizes. Thus it may turn out that our understanding of the biology of these phages and of the roles of exceptionally large phages in the environment could expand substantially in the near future.

This article is organized starting with what has been learned about the evolution and genomic relationships among the smaller tailed phages, those with genomes in the vicinity of 50 kbp, where we have the clearest understanding of these issues. We will then ask how the things we see in these medium-sized phages compare to what is seen in the group of bigger phages with genomes in the 100- to 200-kbp range, and finally we will ask what additional features are found in the jumbo phages with genomes above 200 kbp. Implicit in this organization of the data is the suggestion that the evolution of phages might have followed a similar path from smaller to larger genomes. This in not an uncontroversial idea (Claverie 2006), but at least in the case of the tailed phages it appears to be the most uncomplicated way to reconcile the differences observed among the phages of different sized genomes. This point will be considered in more detail in the following sections.

## Genome Structure and Evolution of Medium Phages

The genomes of the medium-sized tailed phages share several salient features across the entire group of sequenced genomes (Juhala et al. 2000; Hendrix 2002; Hatfull et al. 2006). These include the following:

1. The genomic real estate is very efficiently used. Typically more than 90% of the genome is occupied by protein-coding genes, frequently organized into operons and often with little or no space between the coding regions. Noncoding regions between genes often contain transcription-regulating sequences such as promoters and operators.
2. There can be great divergence of sequence – nucleotide or encoded protein sequence – between homologous genes. In some pair-wise comparisons, a protein encoded by one phage may be very similar or even identical in sequence to its homolog in the other phage, but more often homologous sequences are highly diverged; frequently two proteins that are thought on independent grounds to be homologs (i.e., to share common ancestry) will have no detectable sequence similarity remaining at all.
3. The genomes are genetic mosaics with respect to each other. Thus in a pair-wise comparison a particular region of sequence may be very similar between the two phages but with a sudden transition to an adjacent region of lesser or, more typically, no sequence similarity. This is taken as evidence of a nonhomologous (illegitimate) recombination event in the ancestry of one of the two phages. The overall result of such nonhomologous recombination among the members of a phage population is horizontal transfer of genes between members of the population. The modules of mosaicism defined and punctuated by sites of nonhomologous

recombination are most often individual genes, but they can also more rarely correspond to parts of genes. In some cases, groups of genes travel together through evolution, that is, there is little or no evidence of past nonhomologous recombination within the group. Such a lack of exchange among lineages is seen most dramatically in the structural genes, especially those encoding proteins involved in formation of the head of the virion. A similar lack of exchange is seen, for example, between a repressor gene and the operators that the encoded repressor binds to. In general, genetic elements that stay together over evolutionary time in this way appear to be ones that encode proteins (or a protein and its binding site) that interact intimately during the life cycle of the phage.

4. The genetic diversity across the population is extremely large. Much of this is due to the high level of mutational drift in the sequences of homologous genes encoding essential functions, to analogous but nonhomologous substitutions of essential genes, and to reassortment of these different versions of essential genes into new combinations by the recombinational processes leading to genomic mosaicism. However, there are also a few genes in these phages that are unusually variable in regard to their presence or absence in otherwise very similar genomes. These genes can typically be deleted without noticeable effect on growth of the phage in laboratory conditions. A possibly familiar example of such genes is those in the center of the phage lambda genome, in the region classically known as the b2 region (Juhala et al. 2000). Other closely related phages typically also have genes in this region but they are usually different from those in lambda.

According to the model that has been developed to account for these observations, diversity in the phage population is developed by a combination of point mutation, nonhomologous (illegitimate) recombination, and homologous recombination. Of these, nonhomologous recombination is the most evolutionarily creative force because it generates novel joints in the sequence by joining together two sequences that were not previously together. This can create not only rearrangements and analogous substitutions in an existing genome, but also deletions or insertions of novel sequences. Nonhomologous recombination is also expected to be the least often successful in producing a viable phage as the product of recombination, since the vast majority of the recombinants are expected to be the inviable results of inappropriate joining of mismatched parts of different coding regions or nonfunctional genome rearrangements, etc. Homologous recombination occurs many orders of magnitude more frequently than nonhomologous recombination. By definition, it does not create any novel joints in the genome sequence, but it can still serve as a source of evolutionary innovation by creating new combinations of the novel joints and other features flanking the site of recombination.

The combination of these events creates diversity in the population. For recombination, the recombining partners may be co-infecting phages, the prophage DNA found in almost all bacterial cells, and in the case of deletions, duplications, and inversions, the phage's own DNA. The amount of diversity generated in this way, integrated over evolutionary time, must be staggeringly large in order to account for the tremendous diversity seen in the observable population and for the expectation that all but a tiny fraction of the nonhomologous recombinants will be functionally compromised and rapidly eliminated by natural selection. The recombinants that do survive are those that have not compromised function, for example, those at gene boundaries. Despite the essentially random nature of the mutational processes involved, the net result is to produce genomes that are neatly laid out and appear designed for efficiency of function.

The requirement for extremely large numbers of intrinsically improbable nonhomologous recombination events to explain the observations from genomic comparisons begins to seem reasonable when we consider recent information regarding the size and dynamics of the global phage population. Current estimates place the number of individual tailed phages on the planet at about $10^{31}$ (Hendrix 2002). The population turns over every few days (at least in the oceans where the measurement has been made; Suttle 2007), as members of the population find new cells to infect, fall victim to predation, or are inactivated by environmental conditions such as UV light or drying. Replacing a population of $10^{31}$ every few days requires $10^{24}$–$10^{25}$ productive infections per second. If we imagine, somewhat arbitrarily, that there is one nonhomologous recombination every $10^6$ infections, and that one in $10^9$ of those recombinants yields a fully functional phage, there will be $10^9$–$10^{10}$ fully functional novel phage variants produced every second on a global scale.

The genomes of the phages in the medium size range are typically highly organized by function. In the well-studied phage lambda, there are two diverging operons of early genes, including genes for replication and recombination functions and genes for regulating the temporal progression of transcription, and a single late operon encoding the lysis functions and the structural genes for the head and tail. Other phages of this size class are similarly organized, with functionally related genes clustered in operons, although the arrangements of the operons relative to each other can vary. One of the most highly conserved features of genome organization across all these phages is the organization of the head genes. These are almost always arranged in the same stereotypical order along the genome, even when the primary sequences of the encoded proteins have diverged past recognizability, and there are only rarely small extra genes inserted between the canonical set of head genes.

## Big and Bigger Phages

There are two groups of phages in the over-100-kbp genome size range for which there are enough genome sequences available to begin serious comparative analysis. These are the group of phages infecting Gram-positive bacteria and typified by *B. subtilis* phage SPO1 and the group of phages infecting Gram-negative hosts typified by *E. coli* phage T4. There are currently six genome sequences available for the SPO1-like group, ranging in size from 130 to 139 kbp and approximately ten sequences in the T4-like group, ranging from 166 to 341 kbp.

Compared to the medium-sized phages, these big phages have a larger number of genes for which biochemical functions can be deduced from the sequence (though also a higher proportion of genes with unknown functions). Some of these functionally identified genes encode virion structural components, corresponding to the slightly greater structural complexity of the virions of the bigger phages. However, most of the new genes in this category are concerned with DNA replication and nucleotide metabolism. Medium phage lambda, for example, has two genes encoding proteins that recruit the host replication machinery to the phage DNA, while big phage T4 encodes a complete DNA replication machine. Curiously, *Pseudomonas* phage KZ (280 kbp) breaks this general pattern by having no recognizable DNA polymerase or other replication-associated genes.

The concept of core genes has been proposed for the T4-like group to describe the genes that are shared across the whole group of T4-like phages (Filee et al. 2006). These comprise, to a close approximation, the structural genes encoding head and tail proteins and the DNA replication and nucleotide metabolism genes. The SPO1-like group also has a group of genes that could similarly be described as core genes, though there seems to be less complete coverage than in T4 of the DNA replication and nucleotide metabolism genes. A striking and somewhat unexpected feature of the core genes of the T4 group (Filee et al. 2006), and probably also the SPO1 group (C. Stewart et al., unpublished observations), is that there is almost no evidence of horizontal exchange of these genes among the phages in the group. This is similar to the coherence among the head genes described above for the medium-sized phages, but in the case of the T4 group, the cohering genes include all the core genes, a much more extensive group. We propose a similar explanation for the lack of mixing of lineages of core genes in the big phages as for the lack of mixing of head gene lineages in the medium phages; that is, all the proteins encoded by the core genes have co-evolved to function together and cannot accept substitutions from differently co-evolved lineages.

The core genes account for somewhat less than half the genes in phage T4; the rest are non-core genes, which by definition are found in only a subset of the T4-like phages. In a comparison of eight T4-like genomes, 43% of the T4 genes were either unique to T4 or matched only one other phage in the set of eight, and 40%, corresponding roughly to the core genes, were found in all of the phages in the comparison (R.W. Hendrix, unpublished data). In a comparison of six SPO1-like phages, 30% of the SPO1 genes had at least one homolog in one of the other phages (Stewart et al., unpublished observations). Because the core genes are on average bigger than the non-core genes, the 30% of the genes with homologs account for roughly half of the protein-coding capacity of the SPO1 genome.

For the most part, the functions of the non-core genes are not known. In T4, where most of the work on these genes has been done, a number of them have been shown to have functions that are not absolutely essential for phage growth but may adapt the phage for growth in a particular ecological situation. A good example of such a gene is the *psbA* gene encoding a component of the photosynthesis reaction center of some of the large T4-like cyanophages (Mann et al. 2003). It is commonly assumed that all of the genes in a phage genome must provide a selective benefit to the phage or they would be replaced by genes that did provide such a benefit. However, the constraints that the capsid imposes on genome size may mean that some genes can be tolerated in the genome even if they provide no selective benefit, as discussed in the next section.

If we compare two similar phage genomes, the genes that are present in one genome and not the other must be genes that have entered or departed one of the genomes since the phages' last common ancestor. This means that the non-core genes must be in rapid flux into and out of the phage genomes, at the same time that the core genes are not moving between genomes at an appreciable rate. We propose that this situation can be explained by invoking the same model for genome evolution as we described above for the medium-sized phages. That is, diversity is generated across the genome by nonhomologous recombination, and any recombinants that disrupt the integrity of the core gene set are rapidly eliminated from the population. Numerous recombinants that reassort the non-core genes are evidently tolerated, and this is the source of the diversity we observe among the catalogs of non-core genes found in different phages.

The gene order for the head and tail genes, which is seen to be strongly conserved in the medium phages, is largely but not absolutely preserved in the T4-like group. Considering all the core genes together, there is evidence for some DNA rearrangements within the group, possibly associated in some cases with recombination between copies of homing endonuclease genes that are typically present in multiple copies in these phages. However, the most striking difference in genome organization between the medium and big phages is in the much larger number of non-core genes, which are generally found in clusters, with those clusters interspersed among clusters of core genes, across the entire genome (Comeau et al. 2007) (see Fig. 1).

## Capsid Size and Genome Size

In construction of the virion of all the tailed phages, an empty protein capsid is first assembled and then DNA is pumped into it. This clearly puts an upper size limit on the genome, and in fact the DNA is usually packed nearly as tightly as is physically possible. Different phages have different ways to determine where in the genome sequence the physical ends of the packaged DNA lie. Here we will only consider the headful packaging mechanism, which applies to phage T4 (Black et al. 1994) and appears to apply to bigger phages as well. In phages that use this mechanism, the product of DNA replication is a concatemer, that is, a tandem array of multiple copies of the genome sequence. The packaging machinery cuts the DNA at an arbitrary position in the sequence and pumps DNA into the prohead until it is full and makes the second cut to define the second end of the packaged DNA. In the case of T4, the headful of DNA corresponds to approximately 102% of the genome sequence, giving some terminal redundancy that provides for recombination following infection to allow reconstitution of circles and concatemers.

Most and possibly all of the major capsid proteins of tailed phages have common ancestry (Fokine et al. 2005), and so the structures of the capsids must also have common ancestry. The fact that not all phages have the same capsid size implies that they have changed size over evolutionary time, and we can ask what sorts of

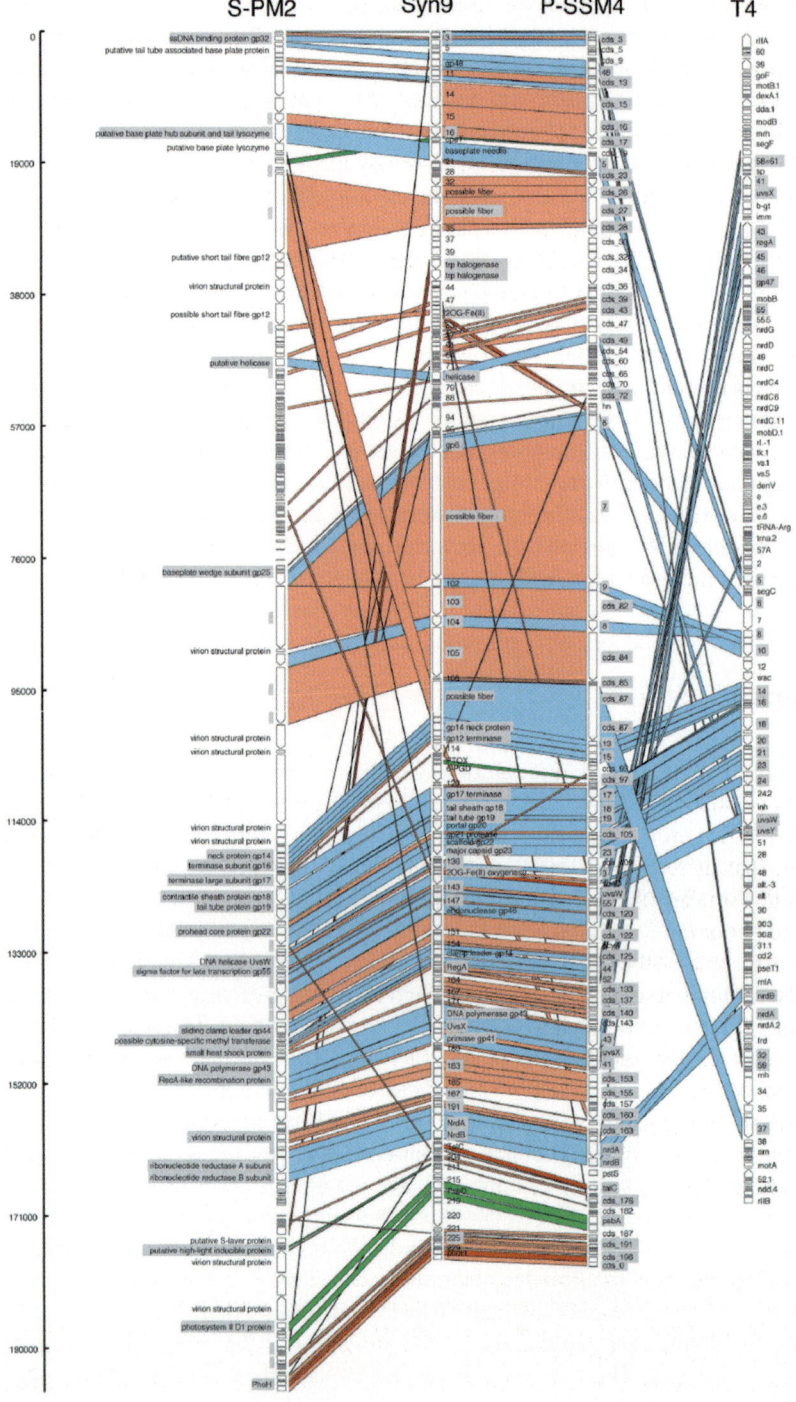

changes are plausible, given our understanding of the restrictions placed on genome size by capsid size and the quantized nature of possible changes in capsid size (Caspar and Klug 1962). Taking T4 as an example, we can ask if a smaller genome could plausibly evolve. T4 could probably fit all of its essential core genes into a smaller capsid – say an isometric T=13 shell rather than its current T=13 prolate shell – but to do so it would have to delete all the non-core genes before the capsid could become smaller. Because the non-core genes are interspersed with the core genes, this would require several separate deletions of non-core genes before a mutant with a smaller capsid would be viable. To the extent that the non-core genes provide a selective benefit to the phage, those deletions would be counter-selected, and moreover the only likely benefit of a smaller head – less DNA to replicate to fill a virion – could not be achieved until all the deletions were in place.

Increasing the size of the capsid appears to be considerably more likely. An increase in capsid size is immediately accommodated simply by packaging more DNA, with a correspondingly large terminal redundancy. Novel DNA can now enter the genome without displacing any essential genes. However, once such DNA has come in, the pathway back to the original capsid size is blocked. Much of the novel DNA that enters the genome will likely provide no selective benefit to the phage, but over time genes that do provide a selective benefit should be preferentially retained.

# Jumbo Phages

The jumbo phages, including the biggest, phage G (Fig. 2), have for the most part not been analyzed in as much detail as the smaller phages. This is largely because there are few enough of them sequenced, and they are different enough from each other, that there are fewer possibilities for meaningful comparative analysis than with the smaller phages. With that caveat, we can ask what we can say about the jumbo genomes.

First, some of the jumbo phages – N3, KPV40, 121Q – are in the T4-like group by the criterion that they have T4-like head and tail genes. It would appear that they are like T4 except for having a larger fraction of non-core genes, which in this case have to be judged to be in this class by their lack of matches to the group of smaller T4-like phages.

All of the big phages (with the exception of KZ, cited above) have genes recognizable as DNA replication and nucleotide metabolism genes. As has been noted elsewhere, this is a feature of large viruses in general and not just large phages.

---

**Fig. 1** Comparison of the genome content and organization of four T4-type phages. S-PM2, Syn9, and P-SSM4 are cyanophages, closely related to each other and distantly related to coliphage T4. The *filled trapezoids* indicate correspondences between genes that are homologous, as judged by detectable sequence similarity between their encoded proteins. *Blue trapezoids* connect genes with a homolog in T4; *orange trapezoids* connect homologs found only in the cyanophages. (From Weigele et al. 2007, with permission)

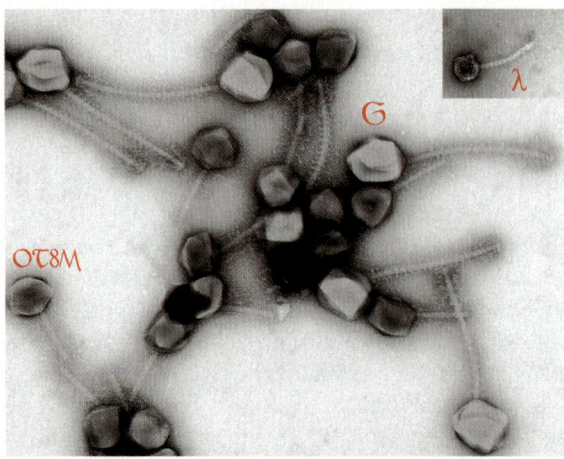

**Fig. 2** Electron micrograph of a mixture of *Bacillus* phage G and *Serratia* phage OT8M. The *inset* shows coliphage lambda to the same scale. Lambda has a head with a diameter of 60 nm and a genome approximately 10% the size of the phage G genome

A feature that sets the jumbo phages apart from their smaller brethren is that they are much more likely to have families of paralogs within the genome. This reaches an extreme in phage G, in which more than 10% of its 682 predicted protein coding genes are members of such families. To date none of these genes has made a database match that would suggest a function.

In phage G we see a feature that has been documented in the big viruses of eukaryotic hosts as well, notably in Mimivirus: it encodes a number of components of the translation system. Phage G encodes 17 tRNAs covering 14 codon specificities and an apparent homolog of a serine aminoacyl tRNA synthetase. tRNAs have been known in phages for decades; phage T4 has seven, even some of the medium-sized phages have one or two, and some mycobacteriophages, as well as vibriophage KVP40, have as many as 30 (Lee et al. 2004; Hertveldt et al. 2005). What function they serve is less clear. One possibility that has been widely considered is that they read codons that are more abundant in the phage than in the host, especially in the highly translated structural genes, and so increase the efficiency of phage-specific translation. This is an attractive idea with statistical support in some cases, and it seems likely to be a correct explanation for at least some of the phage-encoded tRNAs. However, the presence of a tRNA synthetase in phage G (and four in Mimivirus) suggests that the translational components of phages may have additional functions. An attractive possibility is that phages incorporate nonstandard amino acids into their proteins, perhaps by mechanisms analogous to the mechanisms by which cells incorporate selenocysteine or pyrrolysine, or how bacteria lacking glutamyl tRNA synthetase incorporate glutamine (Cathopoulis et al. 2007). This currently entirely hypothetical possibility should be susceptible to experimental test.

The most unexpected feature of phage G at this stage of analysis is that the chromosome – i.e., the DNA packaged in the virion –is roughly 35% longer than the genome sequence. The genome sequence is 497,513 bp, but the chromosome is approximately 670,000 bp (Sun and Serwer 1997). This must mean that each virion has a terminal redundancy about the size of the entire genome of phage T4. We have considered two alternative explanations for this interesting discrepancy. First, it may be that phage G had, in the recent past, a genome of roughly 670 kbp but has suffered a large deletion, of approximately 170 kbp, perhaps during passage in the laboratory. This seems somewhat unlikely because it would mean that the deleted 170 kbp contained no genes essential for the growth of the phage, but sequences of independently isolated phages closely related to G will be needed before this possibility can be evaluated.

The alternative explanation of the genome size discrepancy is that phage G in its recent evolutionary past has had an increase in its capsid size. In this scenario the phage had a genome size perhaps similar to the current 498 kbp, which it packaged with only a small terminal redundancy. The hypothetical increase in capsid size would lead to a large increase in terminal redundancy, and we would have to postulate that that event was recent enough that genetic space specified by the new capsid size has not yet filled up with newly acquired genes. As with the first explanation, this will require new sequences from phage G relatives before it can be fully assessed.

## Conclusions

The giant bacteriophages can be seen as derivatives of smaller phages that have acquired novel genetic functions and increased genome functions over evolutionary time. The implied increases in genome size are coordinated with and perhaps driven by increases in capsid size. The novel genetic functions in the giant phages are largely unidentified and are a testament to the remarkable degree of genetic diversity in the phage universe and an inspiration for continuing work on these fascinating viruses.

**Acknowledgements** Work in the author's laboratory is supported by NIH grants GM51975 and GM47795 and NSF grant 0333112.

## References

Black LW, Showe MK, Steven AC (1994) Morphogenesis of the T4 head. In: Karam JD (ed) Molecular biology of Bacteriophage T4. ASM, Washington, DC, pp 218–258

Caspar DLD, Klug A (1962) Physical principles in the construction of regular viruses. Cold Spring Harbor Symp Quant Biol 27:1–24

Cathopoulis T, Chuawong P, Hendrickson TL (2007) Novel tRNA aminoacylation mechanisms. Mol Biosyst 3:408–418

Claverie JM (2006) Viruses take center stage in cellular evolution. Genome Biol 7:110

Comeau AM, Bertrand C, Letarov A, Tetart F, Krisch HM (2007) Modular architecture of the T4 phage superfamily: a conserved core genome and a plastic periphery. Virology 362:384–396

Filee J, Bapteste E, Susko E, Krisch HM (2006) A selective barrier to horizontal gene transfer in the T4-type bacteriophages that has preserved a core genome with the viral replication and structural genes. Mol Biol Evol 23:1688–1696

Fokine A, Leiman PG, Shneider MM, Ahvazi B, Boeshans KM, Steven AC et al (2005) Structural and functional similarities between the capsid proteins of bacteriophages T4 and HK97 point to a common ancestry. Proc Natl Acad Sci U S A 102:7163–7168

Hatfull GF, Pedulla ML, Jacobs-Sera D, Cichon PM, Foley A, Ford ME et al (2006) Exploring the mycobacteriophage metaproteome: phage genomics as an educational platform. PLoS Genet 2:e92

Hendrix RW (2002) Bacteriophages: evolution of the majority. Theor Popul Biol 61:471–480

Hertveldt K, Lavigne R, Pleteneva E, Sernova N, Kurochkina L, Korchevskii R et al (2005) Genome comparison of *Pseudomonas aeruginosa* large phages. J Mol Biol 354:536–545

Juhala RJ, Ford ME, Duda RL, Youlton A, Hatfull GF, Hendrix RW (2000) Genomic sequences of bacteriophages HK97 and HK022: pervasive genetic mosaicism in the lambdoid bacteriophages. J Mol Biol 299:27–51

Lee S, Kriakov J, Vilcheze C, Dai Z, Hatfull GF, Jacobs WR Jr (2004) Bxz1, a new generalized transducing phage for mycobacteria. FEMS Microbiol Lett 241:271–276

Mann NH, Cook A, Millard A, Bailey S, Clokie M (2003) Marine ecosystems: bacterial photosynthesis genes in a virus. Nature 424:741

Serwer P, Hayes SJ, Thomas JA, Hardies SC (2007) Propagating the missing bacteriophages: a large bacteriophage in a new class. Virol J 4:21

Sun M, Serwer P (1997) The conformation of DNA packaged in bacteriophage G. Biophys J 72:958–963

Suttle CA (2005) Viruses in the sea. Nature 437:356–361

Suttle CA (2007) Marine viruses – major players in the global ecosystem. Nat Rev Microbiol 5:801–812

Weigele PR, Pope WH, Pedulla ML, Houtz JM, Smith AL, Conway JF et al (2007) Genomic and structural analysis of Syn9, a cyanophage infecting marine Prochlorococcus and Synechococcus. Environ Microbiol 9:1675–1695

# Index

# Current Topics in Microbiology and Immunology

## Volumes published since 2002

Vol. 295: **Sullivan, David J.; Krishna Sanjeew (Eds.):** Malaria: Drugs, Disease and Post-genomic Biology. 2005. 40 figs., XI, 446 pp. ISBN 3-540-25363-7

Vol. 296: **Oldstone, Michael B. A. (Ed.):** Molecular Mimicry: Infection Induced Autoimmune Disease. 2005. 28 figs., VIII, 167 pp. ISBN 3-540-25597-4

Vol. 297: **Langhorne, Jean (Ed.):** Immunology and Immunopathogenesis of Malaria. 2005. 8 figs., XII, 236 pp. ISBN 3-540-25718-7

Vol. 298: **Vivier, Eric; Colonna, Marco (Eds.):** Immunobiology of Natural Killer Cell Receptors. 2005. 27 figs., VIII, 286 pp. ISBN 3-540-26083-8

Vol. 299: **Domingo, Esteban (Ed.):** Quasispecies: Concept and Implications. 2006. 44 figs., XII, 401 pp. ISBN 3-540-26395-0

Vol. 300: **Wiertz, Emmanuel J.H.J.; Kikkert, Marjolein (Eds.):** Dislocation and Degradation of Proteins from the Endoplasmic Reticulum. 2006. 19 figs., VIII, 168 pp. ISBN 3-540-28006-5

Vol. 301: **Doerfler, Walter; Böhm, Petra (Eds.):** DNA Methylation: Basic Mechanisms. 2006. 24 figs., VIII, 324 pp. ISBN 3-540-29114-8

Vol. 302: **Robert N. Eisenman (Ed.):** The Myc/Max/Mad Transcription Factor Network. 2006. 28 figs., XII, 278 pp. ISBN 3-540-23968-5

Vol. 303: **Thomas E. Lane (Ed.):** Chemokines and Viral Infection. 2006. 14 figs. XII, 154 pp. ISBN 3-540-29207-1

Vol. 304: **Stanley A. Plotkin (Ed.):** Mass Vaccination: Global Aspects – Progress and Obstacles. 2006. 40 figs. X, 270 pp. ISBN 3-540-29382-5

Vol. 305: **Radbruch, Andreas; Lipsky, Peter E. (Eds.):** Current Concepts in Autoimmunity. 2006. 29 figs. IIX, 276 pp. ISBN 3-540-29713-8

Vol. 306: **William M. Shafer (Ed.):** Antimicrobial Peptides and Human Disease. 2006. 12 figs. XII, 262 pp. ISBN 3-540-29915-7

Vol. 307: **John L. Casey (Ed.):** Hepatitis Delta Virus. 2006. 22 figs. XII, 228 pp. ISBN 3-540-29801-0

Vol. 308: **Honjo, Tasuku; Melchers, Fritz (Eds.):** Gut-Associated Lymphoid Tissues. 2006. 24 figs. XII, 204 pp. ISBN 3-540-30656-0

Vol. 309: **Polly Roy (Ed.):** Reoviruses: Entry, Assembly and Morphogenesis. 2006. 43 figs. XX, 261 pp. ISBN 3-540-30772-9

Vol. 310: **Doerfler, Walter; Böhm, Petra (Eds.):** DNA Methylation: Development, Genetic Disease and Cancer. 2006. 25 figs. X, 284 pp. ISBN 3-540-31180-7

Vol. 311: **Pulendran, Bali; Ahmed, Rafi (Eds.):** From Innate Immunity to Immunological Memory. 2006. 13 figs. X, 177 pp. ISBN 3-540-32635-9

Vol. 312: **Boshoff, Chris; Weiss, Robin A. (Eds.):** Kaposi Sarcoma Herpesvirus: New Perspectives. 2006. 29 figs. XVI, 330 pp. ISBN 3-540-34343-1

Vol. 313: **Pandolfi, Pier P.; Vogt, Peter K. (Eds.):** Acute Promyelocytic Leukemia. 2007. 16 figs. VIII, 273 pp. ISBN 3-540-34592-2

Vol. 314: **Moody, Branch D. (Ed.):** T Cell Activation by CD1 and Lipid Antigens, 2007, 25 figs. VIII, 348 pp. ISBN 978-3-540-69510-3

Vol. 315: **Childs, James, E.; Mackenzie, John S.; Richt, Jürgen A. (Eds.):** Wildlife and Emerging Zoonotic Diseases: The Biology, Circumstances and Consequences of Cross-Species Transmission. 2007. 49 figs. VII, 524 pp. ISBN 978-3-540-70961-9

Vol. 316: **Pitha, Paula M. (Ed.):** Interferon: The 50th Anniversary. 2007. VII, 391 pp. ISBN 978-3-540-71328-9

Vol. 317: **Dessain, Scott K. (Ed.):** Human Antibody Therapeutics for Viral Disease. 2007. XI, 202 pp. ISBN 978-3-540-72144-4

Vol. 318: **Rodriguez, Moses (Ed.):** Advances in Multiple Sclerosis and Experimental Demyelinating Diseases. 2008. XIV, 376 pp. ISBN 978-3-540-73679-9

Vol. 319: **Manser, Tim (Ed.):** Specialization and Complementation of Humoral Immune Responses to Infection. 2008. XII, 174 pp. ISBN 978-3-540-73899-2

Vol. 320: **Paddison, Patrick J.; Vogt, Peter K. (Eds.):** RNA Interference. 2008. VIII, 273 pp. ISBN 978-3-540-75156-4

Printing: Krips bv, Meppel, The Netherlands
Binding: Stürtz, Würzburg, Germany